T0328282

Kidney Biomarkers

Clinical Aspects and Laboratory Determination of Biomarkers Series

Series Editor: Amitava Dasgupta

Volume 1

Alcohol and Its Biomarkers: Clinical Aspects and Laboratory Determination

Volume 2

Biomarkers in Inborn Errors of Metabolism: Clinical Aspects and Laboratory Determination

Volume 3

Endocrine Biomarkers: Clinical Aspects and Laboratory Determination

Volume 4

Kidney Biomarkers: Clinical Aspects and Laboratory Determination

Kidney Biomarkers

Clinical Aspects and Laboratory Determination

Edited by

Seema S. Ahuja

Brian Castillo

Elsevier
Radarweg 29, PO Box 211, 1000 AE Amsterdam, Netherlands
The Boulevard, Langford Lane, Kidlington, Oxford OX5 1GB, United Kingdom
50 Hampshire Street, 5th Floor, Cambridge, MA 02139, United States

Library of Congress Cataloging-in-Publication Data
A catalog record for this book is available from the Library of Congress

British Library Cataloguing-in-Publication Data
A catalogue record for this book is available from the British Library

ISBN 978-0-12-815923-1

For information on all Elsevier Press publications
visit our website at https://www.elsevier.com/books-and-journals

Publisher: Stacy Masucci
Acquisitions Editor: Tari K. Broderick
Editorial Project Manager: Megan Ashdown
Production Project Manager: Omer Mukthar
Cover Designer: Matthew Limbert

Typeset by SPi Global, India

Contents

Contributors

Soraya Arzhan Kidney Institute of New Mexico, Albuquerque, NM, United States
Brian Castillo Houston Methodist Hospital, Houston, TX, United States
Sara Faiz Baylor College of Medicine, Houston, TX, United States
Melissa Fang University of New Mexico School of Medicine, Albuquerque, NM, United States
Kavitha Ganta University of New Mexico Health Science Center; New Mexico Veterans Administration Health Care System, Albuquerque, NM, United States
Asadullah Khan Kidney & Hypertension Consultants; Clinical Assistant Professor, Department of Internal Medicine, ACTAT, McGovern Medical School; Clinical Assistant Professor of Medicine, Weill Cornell Medical College-Houston Methodist, Houston, TX, United States
Vanessa Moreno Arkana Laboratories, Little Rock, AR, United States
Parisa Mortaji Department of Medicine, University of Colorado, Aurora, CO, United States
Smita Pattanaik Additional Professor of Clinical Pharmacology, Post Graduate Institute of Medical Education and Research, Chandigarh, India
Brent Wagner University of New Mexico Health Science Center; New Mexico Veterans Administration Health Care System; Kidney Institute of New Mexico, Albuquerque, NM, United States
Haiyan Zhang UC San Diego Health, Anatomic Pathology, La Jolla, CA, United States

CHAPTER 1

Ideal biomarkers of acute kidney injury

Asadullah Khan, M.D.

Kidney & Hypertension Consultants, Houston, TX, United States; Clinical Assistant Professor, Department of Internal Medicine, ACTAT, McGovern Medical School, Houston, TX, United States; Clinical Assistant Professor of Medicine, Weill Cornell Medical College-Houston Methodist, Houston, TX, United States

Acronym key

ADPKD	adult polycystic kidney disease
ADQI	adult dialysis quality initiative
AKI	acute kidney injury
AKIN	acute kidney injury network
ATI	acute tubular injury
ATN	acute tubular necrosis
AUCROCC	area under the curve receiver operating characteristic curve
BUN	blood urea nitrogen
CIN	contrast induced nephropathy
CKD	chronic kidney disease
COPD	chronic obstructive pulmonary disease
CPB	cardiopulmonary bypass
CRP	C-reactive protein
D+HUS	diarrhea positive hemolytic uremic syndrome
Da, kDa	dalton, kilodalton
DCD	donation after cardiac death
DGF	delayed graft function
ED	emergency department
ELISA	enzyme linked immunosorbent assay
FeNa	fractional excretion of sodium
GFR	glomerular filtration rate
GI	gastrointestinal
HIVAN	human immunodeficiency virus associated nephropathy
HRS	hepatorenal syndrome
IGFBP-7	insulin-like growth factor-binding protein-7
KDIGO	Kidney Disease Improving Global Outcomes
L-FABP	L type fatty acid binding protein
NAG	*N*-acetyl-β-D-glucosaminidase
PICU	pediatric intensive care unit

Seema S. Ahuja and Brian Castillo: Kidney Biomarkers. https://doi.org/10.1016/B978-0-12-815923-1.00001-8

p-NGAL	plasma neutrophil gelatinase-associated lipocalin
PRA	prerenal azotemia
RIFLE	Risk, Injury, Failure, Loss, End stage kidney disease
RRT	renal replacement therapy
SCr	serum creatinine
SIRS	systemic inflammatory response syndrome
SLE	systemic lupus erythematosus
s-NGAL	serum neutrophil gelatinase-associated lipocalin
TIMP-2	tissue inhibitor metalloproteinase-2
TRIBE-AKI	Translational Research Investigating Biomarker Endpoints
U-KIM-1	urine kidney injury molecule 1
u-NGAL	urinary neutrophil gelatinase-associated lipocalin
UTI	urinary tract infection

Introduction

Over the last 15 years, there have been extensive efforts in examining the utility of new serum and urinary biomarkers to diagnose and prognosticate acute kidney injury.

The search for the ideal biomarker of AKI is still very much "on," despite multiple discoveries regarding a myriad of substances that spike in the setting of AKI.

To date, serum creatinine has been the mainstream marker for diagnosing and managing AKI, but it has several pitfalls. It has poor sensitivity; its ascent is significantly delayed after the actual renal injury; it is not specific to the location or cause of the injury; and it does not differentiate hemodynamic change in glomerular filtration rate (GFR) from intrinsic or obstructive etiologies. It does not reflect the actual GFR unless it has reached steady state, hence true fluctuations in GFR are not detectable in creatinine changes. Another drawback is that tubular secretion of creatinine makes urinary measurements of GFR inaccurate, and certain drugs alter tubular secretion of creatinine, thereby influencing it without an actual change in GFR.

Three major classification systems have been developed for acute kidney injury, namely the KDIGO, AKIN, and RIFLE systems; the latter two involve gradations based on the severity of AKI. However, these systems still utilize serum creatinine and urine output criteria, and are still not representative of a paradigm shift in diagnosis.

It is well known that the event of renal injury predates the biochemical rise in creatinine, and this is where a better biomarker could create opportunities to intervene.

Several additional renal biomarkers have emerged, but none seem to fit the bill as an effective replacement for the status quo. They have been looked at through several angles and lenses, some of which include focused clinical settings, serum vs urinary substances, diagnostic or prognostic features, but the ideal biomarker remains an enigma.

So what qualities should an ideal biomarker possess?

It should be sensitive. It should be an early predictor of kidney disease. It should be rapidly altered after injury occurs. It should remain elevated to allow a wide diagnostic window. It should also be specific, providing insight into the etiology and location of an injury.

It should not be affected by patient characteristics or clinical status. It should have prognostic utility, provide the ability to monitor therapeutic responses, and predict hard clinical outcomes, such as the need for renal replacement therapy, and potential for renal recovery. The sample for the assay should be easy to obtain in a noninvasive manner and results should be available promptly. It should be cost-effective and utilize clinically available techniques. Additionally, it should have the ability to differentiate de novo AKI vs AKI on top of underlying preexisting CKD.

Serum creatinine

Even though serum creatinine (SCr) is not sensitive or specific, it has somehow withstood the test of time, perhaps due to inherent limitations associated with the newer biomarkers. Therefore, it still reserves its spot as the first for discussion and provides a window into the landscape for the rest.

Creatinine is a 113 Da molecule, generated in muscle from the conversion of creatine and phosphocreatine. The Jaffe reaction is used to detect creatinine, but it also detects other chromogens in the serum. Various substances can partially interfere with this reaction, mainly glucose, uric acid, ketones, cephalosporins, furosemide, hemoglobin, bilirubin, and paraproteins, influencing the red-orange color change that indicates creatinine.

SCr is also affected by demographic, physiologic, pharmacologic, and clinical factors. Some examples are strenuous physical activity, protein supplements, dietary protein intake or lack thereof, total parenteral nutrition, infections, cimetidine, trimethoprim, and salicylates [1].

With such an extensive list of factors that can influence SCr, one can only imagine the gap between the current gold standard and the desirable gold standard.

The blood urea nitrogen that is reported with the SCr carries far less specificity. It is heavily influenced by protein intake, catabolic state, steroids, gastrointestinal bleeding, and volume status, making it one of the least desirable biomarkers for AKI [2].

Trials in AKI involving anaritide and fenoldopam were unsuccessful, most probably because the utilization of BUN and SCr for diagnosis resulted in delayed timing of treatment initiation relative to actual renal injury [3–5].

Neutrophil gelatinase-associated lipocalin

NGAL is a 23-kDa protein involved in the immune response to bacterial infections. It is patho-physiologically expressed in neutrophils, hepatocytes, and renal tubular cells, and has bacteriostatic activity. It is protease resistant and therefore easily detectable in urine. It has a brisk quantitative increase in renal tubules and in the urine itself, immediately after ischemic AKI induced in murine models, preceding other urinary biomarkers. Additionally, experimental models of cisplatin-induced tubular injury also demonstrated NGAL in urine of mice at the early stage of injury.

The source of u-NGAL seems to be distal tubular cells, based on imaging.

u-NGAL has demonstrated features of an early biomarker of ischemic and nephrotoxic injury. Serum NGAL is less specific compared to u-NGAL. It is elevated in COPD, critically ill patients, bacterial infections, acute pancreatitis, inflammatory bowel disease, and CKD. A few studies representing the behavior and potential utility of NGAL are as follows:

- 143 pediatric SIRS and septic shock patients over 15 PICUs, against 25 healthy controls concluded s-NGAL to be sensitive but not specific as a predictor of AKI [6];
- 151 adult patients; NGAL in Sepsis-AKI, Sepsis-non AKI and non-sepsis-non AKI were looked at; u-NGAL concluded to be a better predictor of AKI because s-NGAL was elevated in non-AKI patients as well [7]; and
- 65 patients in a study had a similar conclusion of exercising caution when interpreting s-NGAL due to its rise in SIRS and septic shock even in the absence of AKI [8];

It is shown to be elevated in adult polycystic kidney disease (ADPKD) as well.

Twenty-six ambulatory patients with ADPKD were studied and both s-NGAL and u-NGAL were found to be elevated. NGAL demonstrated correlation with cyst counts, with a statistically significant difference in low vs high cyst count groups, and inverse relation with residual renal function. This may warrant further investigation to elucidate a potential pathophysiologic correlation [9].

u-NGAL has been looked at in terms of prognosticating RRT in pediatric diarrhea positive hemolytic uremic syndrome (D + HUS) in a cohort of 34 patients. Normal levels in the acute phase carried a negative predictive value for subsequent need for RRT, but higher levels did not correlate reliably with RRT. u-NGAL also helped to determine that about 60% of D + HUS patients showed acute tubular injury (ATI) [10].

In human immunodeficiency virus associated nephropathy (HIVAN), the presence of iron in a subject's urine pointed toward iron-associated proteins, including u-NGAL (given its role in iron trafficking), indicating potential targets for further trials [11].

A pediatric trial of 85 patients determined elevated u-NGAL to be a reliable biomarker of renal damage in SLE. The same could not be reliably demonstrated for s-NGAL [12].

A single measure of u-NGAL in 635 patients in the emergency department (ED) admitted for AKI was tested among other biomarkers which included NAG, A1-microglubulin, A1 acid glycoprotein, fractional excretion of sodium (FeNa), and SCr. u-NGAL was determined to have superior sensitivity and specificity at a cutoff value of 130 μg/g creatinine, for determining AKI as well as predictive values for outcomes including renal consultation and dialysis, compared with the others [13].

In 55 decompensated cirrhosis patients with AKI, u-NGAL, among other biomarkers, demonstrated promise in distinguishing ATN from functional or prerenal types of AKI, for example, HRS [14].

u-NGAL was found to be an effective early biomarker in adult cardiac surgery, correlating well with subsequent development of AKI. This was demonstrated in a trial with 81 cardiac surgery adult patients. The marker increased early in the trial, and stayed elevated from 3 to 18 h after surgery [15].

p-NGAL demonstrated early predictive biomarker activity, morbidity, and mortality in a prospective, uncontrolled trial of 45 AKI patients [16]. Not only was it shown to be superior to BUN and Cr as an early biomarker of AKI, but also it was an independent predictor of duration and severity of AKI and duration of ICU stay, in a trial of 100 patients after cardiac surgery [17, 18].

Another study of 50 adult patients cited u-NGAL and s-NGAL as predictive biomarkers of AKI as early as 2 h after cardiopulmonary bypass.

NGAL was studied in coronary angiography and percutaneous coronary intervention (PCI) cases. p-NGAL had a significant rise at 2 and 4 h and u-NGAL at 4 and 12 h after contrast. These subjects did not develop contrast nephropathy so further investigation would be needed; however, NGAL may have a potential role in contrast induced nephropathy (CIN) [19].

In terms of CIN, a pediatric study of 91 patients revealed p-NGAL and u-NGAL to be sensitive markers, demonstrating development of contrast injury in 11 of the patients, indicating a rise in SCr [20].

p-NGAL had a high sensitivity, specificity, and area under the curve receiver operating characteristic curve (AUCROCC) in a study of 88 adult critically ill patients compared to controls [21].

In a metaanalysis of 58 manuscripts spanning 16,500 patients, NGAL predicted AKI with a high AUCROCC in three areas, which included cardiac surgery patients, critically ill patients, and renal transplant patients with DGF [22].

Diabetics carried significantly higher s-NGAL and u-NGAL values than nondiabetics.

NGAL had an AUCROCC of 0.98 in adult poly trauma, and has predicted DGF after renal transplant. It has robust correlation, AUCROCC of 0.95 in terms of renal consult, dialysis, and ICU admission. u-NGAL shows promise as an early biomarker in cardiopulmonary bypass, critically ill, ED, trauma, and contrast injury settings.

Caution is advised in interpretation in patients with urinary tract infections, leucocyturia, and proteinuria whereby u-NGAL levels are significantly higher, hence limiting its diagnostic accuracy in these cases.

Cystatin C

Cystatin C is a 13 kDa protein produced in nucleated cells, discovered in 1961, when Butler et al. studied urine proteins of 223 individuals by starch-gel electrophoresis [23].

Freely filtered at the glomerulus, it is completely reabsorbed at the proximal convoluted tubule, thus absent in urine under normal circumstances. It has shown better correlation with GFR at levels 10–60 mL/min [24].

Among older age groups, it is less influenced by loss of muscle mass compared with SCr. However, like SCr, it is influenced by gender and lean muscle mass [25–27].

It was at least as good as SCr as a marker of GFR, and especially in children and the elderly, where decreased muscle mass could be an issue, it may have a role in measuring GFR more reliably [28].

A rise in cystatin-C occurred 1 day after unilateral nephrectomy, in living kidney donors, compared with SCr, which increased after 2 days [29].

A rise in cystatin-C precedes SCr in AKI, as shown in an 85-patient study using RIFLE criteria for AKI diagnosis, the lead time being 0.6 day. It showed favorable AUCROCC [30] 2 days before the R phase in RIFLE criteria [31].

In a pediatric ICU setting, a study of 25 patients showed a better AUROCC compared with SCr at Cr Cl under 80 mL/min [32].

In elective cardiac surgery involving 72 adults, u-Cystatin C was found to be effective for early diagnosis of AKI, where it predicted AKI as early as 6 h earlier compared with p-Cystatin C and NGAL [33].

In children, in a prospective trial of 288 patients undergoing cardiac surgery, preoperative s-Cystatin C did not show an association with AKI but postoperative s-Cystatin-C measured at 6 h predicted Stage 1 and 2 AKI. In the postoperative setting, it also independently predicted longer ventilator and ICU days [34].

s-Cystatin C clearance during continuous veno-venous hemofiltration, using dialysate flow of 2 L/h, was found to be less than 30% of its production, in critically ill patients with acute kidney injury, in a study of 18 patients. Hence, if the serum concentration of s-Cystatin C is not significantly influenced with dialysis clearance during continuous veno-venous hemofiltration, there could be a potential utility in monitoring residual renal function during continuous dialysis modalities [35].

Cystatin C levels can be influenced by thyroid function abnormalities, steroid use, inflammation, and C-reactive protein, reducing its specificity.

The TRIBE-AKI Consortium found s-Cystatin-C to be less sensitive than SCr for AKI detection. However, combined s-Cystatin-C and SCr identified subsets with higher risk of adverse outcomes, like dialysis and mortality [36].

Kidney injury molecule-1

KIM-1 is a transmembrane glycoprotein with an immunoglobulin domain expressed in epithelial cells. Its mRNA expression is minimal normally, but increases significantly in postischemic renal injury.

In rats, u-KIM-1 is an ELISA-based rapid and sensitive biomarker which can detect injury from ischemic and cisplatin nephrotoxicity relatively early. After cisplatin, it demonstrated a 3- to 5-fold rise, and after ischemia, a 10-fold increase after bilateral renal pedicle clamping. It also correlated with proteinuria and interstitial damage.

There is a rapid assay in humans for u-KIM-1.

Urine samples were collected from 32 patients with different forms of AKI and also with CKD against eight controls. Patients with biopsy-confirmed ATN had shown significantly elevated u-KIM-1 levels. Ischemic injury was associated with higher levels compared to other forms of acute renal failure and CKD. Both tissue and u-KIM-1 expression were elevated in focal segmental glomerulosclerosis, IgA nephropathy, membranoproliferative glomerulonephritis, membranous glomerulopathy, acute rejection, systemic lupus, diabetic nephropathy, hypertension, and granulomatosis with polyangiitis [37, 38].

A metaanalysis examined the utility of KIM-1 spanning 2979 patients over 11 trials, and found an approximate sensitivity of 74% and specificity of 86% with AUCROCC of 0.86 for AKI diagnosis.

KIM-1 has been studied in cardiac surgery, septic shock with AKI, ICU patients, and non-ICU patients, but larger trials are needed to identify its proper place in the spectrum of biomarkers.

Interleuken-18

The activity of IL-18, a pro-inflammatory cytokine, in ischemic AKI was demonstrated in murine models. Caspase-1 deficient mice did not convert precursor to active IL-18, as compared with wild-type mice. Various biochemical analyses have since been performed, for example, electrochemiluminescence, immunoblot analysis, IL-18 neutralizing antisera, IL-18 binding proteins, and immunohistochemistry. These findings have led to the detection of IL-18 in the urine of human kidneys that may have sustained AKI, therefore identifying its role as a biomarker as well as a potential target for therapy.

The cytokine is not specific to the kidneys, being found in leukocytes, keratinocytes, the bowel, and dendritic cells. Exogenous IL-18 binding protein in mice has shown protection from liver necrosis and certain forms of arthritis, indicating IL-18 as a potential disease mediator in nonrenal tissue as well [39–48].

The role of urine IL-18 as a marker of established tubular injury in humans was demonstrated in a study of 72 patients where the controls included normal subjects and patients with prerenal azotemia, UTIs, nephrotic syndrome, or CKD.

This was also reproduced in pediatric trials in nonseptic, critically ill patients, and it correlated with severity and mortality [49–51].

In pediatric cardiopulmonary bypass patients, IL-18 was found to be a predictive biomarker of AKI, with an increase at 4–6 h, a peak at 12 h, and sustained elevation for up to 48 h [52].

Along with NGAL, IL-18 predicted AKI after cardiac surgery in adults, with elevation in levels as early as 2–4 h postoperatively [53].

However, in an observational study of IL-18 alone, in adults post-CPB, this finding was not observed [54].

The TRIBE-AKI study found that urine IL-18, urine, and plasma NGAL together increased the AUROCC, for early prediction of AKI as well as with longer length of hospital stay, longer ICU stay, and higher risk for dialysis or death [55].

A multicenter, prospective cohort study of 311 children undergoing pediatric cardiac surgery for congenital diseases, conducted by the TRIBE-AKI Consortium, also demonstrated u-IL-18 and u-NGAL predictive of AKI and poor outcomes for hospital stay, ICU stay, and ventilator days [52].

IL-18 and NGAL may have a role in predicting CIN. Thirteen of 150 patients who were diagnosed with CIN had elevated urine IL-18 and NGAL levels 24 h after the procedure compared to non-CIN patients.

In a study of 72 patients, urinary IL-18 demonstrated an AUCROCC of 0.95, for ATN, compared with with optimal sensitivity and specificity using a cutoff of 500 pg/mg. The study also included 22 patients with significantly elevated urinary IL-18 levels measured during the first 24 hours after transplant, among patients who developed delayed graft function compared with patients with prompt renal function after transplant [55a].

L type-fatty acid binding protein

Fatty acid binding proteins (FABPs) are transporter proteins for fatty acids across intra- and extracellular membranes. L-FABP (L Type-FABP) further transports fatty acids to mitochondria for oxidative energy production for epithelial cells. Other than in the kidneys, they have a presence in several other tissues, such as the skin, brain, muscle, liver, gut, and adipocytes [56].

Experimental association of elevated L-FABP has been observed in tubulointerstitial damage, ischemic AKI, and nephrotoxic AKI models [57, 58].

Another area is renal ischemia in renal transplant, where direct correlation was found between urine L-FABP and ischemic time of transplanted kidney, hence demonstrating a potential role as a biomarker for ischemia-reperfusion injury [59].

However, L-FABP in urine and serum has been demonstrated to be elevated in sepsis even in the absence of AKI. In a study of 80 critically ill patients, urine L-FABP was found to be significantly elevated in septic shock as opposed to sepsis without shock, including patients with AKI; this difference was not observed with serum L-FABP [60].

Urinary L-FABP studied in 121 patients with biopsy-proven chronic glomerulonephritis revealed elevated levels predicting disease progression over a period of 5 years [61].

In a study of 40 pediatric cardiac surgery patients, 21 of whom developed AKI, u-L-FABP demonstrated sensitive and predictive biomarker activity at 4 and 12 h postoperatively [62].

In critically ill patients, L-FABP demonstrated an AUCROCC of 0.82 in a 152-patient study, for predicting doubling of SCr, dialysis, or death within 7 days [63].

Additionally, in a 145-patient medical and surgical ICU study, L-FABP was an independent predictor for 90-day mortality [64].

N-Acetyl-β-D-glucosaminidase

NAG has a high molecular weight of 130 kDa, therefore its filtration into the urine suggests tubular epithelial injury which is where it is produced as a lysosomal brush border enzyme. Elevated urine concentrations have been found in several renal diseases, namely AKI, GN, DN, CIN, sepsis, and CPB. Along with KIM-1, there has been suggestion of predictive ability for outcomes like dialysis and mortality. However, diseases such as rheumatoid arthritis and hyperthyroidism have revealed elevation of u-NAG, questioning its specificity. Inhibition of the enzyme in the presence of urea and metallic anions in urine has raised concerns regarding its reliable measurement [65–70].

TIMP-2 and IGFBP-7

A brief mention of these biomarkers is important due to their combined AUCROCC of 0.8 for AKI prediction in large numbers of patients in the setting of shock, sepsis, major surgery, and trauma. One study involved the KDIGO stage 2–3 criteria and the other involved the RIFLE criteria. The limitation was lack of comparison with other biomarkers. These biomarkers play a role in cell cycle arrest. Further validation will be required to determine their true utility as AKI biomarkers [71].

Past, present, and future

Acute renal failure was originally defined in 1964 by Homer Smith. By 2002, there were more than 30 definitions of AKI, and the adult dialysis quality initiative (ADQI) reached a consensus definition in 2004, leading to publication of the RIFLE criteria (Table 1). This was further modified in 2009 when the AKIN criteria were released, and then most recently in KDIGO (Tables 2 and 3). However, the reliance of SCr has not yet been broken.

In 2006, the Food and Drug Administration released an initiative to study biomarkers for further characterization of AKI, which led to acceleration in the search for biomarkers.

Table 1 Risk, injury, failure, loss, and ESKD classification.

	GFR criteria	Urine output criteria
Risk	Increase in SCr > or = 1.5 × baseline or decrease in GFR > or = 25%	UO < 0.5 mL/kg/h × 6 h
Injury	Increase in SCr > or = 2 × baseline or decrease in GFR > or = 50%	UO < 0.5 mL/kg/h × 12 h
Failure	Increase in SCr > or = 3 × baseline or SCr > or = 4.0 mg/dL or decrease in GFR > or = 75%	UO < 0.3 mL/kg/h × 24 h or anuria × 12 h
Loss	Complete loss of kidney function > 4 weeks	
ESKD	End Stage Kidney Disease (>3 months)	

Adapted from Chan-Yu Lin, et al. AKI classification. AKIN and RIFLE criteria. World J Crit Care Med 2012;1(2):40–45.

Table 2 Acute kidney injury network classification.

Stage 1	Increase in SCr > or = 1.5 × baseline or decrease in GFR > or = 25% + increase in SCr > or = 0.3 mg/dL	UO < 0.5 mL/kg/h × 6 h
Stage 2	Increase in SCr > or = 2 × baseline or decrease in GFR > or = 50%	UO < 0.5 mL/kg/h × 12 h
Stage 3	Increase in SCr > or = 3 × baseline or SCr > or = 4.0 mg/dL or decrease in GFR > or = 75% + initiation of RRT	UO < 0.3 mL/kg/h × 24 h or anuria × 12 h

Adapted from Chan-Yu Lin, et al. AKI classification. AKIN and RIFLE criteria. World J Crit Care Med 2012;1(2):40–45.

Table 3 Kidney disease improving global outcomes classification.

Stage	Serum creatinine	Urine output
Stage 1	1.5–1.9 × baseline or > or = 0.3 mg/dL increase	UO < 0.5 mL/kg/h for 6–12 h
Stage 2	2.0–2.9 × baseline	UO < 0.5 mL/kg/h for > or = 12 h
Stage 3	3.0 × baseline or increase in SCr > or = 4.0 mg/dL or initiation of RRT or in patients < 18 years, decrease in eGFR to <35 mL/min/1.73 m^2	UO < 0.3 mL/kg/h for > or = 24 h or anuria for > or = 12 h

Adapted from Brenner and rector's the kidney; [Chapter 30, Section IV, Page 932].

Placing some of these findings in perspective, the following areas in AKI so far have the potential of a shift in clinical approach: prerenal vs ATI, cirrhosis, cardiorenal, and DCD allocations [72].

Prerenal vs ATI

The most widely used biochemical approach to date has been the use of FeNa, where <1% is regarded as prerenal and >1% ATI.

However, FeNa can be confounded by salt loading, diuretics, preexisting CKD and overlap of the FeNa itself in either clinical scenarios.

This is followed by urine microscopy, which is itself limited by bedside availability and specialty training. In this area, among the new biomarkers, u-IL-18 and u-NGAL have the ability to aid in discriminating between prerenal and ATI [73, 74].

These are followed by KIM-1 and L-FABP as potential markers in ATN.

A combination of these, with clinical and information and FeNa, along with a urine microscopy score may have a potential of increasing precision [75].

Cirrhosis

This is an important clinical scenario where differentiation between HRS and ATI can have major therapeutic implications, for example, liver transplant vs dialysis and dual organ transplantation [76–79].

Current clinical utilization mainly involves SCr, FeNa, and response to hydration [80, 81].

Here, recent studies have demonstrated u-NGAL and IL-18 to have the highest levels in ATI, followed by HRS and finally prerenal.

Combined use of IL-18, u-NGAL, and L-FABP also demonstrated a significant likelihood of ATI, even if one of these biomarkers were elevated.

Hence the above biomarkers could likely complement the exclusion of prerenal azotemia and HRS. However, so far in the case of HRS, FeNa has been reliably the lowest, closer to 0.1% in a fairly uniform fashion, as opposed to prerenal and ATI. Similarly, the use of urinary albumin has yielded higher values in ATI, as opposed to prerenal and HRS cases.

Cardiorenal syndrome

This condition usually poses a complex clinical challenge, where a constant balancing act is needed between optimizing volume with diuretics and finding an acceptable rise in SCr to attain a decongested state [82–85].

In this scenario, u-KIM-1 and *N*-acetyl-β-D-glucosaminidase were elevated despite higher GFR measurements, possibly implying tubular dysfunction not captured by SCr.

Hence the above biomarkers could be studied further in combination with markers of hemoconcentration, like albumin and hematocrit, to form a panel that could elucidate the utility of more aggressive diuretics [86, 87].

Kidney allocation from deceased donors

Approximately 2500 procured kidneys from deceased donors are discarded annually, due to risk of DGF, prolonged hospital stay, and prolonged graft failure. Hence, the kidney donor profile index was revised in 2014 with a score based on creatinine, medical, and demographic factors; however, the scope and reliability remain limited to date. There is significant room for improvement in the tools needed to improve assessment of the donor kidney.

Among the newer biomarkers, NGAL and L-FABP showed incrementally worse 6-month recipient GFR, especially in cases without DGF [88].

On the other hand, preclinical trials on a repair protein YKL-40 have shown urinary levels in donors to have a potential of subsequent renal recovery after transplantation [89, 90].

The NephroCheck

In an attempt to bring biomarker use to the bedside, the FDA approved the first AKI point-of-care biomarker device, the NephroCheck. It involves the measurement of the cell cycle arrest biomarkers metalloproteinase-2 and IGF-binding protein 7. The NephroCheck had a positive predictive value of diagnosing stage 2 or 3 KDIGO AKI of 49%, and a negative predictive value of 97% [91].

It may have the potential of identifying high-risk critically ill patients who can be enrolled in further clinical trials. However, it has been argued that a combination of biomarkers of kidney damage and function rather than an isolated marker limited to a single biologic process is more likely to be superior. This has been demonstrated in an example where Cystatin C and NGAL combination were superior to a change in SCr in predicting the severity and duration of AKI [92].

This led to the establishment of the renal angina index to risk-stratify AKI patients based on the severity of the clinical setting and the change in creatinine clearance. There was some validation of this index in a pediatric study of ICU patients, but this approach requires larger trials to show effectiveness [93].

Moving toward a more sophisticated approach, the NIH has initiated a Kidney Precision Medicine Project, which is an attempt to move away from reliance on SCr as a guide, and to localize biomarkers into panels dedicated to more specific clinical settings. Furthermore, the goal will be to establish a tissue-based atlas of renal pathology founded on biopsies and to understand the renal injury biomarkers in relation to pathologic patterns [94].

Acknowledgment

Rohaan Khan, Senior at Michael E. Debakey High School for Health Professions; for organizing the references used, and structuring the tabular data.

References

[1] Stevens LA, Lafayette RA, Perrone RD, Levey AS. Laboratory evaluation of kidney function. In: Schrier RW, editor. Diseases of the kidney and urinary tract. 8th ed. Philadelphia: Lippincott, Williams and Wilkins; 2007. p. 299–336.

[2] Walser M. Determinants of ureagenesis, with particular reference to renal failure. Kidney Int 1980;17:709–21.

[3] Allgren RL, Marbury TC, Rahman SN, et al. Anaritide in acute tubular necrosis. N Engl J Med 1997;336:828–34.

[4] Lewis J, Salem MM, Chertow GM, et al. Atrial natriuretic factor in oliguric acute renal failure. Anaritide Acute Renal Failure Study Group. Am J Kidney Dis 2000;36:767–74.

[5] Kellum JA. Prophylactic fenoldopam for renal protection? No, thank you, not for me—not yet at least. Crit Care Med 2005;33(11):2681–3.

[6] Wheeler DS, Devarajan P, Ma Q, et al. Serum neutrophil gelatinase-associated lipocalin (NGAL) as a marker of acute kidney injury in critically ill children with septic shock. Crit Care Med 2008;36:1297–303.

[7] Aydogdu M, Gursel G, Sancak B, et al. The use of plasma and urine neutrophil gelatinase associated lipocalin (NGAL) and Cystatin C in early diagnosis of septic acute kidney injury in critically ill patients. Dis Markers 2013;34:237–46.

[8] Martensson J, Bell M, Oldner A, et al. Neutrophil gelatinase-associated lipocalin in adult septic patients with and without acute kidney injury. Intensive Care Med 2010;36:1333–40.

[9] Bolignano D, Coppolino G, Campo S, et al. Neutrophil gelatinase-associated lipocalin in patients with autosomal-dominant polycystic kidney disease. Am J Nephrol 2007;27:373–8.

[10] Trachtman H, Christen E, Cnaan A, et al. Urinary neutrophil gelatinase-associated lipocalcin in D + HUS: a novel marker of renal injury. Pediatr Nephrol 2006;21:989–94.

[11] Soler-Garcia AA, Johnson D, Hathout Y, Ray PE. Iron-related proteins: candidate urine biomarkers in childhood HIV-associated renal diseases. Clin J Am Soc Nephrol 2009; 4(4):763–71.

[12] Suzuki M, Wiers KM, Klein-Gitelman MS, et al. Neutrophil gelatinase- associated lipocalin as a biomarker of disease activity in lupus nephritis. Pediatr Nephrol 2008;23:403–12.

[13] Nickolas TL, O'Rourke MJ, Yang J, et al. Sensitivity and specificity of a single emergency department measurement of urinary neutrophil gelatinase-associated lipocalin for diagnosing acute kidney injury. Ann Intern Med 2008;148:810–9.

[14] Ariza X, Sola E, Elia C, et al. Analysis of a urinary biomarker panel for clinical outcomes assessment in cirrhosis. PLoS ONE 2015;10.

[15] Wagener G, Jan M, Kim M, et al. Association between increases in urinary neutrophil gelatinase-associated lipocalin and acute renal dysfunction after adult cardiac surgery. Anesthesiology 2006;105(3):485–91.

[16] Mishra J, Dent C, Tarabishi R, et al. Neutrophil gelatinase-associated lipocalin (NGAL) as a biomarker for acute renal injury after cardiac surgery. Lancet 2005;365:1231–8.

[17] Haase-Fielitz A, Bellomo R, Devarajan P, et al. Novel and conventional serum biomarkers predicting acute kidney injury in adult cardiac surgery—a prospective cohort study. Crit Care Med 2009;37(2):553–60.

[18] Haase M, Bellomo R, Devarajan P, et al. Novel biomarkers early predict the severity of acute kidney injury after cardiac surgery in adults. Ann Thorac Surg 2009;88(1):124–30.

[19] Bachorzewska-Gajewska H, Malyszko J, Sitniewska E, et al. Neutrophil-gelatinase-associated lipocalin and renal function after percutaneous coronary interventions. Am J Nephrol 2006;26(3):287–92.

[20] Hirsch R, Dent C, Pfriem H, et al. NGAL is an early predictive biomarker of contrast-induced nephropathy in children. Pediatr Nephrol 2007;22:2089–95.

[21] Constantin JM, Futier E, Perbet S, et al. Plasma NGAL is an early marker of acute kidney injury in adult critically ill patients: a prospective study. J Crit Care 2010;25(1):176.

[22] Haase-Fielitz A, Haase M, Devarajan P. Neutrophil gelatinase-associated lipocalin as a biomarker of acute kidney injury: a critical evaluation of current status. Ann Clin Biochem 2014;51:335–51.

[23] Butler FA, Flynn FV. The occurrence of post-gamma protein in urine: a new protein abnormality. J Clin Pathol 1961;14:172–8.

[24] Uzun H, Ozmen KM, Ataman R, et al. Serum cystatin C level as a potentially good marker for impaired kidney function. Clin Biochem 2005;38(9):792–8.

[25] Knight EL, et al. Factors influencing serum cystatin C levels other than renal function and the impact on renal function measurement. Kidney Int 2004;65(4):1416–21.

[26] Groesbeck D, et al. Age, gender, and race effects on cystatin C levels in US adolescents. Clin J Am Soc Nephrol 2008;3(6):1777–85.

[27] Kottgen A, et al. Serum cystatin C in the United States: the Third National Health and Nutrition Examination Survey (NHANES III). Am J Kidney Dis 2008;51(3):385–94.

[28] Filler G, Bokenkamp A, Hofmann W, et al. Cystatin C: as a marker of GFR—history, indications, and future research. Clin Biochem 2005;38(1):1–8.

[29] Herget-Rosenthal S, Pietruck F, Volbracht L, et al. Serum cystatin C—a superior marker of rapidly reduced glomerular filtration after uninephrectomy in kidney donors compared to creatinine. Clin Nephrol 2005;64(1):41–6.

[30] Hajian-Tilaki K. Receiver operating characteristic (ROC) curve analysis for medical diagnostic test evaluation. Caspian J Intern Med 2013;4(2):627.

[31] Herget-Rosenthal S, Marggraf G, Husing J, et al. Early detection of acute renal failure by serum cystatin C. Kidney Int 2004;66:1115–22.

[32] Herrero-Morin JD, Malaga S, Fernandez N, et al. Cystatin C and beta2-microglobulin: markers of glomerular filtration in critically ill children. Crit Care 2007;11:R59.

[33] Koyner JL, Bennet MR, Worcester EM, et al. Urinary cystatin C as an early biomarker of acute kidney injury following adult cardiothoracic surgery. Kidney Int 2008;74(8):1059–69.

[34] Zappitelli M, Krawczeski CD, Devarajan P, et al. Early postoperative serum cystatin C predicts severe acute kidney injury following pediatric cardiac surgery. Kidney Int 2011;80:655–62.

[35] Baas MC, Bouman CS, Hoek FJ, et al. Cystatin C in critically ill patients treated with continuous venovenous hemofiltration. Hemodial Int 2006;10(Suppl. 2):S33–7.

[36] Spahillari A, Parikh CR, Sint K, et al. Serum cystatin C- versus creatinine-based definitions of acute kidney injury following cardiac surgery: a prospective cohort study. Am J Kidney Dis 2012;60:922–9.

[37] van Timmeren MM, van den Heuvel MC, Bailly V, et al. Tubular kidney injury molecule-1 (KIM-1) in human renal disease. J Pathol 2007;212:209–17.

[38] Han WK, Bailly V, Abichandani R, et al. Kidney injury molecule-1 (KIM-1): a novel biomarker for human renal proximal tubule injury. Kidney Int 2002;62:237.

[39] Melnikov VY, Ecder T, Fantuzzi G, et al. Impaired IL-18 processing protects caspase-1-deficient mice from ischemic acute renal failure. J Clin Invest 2001;107:1145–52.

[40] Haq M, Norman J, Saba SR, et al. Role of IL-1 in renal ischemic reperfusion injury. J Am Soc Nephrol 1998;9:614–9.

[41] Melnikov VY, Faubel SG, Siegmund B, et al. Neutrophil-independent mechanisms of caspase-1- and IL-18-mediated ischemic acute tubular necrosis in mice. J Clin Invest 2002;110:1083–91.

[42] He Z, Altmann C, Hoke TS, et al. Interleukin-18 (IL-18) binding protein transgenic mice are protected against ischemic AKI. Am J Physiol Ren Physiol 2008;295:F1414–21.

[43] Wu H, Craft ML, Wang P, et al. IL-18 contributes to renal damage after ischemia-reperfusion. J Am Soc Nephrol 2008;19:2331–41.

[44] Edelstein CL, Hoke TS, Somerset H, et al. Proximal tubules from caspase-1 deficient mice are protected against hypoxia-induced membrane injury. Nephrol Dial Transplant 2007;22:1052–61.

[45] Dinarello CA, Fantuzzi G. Interleukin-18 and host defense against infection. J Infect Dis 2003;187(Suppl. 2):S370–84.

[46] Boraschi D, Dinarello CA. IL-18 in autoimmunity: review. Eur Cytokine Netw 2006;17:224–52.

[47] Faggioni R, Jones-Carson J, Reed DA, et al. Leptin-deficient (ob/ob) mice are protected from T cell-mediated hepatotoxicity: role of tumor necrosis factor alpha and IL-18. Proc Natl Acad Sci U S A 2000;97:2367–72.

[48] Faggioni R, Cattley RC, Guo J, et al. IL-18-binding protein protects against lipopolysaccharide-induced lethality and prevents the development of Fas/Fas ligand-mediated models of liver disease in mice. J Immunol 2001;167:5913–20.

[49] Mehta RL. Urine IL-18 levels as a predictor of acute kidney injury in intensive care patients. Nat Clin Pract Nephrol 2006;2(5):252–3.

[50] Parikh CR, Abraham E, Ancukiewicz M, Edelstein CL. Urine IL-18 is an early diagnostic marker for acute kidney injury and predicts mortality in the ICU. J Am Soc Nephrol 2005;16:3046–52.

[51] Washburn KK, Zapitelli M, Arikan AA, et al. Urinary interleukin-18 as an acute kidney injury biomarker in critically ill children. Nephrol Dial Transplant 2008;23:566–72.

[52] Parikh CR, Devarajan P, Zappitelli M, et al. Postoperative biomarkers predict acute kidney injury and poor outcomes after pediatric cardiac surgery. J Am Soc Nephrol 2011; 22(9):1737–47.

[53] Xin C, Yulong X, Yu C, et al. Urine neutrophil gelatinase-associated lipocalin and interleukin-18 predict acute kidney injury after cardiac surgery. Ren Fail 2008;30(9):904–13.

[54] Haase M, Bellomo R, Story D, et al. Urinary interleukin-18 does not predict acute kidney injury after adult cardiac surgery: a prospective observational cohort study. Crit Care 2008;12(4):R96.

[55] Parikh CR, Devarajan P, Zappitelli M, et al. Postoperative biomarkers predict acute kidney injury and poor outcomes after adult cardiac surgery. J Am Soc Nephrol 2011;22(9): 1748–57.

[55a] Parikh CR, Jani A, Melnikov VY, Faubel S, Edelstein CL. Urinary interleukin-18 is a marker of human acute tubular necrosis. Am J Kidney Dis 2004;43(3):405–14.

[56] Edelstein CL. Biomarkers of kidney disease; 2017. p. 241–315 [Chapter 6].

[57] Yokoyama T, Kamijo-Ikemori A, Sugaya T, et al. Urinary excretion of liver type fatty acid binding protein accurately reflects the degree of tubulointerstitial damage. Am J Pathol 2009; 174(6):2096–106.

[58] Negishi K, Noiri E, Doi K, et al. Monitoring of urinary L-type fatty acid-binding protein predicts histological severity of acute kidney injury. Am J Pathol 2009;174(4):1154–9.

[59] Yamamoto T, Noiri E, Ono Y, et al. Renal L-type fatty acid-binding protein in acute ischemic injury. J Am Soc Nephrol 2007;18(11):2894–902.

[60] Nakamura T, Sugaya T, Koide H. Urinary liver-type fatty acid-binding protein in septic shock: effect of polymyxin B-immobilized fiber hemoperfusion. Shock 2009;31 (5):454–9.

[61] Mou S, Wang Q, Li J, et al. Urinary excretion of liver-type fatty acid-binding protein as a marker of progressive kidney function deterioration in patients with chronic glomerulonephritis. Clin Chim Acta 2012;413:187–91.

[62] Portilla D, Dent C, Sugaya T, et al. Liver fatty acid binding protein as a biomarker of acute kidney injury after cardiac surgery. Kidney Int 2008;73:465–72.

[63] Parr SK, Clark AJ, Bian A, et al. Urinary L-FABP predicts poor outcomes in critically ill patients with early acute kidney injury. Kidney Int 2015;87:640–8.

[64] Cho E, Yang HN, Jo SK, et al. The role of urinary liver-type fatty acid-binding protein in critically ill patients. J Korean Med Sci 2013;28:100–5.

[65] Westhuyzen J, et al. Measurement of tubular enzymuria facilitates early detection of acute renal impairment in the intensive care unit. Nephrol Dial Transplant 2003;18(3):543–51.

[66] Liangos O, et al. Urinary N-acetyl-beta-(D)-glucosaminidase activity and kidney injury molecule-1 level are associated with adverse outcomes in acute renal failure. J Am Soc Nephrol 2007;18(3):904–12.

[67] Katagiri D, et al. Combination of two urinary biomarkers predicts acute kidney injury after adult cardiac surgery. Ann Thorac Surg 2012;93(2):577–83.

[68] Wellwood JM, et al. Urinary N-acetyl- beta-D-glucosaminidase activities in patients with renal disease. Br Med J 1975;3(5980):408–11.

[69] Ren L, et al. Assessment of urinary N-acetyl-β-glucosaminidase as an early marker of contrast-induced nephropathy. J Int Med Res 2011;39(2):647–53.

[70] Bondiou MT, et al. Inhibition of A and B N-acetyl-beta-D-glucosaminidase urinary isoenzymes by urea. Clin Chim Acta 1985;149(1):67–73.

[71] Kashani K, et al. Discovery and validation of cell cycle arrest biomarkers in human acute kidney injury. Crit Care 2013;17(1):R25.

[72] Chirag, et al. Perspective on clinical application of biomarkers in AKI. J Am Soc Nephrol 2017;28:1677–85.

[73] Nejat M, Pickering JW, Devarajan P, Bonventre JV, Edelstein CL, Walker RJ, Endre ZH. Some biomarkers of acute kidney injury are increased in pre-renal acute injury. Kidney Int 2012;81:1254–62.

[74] Singer E, Elger A, Elitok S, Kettritz R, Nickolas TL, Barasch J, Luft FC, Schmidt-Ott KM. Urinary neutrophil gelatinase-associated lipocalin distinguishes pre-renal from intrinsic renal failure and predicts outcomes. Kidney Int 2011;80:405–14.

[75] Hall IE, Coca SG, Perazella MA, Eko UU, Luciano RL, Peter PR, Han WK, Parikh CR. Risk of poor outcomes with novel and traditional biomarkers at clinical AKI diagnosis. Clin J Am Soc Nephrol 2011;6(12):2740–9.

[76] Velez JC, Kadian M, Taburyanskaya M, Bohm NM, Delay TA, Karakala N, Rockey DC, Nietert PJ, Goodwin AJ, Whelan TP. Hepatorenal acute kidney injury and the importance of raising mean arterial pressure. Nephron 2015;131:191–201.

[77] Dobre M, Demirjian S, Sehgal AR, Navaneethan SD. Terlipressin in hepatorenal syndrome: a systematic review and meta-analysis. Int Urol Nephrol 2011;43:175–84.

[78] Nassar Junior AP, Farias AQ, D' Albuquerque LA, Carrilho FJ, Malbouisson LM. Terlipressin versus norepinephrine in the treatment of hepatorenal syndrome: a systematic review and meta-analysis. PLoS ONE 2014;9.

[79] Nadim MK, Genyk YS, Tokin C, Fieber J, Ananthapanyasut W, Ye W, Selby R. Impact of the etiology of acute kidney injury on outcomes following liver transplantation: acute tubular necrosis versus hepatorenal syndrome. Liver Transpl 2012;18:539–48.

[80] Wong F, Nadim MK, Kellum JA, Salerno F, Bellomo R, Gerbes A, Angeli P, Moreau R, Davenport A, Jalan R, Ronco C, Genyk Y, Arroyo V. Working party proposal for a revised classification system of renal dysfunction in patients with cirrhosis. Gut 2011;60:702–9.

[81] Davenport A. Difficulties in assessing renal function in patients with cirrhosis: potential impact on patient treatment. Intensive Care Med 2011;37:930–2.

[82] Krumholz HM, Chen YT, Vaccarino V, Wang Y, Radford MJ, Bradford WD, Horwitz RI. Correlates and impact on outcomes of worsening renal function in patients > or = 65 years of age with heart failure. Am J Cardiol 2000;85:1110–3.

[83] Smith GL, Vaccarino V, Kosiborod M, Lichtman JH, Cheng S, Watnick SG, Krumholz HM. Worsening renal function: what is a clinically meaningful change in creatinine during hospitalization with heart failure? J Card Fail 2003;9:13–25.

[84] Weinfeld MS, Chertow GM, Stevenson LW. Aggravated renal dysfunction during intensive therapy for advanced chronic heart failure. Am Heart J 1999;138:285–90.

[85] Knight EL, Glynn RJ, McIntyre KM, Mogun H, Avorn J. Predictors of decreased renal function in patients with heart failure during angiotensin-converting enzyme inhibitor therapy: results from the studies of left ventricular dysfunction (SOLVD). Am Heart J 1999;138:849–55.

[86] Felker GM, Lee KL, Bull DA, Redfield MM, Stevenson LW, Goldsmith SR, LeWinter MM, Deswal A, Rouleau JL, Ofili EO, Anstrom KJ, Hernandez AF, McNulty SE, Velazquez EJ, Kfoury AG, Chen HH, Givertz MM, Semigran MJ, Bart BA, Mascette AM, Braunwald E, O'Connor CM, NHLBI Heart Failure Clinical Research Network. Diuretic strategies in patients with acute decompensated heart failure. N Engl J Med 2011;364:797–805.

[87] Anand IS, Bishu K, Rector TS, Ishani A, Kuskowski MA, Cohn JN. Proteinuria, chronic kidney disease, and the effect of an angiotensin receptor blocker in addition to an angiotensin-converting enzyme inhibitor in patients with moderate to severe heart failure. Circulation 2009;120:1577–84.

[88] Reese PP, Hall IE, Weng FL, Schröppel B, Doshi MD, Hasz RD, Thiessen-Philbrook H, Ficek J, Rao V, Murray P, Lin H, Parikh CR. Associations between deceased-donor urine injury biomarkers and kidney transplant outcomes. J Am Soc Nephrol 2016;27:1534–43.

[89] Puthumana J, Hall IE, Reese PP, Schröppel B, Weng FL, Thiessen-Philbrook H, Doshi M, Rao V, Lee CG, Elias JA, Cantley LG, Parikh CR. YKL-40 associated with renal recovery in deceased donor kidney transplantation. J Am Soc Nephrol 2017;28:389–93.

[90] Schmidt IM, Hall IE, Kale S, Lee S, He CH, Lee Y, Chupp GL, Moeckel GW, Lee CG, Elias JA, Parikh CR, Cantley LG. Chitinase-like protein Brp-39/YKL-40 modulates the renal response to ischemic injury and predicts delayed allograft function. J Am Soc Nephrol 2013;24:309–19.

[91] Bihorac A, Chawla LS, Shaw AD, Al-Khafaji A, Davison DL, Demuth GE, Fitzgerald R, Gong MN, Graham DD, Gunnerson K, Heung M, Jortani S, Kleerup E, Koyner JL, Krell K, Letourneau J, Lissauer M, Miner J, Nguyen HB, Ortega LM, Self WH, Sellman R, Shi J, Straseski J, Szalados JE, Wilber ST, Walker MG, Wilson J, Wunderink R, Zimmerman J, Kellum JA. Validation of cell-cycle arrest biomarkers for acute kidney injury using clinical adjudication. Am J Respir Crit Care Med 2014;189:932–9.

[92] Basu RK, Wong HR, Krawczeski CD, Wheeler DS, Manning PB, Chawla LS, Devarajan P, Goldstein SL. Combining functional and tubular damage biomarkers improves diagnostic precision for acute kidney injury after cardiac surgery. J Am Coll Cardiol 2014;64:2753–62.

[93] Menon S, Goldstein SL, Mottes T, Fei L, Kaddourah A, Terrell T, Arnold P, Bennett MR, Basu RK. Urinary biomarker incorporation into the renal angina index early in intensive care unit admission optimizes acute kidney injury prediction in critically ill children: a prospective cohort study. Nephrol Dial Transplant 2016;31:586–94.

[94] National Institutes of Diabetes and Digestive and Kidney Diseases. Kidney Precision Medicine Project (KPMP).

Further reading

Brenner and rector's the kidney; n.d. [Chapter 30, Section IV, Page 932].

Lin C-Y, et al. AKI classification. AKIN and RIFLE criteria. World J Crit Care Med 2012; 1(2):40–5.

CHAPTER 2

Utility of the "omics" in kidney disease: Methods of analysis, sampling considerations, and technical approaches in renal biomarkers

Vanessa Moreno
Arkana Laboratories, Little Rock, AR, United States

Introduction

Chronic kidney disease (CKD) affects 10%–20% of the population worldwide, with substantial morbidity and mortality and a high burden in health care spending. As a result, there is substantial interest to identify patients at high risk of rapid loss of renal function and progression to CKD, which will subsequently lead to end stage renal disease (ESRD) [1, 2]. Hence, over the last decade, the search for renal biomarkers and surrogates for specific diseases has been rapidly increasing.

The term "biomarker" refers to "any characteristic that is objectively measured and evaluated as an indicator of normal biologic processes, pathogenic processes or pharmacologic responses to a therapeutic intervention" [3, 4]. Therefore, the ideal renal biomarker, in order to be considered clinically applicable, should have the following characteristics [5]:

a. Noninvasive and easy to perform at the bedside or in a standard clinical laboratory using accessible samples such as blood or urine.
b. Rapid and reliable to measure using standardized assay platforms.
c. Highly sensitive to facilitate early detection with a wide dynamic range and cutoff values that allow for risk stratification.
d. Highly specific for the determined disease.
e. Strong biomarker properties on receiver-operating characteristics (ROC) curves.

The exponential growth in available data and the rise of novel biomarkers of renal disease have become possible due to the major advances in the field of "omics" technologies. The appeal of multiomics studies is compelling because they have provided a deeper and better-informed understanding of important

Seema S. Ahuja and Brian Castillo: Kidney Biomarkers. https://doi.org/10.1016/B978-0-12-815923-1.00002-X

biological processes, pathways, and functional modules that can only be obtained through the combined investigation of the genome, transcriptome, proteome, and metabolome from a dynamic biological system [6].

Conceptually, the "-omic" technologies follow the lifespan of a protein. Genes (genomics) are transcribed (transcriptomics) into proteins (proteomics), which may then undergo posttranslational modification prior to carrying out their designated function. Some proteins are enzymes, which are only active in a specific conformational state (catabolomics), which then release metabolic products (metabolomics). Each step is tightly regulated, so changes in gene expression or transcript levels do not necessarily correlate with downstream effects [7].

Regardless of the "-omic" technology used, it is important that novel biomarkers be translated into easy diagnostic assays that can rapidly process multiple samples (e.g., high throughput) to facilitate translation from bench to bedside [7].

As novel candidate biomarkers for acute kidney injury (AKI) and CKD emerge, it is important to establish optimal sample handling and storage conditions, ideally, prior to undertaking extensive clinical studies. It is necessary to know that markers are sufficiently stable to be useful for clinical purposes and also to enable clinicians and researchers to rely on findings that arise from studies in which long-term storage has been used [8]. Objective measurement and evaluation require a clear understanding of preanalytical factors that may affect the integrity of the biomarker and stringent adherence to protocols that will maximize such integrity. It has not been unusual for clinical studies to be undertaken before the stability of a biomarker has been established. This is understandable because establishing clinical utility clearly is more exciting than confirming biomarker integrity. However, failure to appreciate the instability of markers can lead to significant losses during sample handling and storage, which in turn contributes to data misinterpretation and noncompatibility of studies [8].

Statistical methods in analysis and interpretation of biomarkers in kidney disease

Novel markers are being largely identified by new and advanced technology in basic research, including genomics, proteomics, and noninvasive imaging [9, 10]. These markers hold the promise of improving the prediction of diagnostic and prognostic outcomes to bring personalized medicine closer [11]. The rapid expansion of the biomarker field has prompted development of analytic strategies to convert the data to actionable knowledge. Recognizing the value of a biomarker in a particular situation requires knowledge of the sensitivity, specificity, positive and negative predictive values of the test, and the thresholds at which this is effective [12].

The development of biomarkers into diagnostic or prognostic tests can be categorized into three broad phases: discovery, performance evaluation, and impact determination when added to existing clinical measures [13]. Each phase requires a unique study design and statistical considerations to accurately accomplish research objectives [13].

The statistical methodology used to assess biomarker performance differs from the classic methods used in epidemiology or therapeutic research. In the biomarker discovery stage, the metrics used are based on association between the biomarker and outcome (e.g., odds ratios and relative risks). Meanwhile, the metrics used in the process of development and performance of a biomarker are based on classification or discrimination; for example, true-positive rates (TPRs) and false-positive rates (FPRs) [9].

At the end of the discovery phase, the biomarker is subsequently advanced to the next phase where it is considered as a potential surrogate for a disease of interest and can be measured by reliable methods or assays [13, 14]. In the final phase of biomarker development, the objective is to determine the additional value of the biomarker when used to expand existing clinical models as well as the impact in clinical care [13, 14]. In this chapter, we will discuss the statistical tests used in evaluating the performance and clinical value once an ideal biomarker is successfully discovered.

Metrics for prediction performance (diagnostic potential)

During the second phase of biomarker development, the focus is to show the biomarker's ability to discriminate between diseased and nondiseased patients better or earlier than the current clinical risk factors, explore clinical covariates associated with the biomarker, and establish scenarios or subgroups in which biomarker screening or testing criteria could be applied (validation) [13, 14].

In general, the first step adopted by most researchers is to quantify the classification performance with TPRs, FPRs, and ROC curves. In the medical literature, these rates are also referred to as sensitivity, which is also known as TPR; and specificity, which is the true-negative rate (TNR) and calculated as $1 - \text{FPR}$ [15].

Quantification of prediction performance

True positive rate and false positive rate

If we compare the classification assigned by the biomarker with the true disease status, the results can be categorized as a true positive (TP), false positive

Table 1 Biomarker prediction performance by disease status

		Disease status	
		Diseased	**Nondiseased**
Biomarker test	Positive (diseased)	TP	FP
	Negative (nondiseased)	FN	TN

TP, true positive; FP, false positive; FN, false negative; TN, true negative
True-positive rate (TPR) = Sensitivity = TP/(TP + FN)
True-negative rate (TNR) = Specificity = TN/(TN + FP) = 1 − FPR
False-positive rate (FPR) = 1 − Specificity = FP/(FP + TN)
False-negative rate (FNR) = 1 − Sensitivity = FN/(FN + TP)

(FP), true negative (TN), or false negative (FN). A TP occurs when the biomarker correctly classifies the patient as a diseased patient; similarly, a TN result occurs when the biomarker correctly classifies the patient as a non-diseased patient. On the other hand, an FP occurs when a biomarker incorrectly classifies a nondiseased patient as a diseased patient; while the FN incorrectly classifies the diseased patient as a nondiseased patient [13, 14] (see Table 1).

The TPR is the proportion of diseased patients that the biomarker correctly classifies as diseased patients. The FPR is the proportion of nondiseased patients that the biomarker incorrectly classifies as diseased patients. The range of possible values for both the TPR and FPR is between zero and one. A good biomarker has high TPR and low FPR [14, 16].

Traditional epidemiologic metrics, such as odd ratios, quantify the association between the biomarker and outcome but not the discriminatory ability of the biomarker to separate cases from controls, because odds ratios are not directly linked to TPR and FPR levels. Prediction performance can differ even if the odds ratio remains the same. For studies of prediction, TPR and FPR should be used instead of metrics based on association (e.g., odds ratio) [14, 17].

Receiver-operating characteristic curve

The ROC curve is the most widely used methodology for describing the intrinsic performance of diagnostic tests [5, 18, 19]. An ROC curve is a graphical plot of the sensitivity (TPR) on the "y-axis" versus the FPR (1 − specificity) on the "x-axis" for a binary classifier system, as its discrimination threshold is varied [5] (Fig. 1). For biomarker analysis, the ROC curve provides a complete description of the biomarker classification performance as the disease-positive cutoff changes. ROC curves can, thus, guide the selection of cutoffs for diagnosis of a disease [16, 19].

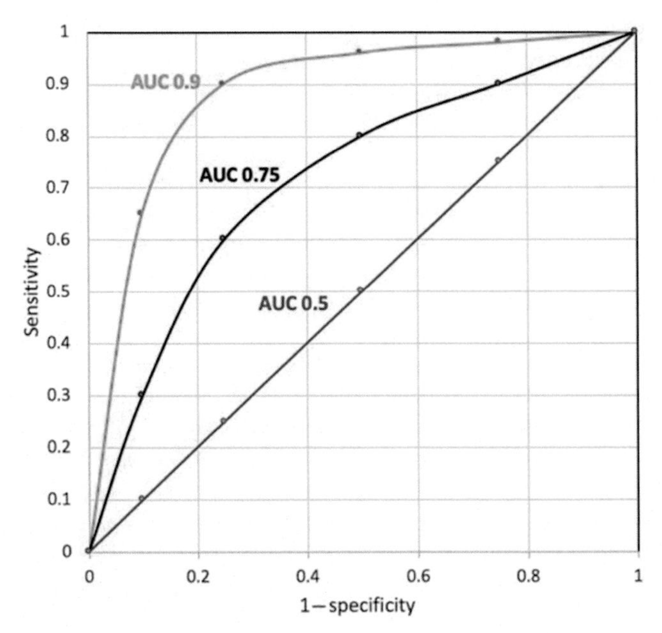

FIG. 1
Receiver-operating characteristic curve (ROC) with three hypothetical biomarkers demonstrating their corresponding area under the curve (AUC). The blue straight line represents a biomarker with an AUC of 0.5, which means a result that is no better than expected by random chance. The red curve shows a biomarker with an AUC of 0.75, which is considered a good biomarker. The green curve depicts a biomarker with an AUC of approximately 0.9, which represents an excellent biomarker.

Characteristically, ROC curves are generated for various cutoff points for the biomarker concentration under consideration. For instance, a perfect biomarker that accurately discriminates all diseased and nondiseased patients would have an ROC curve along the left side of the graph and along the top of the graph representing 100% sensitivity (all true positives detected) and 100% specificity (no false positive found). A completely random biomarker would result in a straight line at a 45° angle from bottom left to top right (the diagonal line of no discrimination) to demonstrate the performance of a biomarker purely due to chance (Fig. 1). If the entire ROC curve lies below the diagonal line, this indicates that the distribution of the biomarker is opposite of usual convention (e.g., lower values of the biomarker are associated with diseased patients) [5].

These analyses are especially valuable for comparing the costs and benefits of single test measures with panels of tests that include more than one diagnostic measure or test. Thus, ROC curves can be used to interpret the interplay of the sensitivity and specificity of each candidate biomarker in isolation, and even more informatively, together with others in the sample plot and the absolute areas under each curve are compared to determine which test or combination of tests has the better diagnostic performance [20].

Summary indices

Area under the ROC curve (AUC)

A commonly derived statistic from the ROC curve is the area under the curve (AUC) or area under the ROC curve (AUC). The AUC is considered the primary tool to report diagnostic potential and is probably the most widely used summary index. The AUC ranges from 0.5 to 1 [14].

The AUC is easy to interpret, where the value of 1 represents perfect discrimination between the diseased and nondiseased patients, which means that all the patients are correctly classified by the test (a perfect biomarker). In contrast, an AUC value of 0.5 represents no discrimination at all, which means that all patients are correctly classified no more frequently than can be attributed to by chance [5, 15]. An AUC of 0.75 or above is generally considered a good biomarker, and an AUC of 0.9 or above would represent an excellent biomarker [5].

The AUC can be interpreted as the probability of the biomarker value being higher in a diseased patient compared with a nondiseased patient if the diseased and nondiseased pair of patients is randomly chosen [14].

ROC curves and AUC can be calculated using most statistical software packages. The AUC can be estimated by the c-index (usually calculated by the trapezoidal rule) or by the Mann-Whitney *U*-statistic [21]. The trapezoidal rule and *U*-statistic are nearly identical when the biomarker is continuous [22], but if the biomarker only has a few distinct values [17, 23], the trapezoidal rule systematically underestimates the true area [24].

Optimal prediction threshold

The optimal classification or prediction threshold is another summary index, which is defined as the cut point with maximum difference between the TPR and FPR [e.g., the Youden index calculated as maximum (TPR – FPR) or equivalently, maximum (sensitivity + specificity – 1)] [14].

This definition may not be optimal threshold, depending on the clinical scenario. For example, for a biomarker to be accepted in clinical practice, it must have a better prediction performance than the existing test, which has an FPR of 10%. Thus, the optimal threshold in this new clinical context would be defined as the maximum TPR for an FPR of at least 5%. TPR and FPR must be reported together, and there is always a tradeoff in the selection of TPR versus FPR [14].

Partial area under the curve

The partial area under the ROC curve (pAUC) is an alternative summary index focusing on the range of practical/clinical relevance. One of the major drawbacks of the AUC as an index of diagnostic performance is that it summarizes the entire ROC curve, including regions that frequently are not relevant to practical applications (e.g., regions with low levels of specificity). Therefore, in order to alleviate this deficiency the pAUC can be used to summarize a portion of the ROC curve over a prespecified range of interest. Occasionally, the pAUC can be used to describe the classification performance within a range of FPR values. For example, certain settings in which treatment is harmful may require very low FPR values (e.g., ≤ 0.2); hence, only the AUC between FPR values of 0 and 0.2 would be of interest [14]. However, the same features that increase the practical relevance of the pAUC introduce some difficulties to resolve issues related to the arbitrariness of specifying the range of interest. Another limitation is that it would require the use of larger sample sizes due to the use of less information resulting in loss of statistical precision, as compared with inferences based on the full AUC.

Limitations of statistical metrics used for prediction performance

Although ROC curves and their summary measures are widely used, there are several limitations:

- The interpretation of the AUC is not directly of clinical relevance because patients do not present as pairs of randomly selected cases and controls [14].
- ROC curves are well established for continuous values of biomarkers and binary outcomes, but the statistical methodology for ROC curves is still evolving for continuous outcomes (e.g., Δcreatinine), ordinal outcomes (e.g., AKI network stages) [10], and time to event outcomes (e.g., months to ESRD) [25, 26].
- The AUC of a new biomarker is highly dependent on its comparison with the gold standard. Therefore, in the presence of an imperfect gold standard, such as serum creatinine for the cases of AKI and CKD, the classification potential of the new biomarker may be falsely diminished [27, 28].
- The use of AUC can be troublesome in conditions with rather low prevalence; therefore, positive predictive value (PPV) and negative predictive value (NPV) should also be provided to allow a good judgment on the accuracy of the biomarker. The use of odd ratios should be avoided in cases with low prevalence [29].

A recent AKI biomarker review raised concerns around the inadequacy of the AUC [30]. Perhaps, the most serious issue with the AUC is that it is an insensitive measure of the ability of a new marker to add value to a preexisting risk prediction model [31]. The disadvantage of the AUC is that from a clinical perspective, it does not provide good information on whether adding this biomarker to the other relevant diagnostic information will accurately identify individual risk [15].

Metrics to evaluating incremental diagnostic value

Often, the classification potential of a biomarker is not adequate alone, which is especially true in settings in which clinical measures or clinical risk models are already in use to facilitate clinical decisions. In such scenarios, it is of interest to determine the contribution of the biomarker to an existing multivariable clinical risk model. Also, if the marker will be used predominantly for predictive purposes, it is of interest to determine the potential of improvement in the clinical risk prediction model with the addition of a novel biomarker [14].

There are several methods to assess the contribution of the new marker. Here we will discuss some of them assuming that we are evaluating the incremental value of a biomarker as an extension of a clinical risk prediction model.

Incremental value: How to evaluate?

First of all, it is essential that the underlying clinical risk prediction model is well calibrated before quantifying the incremental performance value of a biomarker [14, 32]. Good calibration means that risk prediction model-based event rates correspond to those rates observed in clinical settings, which can be assessed using plots (scatter plot of observed versus predicted risk). The most fundamental requirement for a new marker is independent relation to the outcome of the study after adjusting for existing variables in the risk prediction model. In some instances, the biomarker may be related to one or more clinical factors, and its independent association may be diminished in the presence of that clinical factor. For some biomarkers, such as plasma neutrophil gelatinase-associated lipocalin (NGAL), the association with the outcome of AKI diminishes markedly after the addition of postoperative change in serum creatinine [33].

In a logistic regression model, this finding means looking at the coefficient or β and the *P* value for the biomarker in the multivariable clinical risk model. Statistical significance may be inferred from the *P* value, and the strength of clinical association can be measured by the effect size [34]. The interpretation of the magnitude and direction of the effect size should take into consideration several factors such as study design, clinical setting, and clinical relevance. In large studies, a biomarker may have a significant *P* value but a small effect size that is not clinically significant. Therefore, it is recommended to balance the interpretation of statistical and clinical significance by considering the effect size of the biomarker association with the outcome and the *P* value after adjusting for existing clinical measures [14].

Effect size is usually presented as metrics of odds ratios, relative risks, hazard ratios, or absolute risk difference. These effect sizes are not linked to discriminatory performance. Hence, researchers have to move beyond associations and explore other measures in order to understand the incremental value of the biomarker in risk prediction. The metrics of improvement in discrimination and risk classification are the two additional aspects that must be evaluated for a new biomarker to understand its contribution to a risk prediction model [14].

It is important to determine the existence of other factors or variables that influence a biomarker's prediction performance and whether they are related to the outcome of interest [12, 35]. Such factors can be determined by examining the distribution of the biomarker in the nondiseased patients. Some of the factors to consider may be related to patient demographics (e.g., age, race, and sex), clinical parameters (e.g., protein in urine, oliguria, and CKD), or sample processing details (e.g., collection time, freezing time, and length of storage). If there are variables associated with biomarker performance, then diagnostic accuracy can be assessed separately (e.g., biomarker performance determined in adults and children separately in the Translational Research Investigating Biomarker Endpoints—TRIBE—AKI consortium cohort), or more sophisticated methods for adjustment can be applied [36, 37].

Improvement in discrimination

The concordance statistic is equal to the AUC. The C-statistic (sometimes called the "concordance" statistic or C-index) is a measure of goodness-of-fit (GOF) for binary outcomes in a logistic regression model [38]. The C-statistic refers to the probability that predicting the outcome is better than chance.

The increment in the C-statistic or change in AUC (ΔAUC) is applied to quantify the added value offered by the new biomarker. The widely used method by DeLong et al. has been designed to nonparametrically compare two correlated ROC curves (clinical model with and without biomarker); however, it has recently been shown that the test may be overly conservative and may occasionally produce incorrect estimates [24]. Begg et al. have used simulations to show that the use of same risk predictors from nested models while comparing AUCs with and without risk factors leads to grossly invalid inferences [39]. Their simulations reveal that the data elements are strongly correlated from case to case and the model that includes the additional marker has a tendency to interpret predictive contributions as positive information, regardless of whether the observed effect of the marker is negative or positive. Both of these phenomena lead to profound bias in the test [14]. It is also recommended not to pursue additional hypothesis testing on the ΔAUC after showing that the test of the regression coefficient is significant [40, 41].

Researchers have observed that ΔAUC depends on the performance of the underlying clinical model. For example, good clinical models are harder to improve, even with markers that have shown strong association [42]. Hence, Pencina et al. devised alternative metrics for evaluating reclassification with novel biomarkers [16, 43–45].

The proposed new metrics, which include net reclassification improvement (NRI), category-free NRI (cfNRI), and integrated discrimination improvement (IDI) are becoming widely used.

Improvement in reclassification

The NRI, cfNRI, and IDI are the newest metrics that have been recently introduced to assess the added value of a candidate biomarker to preexisting risk prediction models [16, 43–45].

The metrics require a reference risk prediction model that calculates the probability (calculated risk) of a patient having the event of interest (e.g., developing CKD, having AKI, etc.) and then a recalculated probability based on a new model compromising the reference model plus a new biomarker [15].

In the nephrology literature, to date, all risk prediction models are determined from a statistical analysis of risk factors in the studied cohort. Normally, variables with a predetermined low *P* value under univariate analysis are included in a logistic regression model [15]. An alternative approach is to use a model with prespecified variables such as the Framingham risk model for coronary heart disease, as discussed by Kivimaki et al. [46].

This applies to the situation where two models are compared; one of these is the original risk model and the other is the risk model with the biomarker added. If a diseased patient moves "up" in risk classification in the new model, this is seen as an improvement in classification and any "downward movement" is considered as a worse reclassification [15].

The NRI, cfNRI, and IDI each consider separately individuals who develop and who do not develop events. Therefore, they provide additional information not available from the AUC.

Net reclassification improvement. The NRI refers to the overall improvement in reclassification, and is quantified as the sum of two differences [15]:

$$NRI = NRI_{events} + NRI_{nonevents}$$

where each difference represents the proportion of individuals with improved minus worsened reclassification, for those individuals with outcome (NRI_{events}) and for those without outcome ($NRI_{nonevents}$), respectively [15].

Calculation of NRI for those with events [15]:

$$\mathrm{NRI}_{\mathrm{events}} = \frac{\#\,\text{events moving up}}{\#\,\text{events}} - \frac{\#\,\text{events moving down}}{\#\,\text{events}}$$

Calculation of NRI for those without events [15]:

$$\mathrm{NRI}_{\mathrm{nonevents}} = \frac{\#\,\text{nonevents moving up}}{\#\,\text{nonevents}} - \frac{\#\,\text{nonevents moving down}}{\#\,\text{nonevents}}$$

where moving up means that an increase in the calculated risk for the individual moves them to a higher-risk category when a biomarker is added to the model. Similarly, moving down means that a decrease in the calculated risk moves them to a lower-risk category [15].

For the NRI, each individual is assigned to a risk category such as low ($<5\%$), medium (5%–$<20\%$), or high ($\geq 20\%$) based on the event probability calculated by the reference risk prediction model [15].

A second model is constructed by adding the biomarker of interest to the reference model and each individual is reassigned to a risk category. The net proportion of patients with events reassigned to a higher-risk category ($\mathrm{NRI}_{\mathrm{events}}$) and of patients without events reassigned to a lower-risk category ($\mathrm{NRI}_{\mathrm{nonevents}}$) is calculated [15].

The NRI is interpreted as the proportion of patients reclassified to a more appropriate risk category [15]. Of note, it should be remembered that the NRI itself is not a proportion, but rather an index that combines four proportions (upward and downward movement in both event and nonevent groups) [15, 47, 48].

Since the introduction of NRI, there have been various modifications to improve this metric; one of the earliest suggestions was to report NRI separately for events ($\mathrm{NRI}_{\mathrm{events}}$) and nonevents ($\mathrm{NRI}_{\mathrm{nonevents}}$) instead of reporting an overall NRI [49]. This dichotomization has proved to be beneficial, because a biomarker frequently improves reclassification only of participants with the disease or vice versa. The range for both $\mathrm{NRI}_{\mathrm{events}}$ and $\mathrm{NRI}_{\mathrm{nonevents}}$ metrics individually ranges from -100% to 100% [47]. Although the event and nonevent NRIs can be presented as percentages, the overall NRI has no units and should therefore not be presented as a percentage [47].

Among those with the event, if the addition of the biomarker of interest to the model results in more individuals being reclassified to higher-risk categories than to lower ones, then the $\mathrm{NRI}_{\mathrm{events}}$ is positive. Conversely, among those without events, if more are assigned to lower than higher-risk categories, then the $\mathrm{NRI}_{\mathrm{nonevents}}$ is positive [15]. Only those individuals for whom the addition of the new biomarker decreases or increases their calculated risk to the extent that they cross a category threshold contribute to the NRI_{events} [15].

Often, useful information is lost with reporting an overall NRI, and in case of low disease occurrence, the overall NRI would weigh the disease and the non-disease groups equally. Based on the disparate clinical consequence, it would be desirable to report both NRI_{events} and $NRI_{nonevents}$ separately [15].

When there is only one cutoff being evaluated with two risk categories (e.g., low and high), the NRI_{events} equals the improvement in sensitivity. Meanwhile, the $NRI_{nonevents}$ equals the improvement in specificity [16, 50]. Then, the NRI components express the net percentages of persons with or without events correctly reclassified. Negative percentages for the components are interpreted as a net worsening in risk classification [51].

Categorical net reclassification improvement. Categorical NRI is highly dependent on the number of categories. This metric also introduces issues, because higher numbers of categories would lead to increased movement of persons across categories with addition of the new biomarker, thus inflating the NRI value [14].

Weighted net reclassification improvement. This is another metric recommended by some statisticians, where the NRI is weighted by prevalence of events to understand the total value in the population. The weighting extends the NRI_{events} and $NRI_{nonevents}$ interpretation to the whole population [14, 32]. The population weighted NRI can be calculated as follows [32]:

$$\text{Population weighted NRI} = \text{Rho}\,(NRI_{events}) + (1 - \text{Rho})NRI_{nonevents}$$

in which Rho denotes the prevalence of the disease.

However, as with the overall NRI, weighted NRI similarly leads to a loss of information by combining the two groups [14].

Category-free net reclassification improvement. The *cfNRI*, also called the *continuous NRI*, was originally proposed to overcome the problem of selecting categories in applications in which they do not naturally exist [45]. It does not require any risk categorization and considers all changes in predicted risk for all events and nonevents [45, 47]. This has several consequences. First of all, most changes in predicted risk do not translate into changes in clinical management; for example, a patient whose 10-year predicted coronary risk doubles from 1% to 2% will probably not be treated differently [51, 52]. Therefore, the interpretation of continuous NRI is different from that of the category-based NRI [44]. Second, the continuous NRI is often positive for relatively weak markers [44], and it is strongly affected by miscalibration, especially in the setting of external validation [53]. As such, the continuous NRI is less suitable for head-to-head comparisons of competing models, unless these models have been developed from the same data or are correctly calibrated [47]. However, the continuous NRI does provide a consistent message across different models and, therefore, is marker-descriptive rather than model-descriptive; similar to the difference in AUCs between ROC curves [54]. In general, continuous NRI is not recommended to be used and, if so, it should be applied only in special situations and along with other reporting metrics of marker assessment [54, 55].

The cfNRI does not consider the magnitude of the change, it only considers the direction. Each patient is counted as either +1 or −1 depending on whether the change in calculated risk was in the correct direction (higher for those with events, lower for those without events) [15].

The cfNRI is calculated based on the sum of the $cfNRI_{events}$ and $cfNRI_{nonevents}$, where the $cfNRI_{events}$ is the proportion of patients with events who have an increase in calculated risk minus the proportion with a decrease; and the $cfNRI_{nonevents}$ is the proportion of patients without events who have a decrease in calculated risk minus the proportion with an increase [15].

The additional information provided by the $cfNRI_{events}$ and $cfNRI_{nonevents}$ is more revealing than the total cfNRI because it allows for the assessment of the performance of the new model for those with and without events separately. The relative importance of the $cfNRI_{events}$ and $cfNRI_{nonevents}$ will depend on the clinical importance of detecting or excluding an event [15].

Integrated discrimination improvement. The IDI metric is independent of category and separately considers the actual change in calculated risk of each individual for those individuals with and without events [14, 15].

Unlike the cfNRI, IDI does not take into account the direction of change and can be conceptualized as a metric that provides the difference in discrimination slopes or the difference of average probabilities between events and nonevents [15, 16]. Also, unlike NRI, IDI is dependent on calibration of the underlying clinical model. For overall assessment of biomarkers, IDI is a better metric than cfNRI, because it aggregates the magnitude of reclassification. For example, a biomarker receives more weight if it reclassifies risk in someone with an outcome from 55% to 80% than it would from 55% to 60%, although both would be counted as the same increment in continuous or categorical NRI. There are no established criteria for the interpretation of the magnitude of the IDI. As a result, the metric of relative IDI is calculated as the IDI divided by the discrimination slope of the clinical model and may be easier to interpret. If the relative *IDI > 1/number of predictors* in the clinical model, it can be inferred that the biomarker has provided some incremental value beyond existing clinical measures [14].

The IDI for events (IDI_{events}) is the difference between the mean of the new model risk probability for those with the event and the mean of the reference model probability for those with event. Similarly, the IDI for those without events ($IDI_{nonevents}$) is the difference in mean probability for those who do not have the event between the reference and new models [16].

Calculation of IDI for those with events [15]:

$$IDI_{events} = \frac{\Sigma \text{probability of event (new)}}{\# \text{events}} - \frac{\Sigma \text{probability of event (reference)}}{\# \text{events}}$$

Calculation of IDI for those without events [15]:

$$IDI_{nonevents} = \frac{\Sigma \text{probability of event (reference)}}{\# \text{nonevents}} - \frac{\Sigma \text{probability of event (new)}}{\# \text{nonevents}}$$

The overall IDI is the sum of IDI_{events} and $IDI_{nonevents}$. Calculation of the overall IDI is as follows [15]:

$$IDI = IDI_{events} + IDI_{nonevents}$$

The IDI_{events} is also equal to the difference in the average sensitivity (normally termed the *integrated sensitivity* [IS]) of the two models across all risk thresholds and the $IDI_{nonevents}$ is equal to the difference in the average 1 − specificity (*integrated 1 − specificity* [IP]) [15].

The IDI is defined similar to the cfNRI except that instead of adding the value of 1(+1 for moving up or −1 for moving down of category, depending on whether the subject has the event or not, respectively), the actual difference in calculated risk between the models for each individual is added. For example, the cfNRI treats a change of calculated risk of 0.005 and 0.5 identically, whereas the IDI gives more weight to a greater change in calculated risk [15]. The IDI is also the integral of the two-category NRI over all possible thresholds [15].

Decision and clinical use of analytic measures

If a biomarker improves clinical risk prediction, the next important consideration should be its impact on clinical management [56]. For assessment of the potential clinical use of promising markers, decision analytic approaches are needed before a formal cost-effectiveness analysis, which encompasses changes in costs and clinical outcomes in more detail. Decision analytic measures incorporate the prevalence of the disease in the population, the gain in TPRs and FPRs because of the new biomarker, and the benefit and harm related to over- and underdiagnosis [9]. However, the use of such decision analytic measures is limited by the fact that weights for harms and benefits are not firmly established in most fields of medicine, although a range of decision thresholds can be considered in a sensitivity analysis with visualization in a decision curve. One such method of decision curve analysis has easy-to-use software and wide practical application [14].

Overall, there is no one measure that can be used for accepting or refuting a biomarker, because each statistical method has its own strengths and weakness. In addition, different methods have different properties and applicability [14].

Biomarker development is also a phased process, which inherently requires the use of a variety of statistical methods to fulfill different objectives. In the early phases, association assessment using techniques such as logistic regression may be sufficient, because the goal is to advance the promising biomarkers to the next phases. At the later phases of development, the primary purpose is to determine the added discriminatory value and incremental benefit provided by the biomarker to traditional clinical measures. Investigators need to choose methods based on the limitations of the statistical measure, biomarker phase of development, hypothesis being tested, sample size, and clinical question [14].

Although ROC curves may be conservative in terms of discovering a new biomarker, NRI may be too aggressive when the marker may not provide predictive information. As with most summary statistics, the NRI should not be interpreted on its own but in the context of complementary statistical measures. If a marker is not associated with the outcome or does not yield an increase in the AUC, a positive NRI should not be expected [10]. In rare instances in which it does occur, random chances or differences in calibration between the models are the most likely causes. Therefore, biomarker reporting guidelines suggest reporting of multiple metrics for full assessment of a novel biomarker [57]. Investigators should turn away from statistical abstractions, such as the NRI and AUC, and rather, move toward depicting the consequences of using a marker or model in straightforward clinical terms [58].

In addition to prognostic information and improvement in risk prediction, it is also conceivable that the current biomarkers under investigation in renal diseases may be used to provide valuable information as exposure biomarkers or predictors of treatment responsiveness. Testing for other applications of biomarkers may require alternate study designs and statistical methods. Finally, investigators and nephrology community are optimistic that novel biomarkers will have important applications and improve risk prediction models, allowing researchers to design more efficient clinical trials for targeted therapeutic agents and improvement of kidney diseases management [14].

Recommendations for sample collection, storage, and preparation

The identification of new biomarkers, along with their validation and translation into practical clinical applications, requires standardized preanalytical procedures for sample handling, stabilization, transport, and storage [59].

Substandard treatment of samples during the preanalytical phase can add substantial variation or bias that may result in misleading analytical results [60].

The contribution of several studies to the definition of standard operating procedures for specimen collection represents a step forward with respect to the available recommendations for biobanking procedures [61]. Standard operating procedures (SOPs) ensure correct implementation of essential biobanking components (sample donor's anonymization as well as proper acquisition, transport, preparation and analysis process faultlessness, storage conditions, and terms of samples) allowing data comparison worldwide [62]. The SOPs are being implemented at the local da Vinci European Biobank (http://www.davincieuropeanbiobank.org) and proposed for adoption by the European Biobanking and Biomolecular Resources Research Infrastructure (BBMRI; http://www.bbmri.eu) [63].

Biobanks, which are infrastructures devoted to the collection, cataloguing and storing of biological samples in order to make them available for medical and clinical research, represent an irreplaceable support for all those studies in which the impact of the results is linked to the large number of the collected samples. At the same time, they have to guarantee that the quality of the stored biological samples remains as close as possible to the fresh sample for any possible future studies [63], including genomics, transcriptomics, proteomics, and metabolomics.

Tissue samples

Deoxyribonucleic acid (DNA), ribonucleic acid (RNA), proteins, and polar metabolites can be extracted from fresh intact tissue as well as fresh-frozen and formalin-fixed, paraffin-embedded (FFPE) tissues. These types of samples can come from animal experiments or patient biopsies.

Due to metabolic alterations that can occur instantaneously secondary to hypoxia, once the sample is collected, it is recommended to use effective methods to arrest this metabolic process. The most common approach is freeze clamping with lower temperature repositories; after immediate freezing in liquid nitrogen [64, 65], it should be stored at −80°C. Contradictory, placing a warm tissue directly into a −80°C freezer may enable the metabolic profile to change in the time it takes for the sample to freeze, which is not an optimal condition [60]. Also, freeze-thaw cycles should be minimized as much as possible to reduce the changes in metabolic profile. Hence, subaliquoting the tissue samples while collecting will reduce the future freeze-thaw cycles and the effect on metabolic profile [60].

Tissue-specific sampling protocols should be used. It is recommended that the tissue sampling should be consistent throughout the same experiment to avoid region-specific metabolite variance due to tissue heterogeneity [66] (e.g., metabolomic alterations during AKI are different in kidney cortex and kidney medulla). For tumor sample analysis, additional care should be taken in terms of location of sample collection (e.g., oxygenated vs. necrotic areas) [66]. The collection of "normal" tissue is also recommended in order to be used as a control [67, 68]. In addition, to avoid contamination, blood should be removed as much as possible from the tissue samples [69].

The detection of metabolites, DNA, and RNA is higher with fresh intact and fresh-frozen tissue than in FFPE samples [65, 70]. However, due to the long-term stability and widespread availability of FFPE tissues, this type of specimen can be very useful for the acquisition of reliable and broadly informative data leading to the discovery of biomarkers [70]. RNA and DNA can be extracted from FFPE samples and analyzed by next-generation sequencing (NGS) [62]. Some of the limitations associated with FFPE tissue include degradation and attrition of compounds of interest due to formalin fixation and paraffin-embedding processes. For instance, it has been shown that messenger RNA (mRNA) levels are significantly lower in FFPE preparations compared with frozen-tissue extractions, and this could be an analogous situation for the metabolome [70].

Furthermore, there are available commercial kits or buffers such as RNA*later*™ stabilization products that can be used in order to stabilize and protect the integrity of RNA in tissue samples [62]. This is an aqueous, nontoxic tissue storage reagent that rapidly permeates tissues to stabilize and protect cellular RNA [71]. RNA*later*™ solution minimizes the need to immediately process tissue samples or to freeze samples in liquid nitrogen for later processing. Tissue pieces can be harvested and submerged in RNA*later*™ solution for storage without jeopardizing the quality or quantity of RNA obtained after subsequent RNA isolation. Therefore, if tissue sample is obtained from excision/biopsy and the study involves RNA isolation, it should be preserved within an hour either by freezing methods or using 10% buffered formalin. If RNA*later*™ solution is available, then this preservative should be added to the sample instead of freezing or using formalin as fixation techniques. Once the sample is submerged in this type of solution, it can be stored for 72 h at 4°C. For long-term storage, the preservation medium should be discarded and the sample can be stored at −80°C [71].

Biofluids

Blood-based biofluids and urine are the most frequently collected human samples in -omics technology because they provide a global view of metabolism operating in multiple different tissues and organs as well as the feasibility for collection [60].

Blood products. Plasma and serum can be considered as two different biofluids both originating from blood. Plasma is collected using anticoagulants; hence, it preserves most clotting factors. On the other hand, serum is produced using natural blood clotting products. Therefore, it is important to know the type of sample required because the production of serum involves coagulation, whereas plasma does not. In addition, the time and temperature allowed for coagulation are important preanalytical factors to consider and should be standardized for sample processing [60].

Moreover, hemolyzed blood samples should be avoided in -omics studies; therefore, special care should be taken during drawing and handling of samples. The collection of blood products and its processing should be performed according to standard protocols [60]. The time between collection and processing should not exceed 2 h; however, samples can be processed within 24 h but during this time delay samples should be kept at 4°C [63].

Regarding the type of collection tube for **plasma samples:** there are three standard types of anticoagulants that can be applied: lithium (or sodium) heparin, sodium citrate, and potassium ethylenediaminetetraacetic acid (EDTA) [60]. Citrate and EDTA are low-molecular-weight chemicals, which can interfere in the detection of metabolites in nontargeted metabolomic studies and can be detected by most analytical platforms [59]. Another limitation of citrate is that it is an endogenous metabolite that can alter the pH of the sample as well as the extraction conditions [60]. On the other hand, heparin is not detected in a typical nontargeted metabolomics assay, which is advantageous; however, the introduction of lithium into the sample can result in unwanted lithium adducts in mass spectrometry (MS)-based analysis. EDTA-based samples are frequently used and may be more beneficial and helpful if concurrent proteomics is planned with the same sample [60]. For **serum samples**, there are two type of collection tubes, which include gel-free and gel-containing (polymeric) tubes [60]. The quality of these tubes is important, and the use of gel-free tubes removes the possibility of sample contamination from the gel (polymer) [60, 62]. The gel-containing tubes are also known as serum-separating tubes (SST), "tiger-tops," or "gold-topped tubes." They contain a special gel that separates blood cells from serum as well as particles to cause blood to clot quickly. The blood sample may then be centrifuged, allowing to clear serum to be removed for testing [72].

Processing should be undertaken before freezing the samples because freezing whole erythrocytes will result in their lysis and substantial contamination of the plasma metabolome [60]. Whole blood stored at room temperature is much more metabolically active than that stored on ice at approximately 4°C, which will greatly affect the quantitative metabolic profile. If storage is at room temperature, then centrifugation is recommended to be done in a short period of time (<30 min); however, storage at 4°C on ice or in a refrigerator for longer period of time is accepted by some large biobanks. Moreover, centrifugation time and temperature must be consistent according to a standard operating procedure [60]. Some metabolites, such as cysteine and cystine, are unstable even under acceptable processing conditions [60].

For long-term storage, the samples are recommended to always be stored frozen, preferably in liquid nitrogen or at −80°C, when liquid nitrogen is not available, and analyzed within the shortest time window possible after collection. Also, it is recommended storing samples from an individual in multiple 0.5- or 1.0-mL aliquots [60]. This will minimize any need for multiple, potentially damaging, freeze-thaw cycles in case that an individual's sample require multiple analyses [73]. Where freeze-thawing cannot be avoided, the number of freeze-thaw cycles should be recorded and standardized among samples in the same study to avoid bias and should be included as a confounding factor when analyzing the data [60]. Although there is limited research reported in the literature regarding freeze-thaw cycles and the influence on metabolic profile, there is data showing that the metabolic profile is not affected when the sample undergoes up to three freeze-thaw cycles in a targeted assay [74, 75].

All these recommended guidelines for blood derivatives biobanking are also focused on the use of stored samples for future extraction of DNA for genetic analysis. Regarding DNA extraction, this can be obtained from peripheral blood on EDTA tube (lavender top) or with the use of DNA PAXgene® tube. The PAXgene® blood DNA tube is intended to collect, anticoagulate, stabilize, transport, and store a venous whole blood sample for preparation in order to obtain the human DNA with appropriate molecular diagnostic testing methods [62].

Blood specimen for RNA isolation can be collected on PAXgene®/Tempus™ tubes, which are made of silica-membranes and magnetic beads, respectively. These collecting tubes allow instant preservation and stability of RNA from blood specimen for about 3 days/5 days (PAXgene®/Tempus™) at room temperature (18–25°C) and about 5 days/7 days (PAXgene®/Tempus™) at lower temperatures (2–5°C). For optimal performance, blood on RNA isolation tubes should be held at room temperature for a minimum of 2 h or at maximum of 3 days/5 days (PAXgene®/Tempus™) before processing or storage. For longer storage time, specimens may require to be frozen prior to testing. In order to do this, freeze the sample first at −20°C for 24 h, and then transfer the sample to a −80°C freezer. Blood in Tempus tubes may be stored up to 6 years at −80°C [62, 76].

Urine. It can be collected as a single midstream collection at a specific time (i.e., spot urine sample) or as a total urine output in a period of 24 h (i.e., 24-h urine collection). In both cases, it is important to prevent bacterial growth and ensure stability of the metabolites. The addition of sodium azide (NaN_3; 0.01%–0.1%) to limit residual bacterial growth and enzyme activity may be considered [59]. For instance, a recent study assessing the impact on metabolite composition from urine samples and collection conditions demonstrates that bacterial growth can be prevented by adding a preservative; however, the metabolite degradation cannot be avoided [77]. Meanwhile, storage at 4°C inhibits bacterial growth and metabolite degradation. Thus, the results from this study support maintaining samples at 4°C during the collection period [60].

Moreover, food intake has a substantial impact on the urinary metabolic profile; thus, overnight fasting prior to urine collection gives a more stable homeostatic picture of an individual's urinary metabolome. Hence, it is recommended that all urine samples should be collected the morning after overnight fasting to reduce the effects of diurnal variations [78].

Another recommendation is to avoid transporting or storing samples for >8 h in a cool pack or at room temperature. In fact, the Da Vinci biobank recommends storing urine for nuclear magnetic resonance (NMR) for no more than 2 h at 4°C and centrifuging before freezing to avoid cell lysis [79].

Urine samples should be processed and aliquoted within 2 h from the time of collection but preferably faster. Samples must be kept refrigerated at 4°C or in wet ice for up to 24 h before processing, and must not be frozen prior to processing, to avoid possible cell breakage [62, 63]. Before aliquoting and long-term storage, urine samples should be centrifuged at 1000–3000 RCF (5 min at 4°C) and, optionally, filtered using a 0.22-μm filter to remove cells and other particulates [63]. Care is required with filtration because there is a potential loss of metabolites. As a result, many studies process urinary samples using centrifugation only [60].

For long-term storage, urine samples should be stored at −80°C. If possible, for very long-term storage, it is better to use liquid nitrogen vapor [63]. Appropriate labeled cryovials should be used to store urine samples [59]. These recommendations have been supported by recent studies in which urine samples stored for long-term period were evaluated using both targeted liquid chromatography-tandem mass spectrometry (LC-MS) and global profiling approach. Based on these studies, they found that storage at −20°C for short periods of time or up to 20 days was sufficient to ensure the stability of metabolites, whereas storage at −80°C was recommended for longer periods of time [80–82].

Furthermore, when urine sample is frozen, the number of freeze-thaw cycles should be limited to two [60]. This is based on a recent study where Rotter et al. demonstrated that two freeze-thaw cycles did not have significant impact on metabolic concentrations [82]. However, after three freeze-thaw cycles, significant effects were identified [82], leading to the recommendation to limit the number of freeze-thaw cycles [60].

Nontargeted versus targeted analysis

One of the strengths of "omics" technologies is that they allow for nonbiased screening of a large number of gene regulation processes, signal transductions, and metabolic pathways. Nontargeted screening technologies have the goal to capture as many unfiltered and nonbiased data as possible. The molecular entities underlying the recorded signals are often unknown. This dramatically minimizes the chance of missing important data or, in other words, reducing the risk of false-negative results. However, as a consequence, the largest problem with nontargeted screening technologies, such as proteomics and metabolomics, are false-positive results [83].

As of today, nontargeted profiling technologies are challenging in a clinical setting because of complex analyses and software tools required. Combinatorial molecular markers typically consist of 3–10 individual parameters [84]. In general, specific combinatorial biomarker patterns confer significantly more information than a single measurement and can thus be expected to have better specificity and detection power [83].

On the other hand, targeted assays are often multianalyte assays and measure well-defined compounds. Analytical technologies include, but are not limited to, antibody and aptamer arrays, bead immunoassays for proteins, and LC-MS assays (for proteins or metabolites) that allow for the assessment of molecular marker panels ideally in a single run [83]. Although bead- and array-based multianalyte protein assays are available, as antibodies are derived from biological sources, antibody cross-reactivities, manufacturing, and batch-to-batch reproducibility can provide challenges [83].

Targeted assays are usually semiquantitative or quantitative and can be validated. Although the quality of the results is much better understood, these assays are limited in terms of their ability to detect unknown effects. Since only selected compounds are measured, targeted analyses are inherently biased [83].

Proteomics and metabolomics in nephrology

The cells, either directly or indirectly (via extracellular fluid), communicate with body fluids. Metabolites, peptides, proteins, and extracellular membrane vesicles (microparticles) are released from cells or taken up from body fluids by transmembrane diffusion or transport, and throughout death process the cells release all of their contents into body fluids. Thus, at least to a certain extent, biochemical and protein changes in cells and organs are reflected in body fluids [85].

While tissue samples, biopsies, and certain fluids such as urine (kidney), bile (liver), and cerebrospinal fluid (CNS) mainly reflect changes in specific organs and thus are considered "proximal matrices," plasma sample reflects systemic changes that often cannot be traced back to a certain organ [85]. Such changes of metabolites, peptides, and proteins in body fluids, if mechanistically linked to disease processes and drug effects in tissues and organs, have the potential to serve as surrogate markers or biomarkers [83].

Proteomics and metabolomics, can be viewed as ultra-high-throughput clinical chemistry/biochemistry, assessing hundreds and sometimes thousands of proteins or metabolites in a single analytical run [86–88]. As such, the assessment of the proteome and metabolome, unlike genomic analysis, provides a view of biological processes at their level of occurrence. The knowledge gained through this systematic analysis is important not only for a better understanding of renal physiology and pathophysiology but also for the identification of disease markers and the development of new therapies [89]. Hence, proteomics and metabolomics complement genomics and are considered phenotypic molecular markers [85].

Potential impact of proteomics and metabolomics in renal diseases

1. Assessment of molecular mechanisms of disease and drug effects in in vitro, animal and clinical studies. This may lead to better understanding of disease mechanisms, drug effects, and toxicities as well as the identification of new therapeutic targets [83].
2. Drug development. The availability of specific and sensitive molecular markers will impact drug development with better targeted therapy and regimens, as follows [83, 90]:
   - Faster and more efficient preclinical and early clinical development
   - Selection of lead drug candidates with a better therapeutic index
   - Earlier and more sensitive detection of toxic effects
   - Monitoring of pharmacodynamics and toxicodynamics during preclinical and clinical development
   - Development of more efficient and predictive animal models
   - Identification of pharmaco- and toxicodynamic mechanisms
   - Better long-term safety and efficacy
   - Stratification of patient populations during clinical trials
   - Identification of "enriched" populations with better chance of therapeutic efficacy and tolerability
   - Provide new supporting or surrogate study endpoints
   - Diagnostic tools for clinical management of drugs in clinical practice
3. Clinical diagnostics and outcomes. Well-qualified molecular marker protein and/or metabolite patterns yield more detailed and mechanistically relevant information than currently, often used single markers, translating into good specificity. The better the specificity of a

molecular marker pattern, the more this will reduce nonspecific background noise. At the same time, reduced background noise usually results in better sensitivity. This will allow for detection and identification of changes in molecular signatures associated with disease processes and drug effects in body fluids such as plasma and urine before symptoms and irreversible injury occur [83].

Proteomics and the kidney

Proteomics refers to the systematic study of proteomes, which describes the entire protein content on one or all cells of an organism. Some researchers define proteomics as the use of quantitative protein-level measurement of gene expression to characterize biological processes and decipher the mechanisms of gene expression control [91].

Proteomics uses a rapidly evolving group of technologies to identify, quantify, and characterize a global set of proteins [92, 93]. This is a rapidly growing field whose development has been a major spin-off of the Human Genome Project (HGP) and genome sequencing projects for other living species [94].

The study of proteomes is becoming of great interest among the renal community wherein proteomic technologies are embraced at an increasing rate [92]. The reasoning to studying proteins rather than nucleic acids is because they mediate most of the physiologic functions within the cell. Hence, the importance of studying proteins as prime mediators of function in organs, tissues, or cells. Moreover, the analysis of body fluids such as urine [95–97] can only be accomplished by proteomics approaches because nucleic acids play no direct functional role in extracellular fluids. In addition, proteins are regulated in a multiplicity of ways, many of which do not involve changes in mRNA levels [94].

Compared to static gene expression, the proteome shows dynamic properties with protein profiles changing depending on a variety of extra- and intracellular stimuli (i.e., cell cycle, temperature, differentiation stress, apoptotic signals) [98]. The complexity of proteomes is highly influenced by the generation of protein isoforms through the mechanism of alternative splicing of cognate mRNAs, proteolytic processing of translated protein, and subsequently, covalent modifications of amino acids in proteins (i.e., phosphorylation, methylation, acetylation, glycosylation, nitrosylation, nitration, sulfoxidation, arginylation, ubiquitinylation, sumoylation, among others). Hence, all these mechanisms have the potential to regulate protein complex formation, activity, and function [89, 98–100]. However, additional regulatory mechanisms depend on changes in protein abundance not only as a result of altered transcription or mRNA stability but also as a result of direct regulation of translation or regulation of protein half-life. Because protein abundance is regulated by both transcriptional and nontranscriptional mechanisms, studies profiling mRNA levels and protein levels have generally demonstrated only a limited correlation between the two variables [94]. Proteomic methods are under development for the large-scale study of all of these modes of regulation, providing information extending far beyond transcriptional regulation. Therefore, it is clear that most forms of cellular regulation would not be detectable by complementary DNA (cDNA) or oligonucleotide arrays. Thus, the field of proteomics can be viewed as being complementary to the area of functional genomics [94].

The proteomics sample

The renal proteome is considered a mosaic in its origin, being made up of the multiple cell types that comprise the kidney and its extrarenal compartments, including blood. Accordingly, the renal proteome can be viewed as an assembly of subproteomes of lesser complexity than the whole, released by or contained in kidney cell compartments (e.g., plasma membranes, nuclei, cytosol, and mitochondria). As urine can harbor proteins from all kidney subproteomes, and the protein composition of urine is affected by kidney injury or disease, the urine proteome can signal the status of kidney health, and the onset and nature of dysfunctions [100].

Urine

Urine proteomics is becoming a routine tool in research and is at the transition toward clinical implementation. The urine proteome is a particularly appealing source of diagnostic and prognostic biomarkers of kidney disease because it is accessed in a noninvasive fashion providing blood-derived and kidney-derived proteins [100]. It has been estimated that about 70% of urinary protein and peptides are generated from the kidney, while the other 30% derived from plasma [101–103].

Proteins in urine arise from five major sources: 1. filtration and secretion of normal or abnormal plasma proteins (originating extrarenally), 2. secretion by the various segments of healthy and diseased nephrons, 3. proteolytic degradation products of the extracellular matrix, 4. secretion by the lower urinary tract, and 5. physiological and/or pathological cell demise in blood and the urinary tract. Therefore, knowledge of urinary proteins would be expected to yield information pertinent to the functions of both extrarenal organs and renal-resident cells [100].

Urinary proteomics is apparently becoming more valuable to reveal systemic and kidney-specific dysfunctions [100, 104]. Besides, a disease-associated protein can serve not only as a biomarker of disease but also as a signal of molecular events associated with the disease pathogenesis helping to find targets for the development of novel therapies [100]. Moreover, the technologies available today allow the assessment of thousands of peptides and proteins in a urine sample within 60–90 min [105, 106].

The urinary proteome has been the focus on large multidisciplinary investigations and has only been recently performed in large cohorts [107, 108]. It has been characterized extensively using different technological approaches, and nowadays a standard urine sample for comparison of datasets between laboratories and different technologies is available [109]. In particular, capillary electrophoresis coupled with mass spectrometry (MS) appears to be well suited for urine proteome analysis. To date, over 20,000 urine samples have been analyzed using this approach, under identical conditions and with identical instruments, resulting in well comparable datasets [110]. Since each individual analysis contains data on over 1000 urinary peptides and proteins, these data contain an enormous wealth of information that can be exploited for clinical purpose [104].

Most analytical approaches assessing the urine proteome include an initial sample preparation step enriching the proteins of interest. A common strategy used is the removal of high-abundance proteins that confers little diagnostic information using techniques, such as column purification (size exclusion, ion exchanger, affinity columns), selective surfaces, immunodepletion, and equalizer beads [105, 111, 112]. Immunodepletion has the inherent risk of losing proteins of interest by codepletion that may be caused by protein-protein interactions independent of the desired specific antibody interactions [105].

Advantages. Compared with other body fluids, urine has several characteristics that make it a preferred choice for biomarker discovery. First of all, urine can be obtained in large quantities using noninvasive procedures, allowing repeated sampling of the same individual for disease surveillance. The availability of urine also allows easy assessment of reproducibility or improvement in sample preparation protocols [105].

Second, urinary peptides and lower-molecular mass proteins are generally soluble. Therefore, solubilization of these low-molecular-weight proteins and peptides, a process with a major influence on the proteomics analysis of cells or tissues, generally is not an issue. Further, these lower-molecular mass compounds (<30 kDa) can be analyzed in a mass spectrometer without additional manipulation or enzymatic digestion (e.g., tryptic digests) [105].

Third, in general, the urinary protein content is relatively stable probably due to the fact that urine "stagnates" for hours in the bladder; hence, proteolytic degradation by endogenous proteases may be essentially complete by the time of voiding. This is in sharp contrast to blood for which activation of proteases (and consequently generation of an array of proteolytic breakdown products) is inevitably associated with its collection [113, 114].

Finally, as described earlier, the changes in the kidney and genitourinary tract are reflected by changes not only in the urinary proteome but also at more distant sites [100, 104, 105].

Disadvantages. Urine has the disadvantage of having a wide variation in protein and peptide concentrations, mostly, because of differences in the daily intake of fluid. However, this shortcoming can be countered by standardization based on creatinine [115] or peptides generally present in urine [116].

Moreover, definition of disease-specific biomarkers in urine, and mostly likely in other compartments, is complicated by significant changes in the proteome during the day. These changes are likely caused by variations in the diet, metabolic or catabolic processes, circadian rhythms, exercise as well as circulatory levels of various hormones [117]. The reproducibility of any analysis is reduced by these physiological changes even if the analytical method shows high reproducibility. However, these variations appear mostly limited to a fraction of the urinary proteome; a large portion remains unaffected by these processes [118].

In addition to the aforementioned, protein and peptide concentrations can also be affected by sample collection and storage. Midstream urine samples are preferred over first-void urine because the latter has a different composition than the former and is more prone to protein degradation [119]. Although long-term storage of urine at −80°C has been described as a safe approach, when frozen samples are thawed an initial loss of minor protein signals can be observed [120, 121]. Therefore, urine samples for proteomic analysis have been generally found to be stable for at least three freeze-thaw cycles; however, marked losses of proteins have been found if samples are frozen and thawed more often [119, 122, 123]. Overall, it is recommended to avoid freeze-thaw cycles whenever possible [124]. When frozen, the number of freeze-thaw cycles should be limited to two [60].

Kidney tissue and cell culture

The basic principle of tissue sample preparation is that the heterogeneity should be diminished as much as possible and that the sample should be pure and relevant. Two methods are recognized for proteome analysis which include homogenization and removal of contaminants. Homogenization can be divided in five categories: mechanical, ultrasonic, pressure, freeze-thaw, and osmotic or detergent lysis [125–127]. Also, it is important to protect the sample from proteolysis during processing. The most protective measures are protein denaturation and the addition of protease inhibitors [128]. Subsequently, the later steps will include the removal of contaminants such as salts, detergents, abundant proteins, lipids, polysaccharides, nucleic acids, and other contaminants as well as protein enrichment using precipitation, centrifugation, prefractionation, electrophoretic, antibody-based procedures, and/or chromatographic techniques [125–127].

Protein profiles and images can be obtained directly from fresh-frozen tissue blocks or from FFPE tissues. Although there are encountered feelings about using FFPE tissue instead of frozen tissue for proteomic and molecular analysis, both types of tissues can be used. Several studies performed in renal and non-renal tissue have shown that the number of proteins identified in frozen tissue is similar to those found in FFPE tissue [129–132].

Ideally, tissue biopsies or other relevant tissue samples should be frozen immediately after excision/biopsy in liquid nitrogen or isopentane to preserve the sample's morphology and minimize protein degradation through enzymatic proteolysis [133]. Thin sections or, for most applications, 5–20-μm-thick sections are cut on a cryostat at −15°C (exact cutting temperature is tissue dependent) and thaw-mounted on an electrically conductive sample plate. The sections are dried in a desiccator for several minutes before MS [133].

Tissue proteomics has increasingly advanced due to recent technical developments that permit extraction analysis by MS from FFPE tissues [131], which makes analysis of large archives of kidney biopsy material feasible and allows assembly of sufficient sample sizes for more robust validation of biomarker discoveries in glomerular diseases [134]. Hence, an important step in protein MS analysis is sample enrichment. The simple idea here is that if a disease process is specific to a particular cell type, then biomarkers are more likely to be found when the sample is processed to enrich the cell type of interest. A convenient modality for cell-type enrichment is laser capture dissection [135, 136]. The value of laser capture microscopy is that allow us to enrich for structures in the kidney that are only affected by specific processes. Alternatively, proteins can be enriched by simple resolution on acrylamide gels. All of the above mentioned coupled with the dramatic increases in the overall sensitivity of the mass spectrometers used for protein MS, it has been possible to analyze smaller and smaller samples. A few thousand cells may be adequate to identify a few thousand proteins, allowing deep analysis of kidney biopsy samples [137, 138].

Laser capture microdissection (LCM) or laser microdissection (LMD) is a more recent developed technology that permits the isolation of single cells or single populations of cells from thin tissue sections (typically 5–10 μm in thickness) mounted on a glass slide [133, 135, 139, 140]. Proteins are extracted from target tissues and protein signatures are determined by MS after peptide separation using both gel-free and gel-based technologies [132, 134]. Using this technique (i.e., LMD/MS) in kidney biopsies allows us to obtain the protein profile of interest of a certain disease affecting a specific renal compartment such as glomeruli, interstitium, and tubules as well as to compare the results with the protein profile of the normal counterpart.

LMD/MS analysis is useful because it uses FFPE tissue instead of fresh, frozen, or other specifically stored tissue samples. Some examples wherein this type of technology can be used include problematic and challenging biopsies, such as those with equivocal Congo red staining, equivocal or negative light-chain staining on immunofluorescence studies, heavy-chain component, and less common forms of amyloidosis that cannot be accurately typed by immunohistochemical staining such as fibrinogen Aα, transthyretin (TTR), leukocyte chemotactic factor 2 (LECT-2), apolipoprotein A-I and A-IV, gelsolin, lysozyme, and β-2 microglobulin amyloidosis [139, 140]. Currently, LMD/MS is used routinely as an ancillary test for the diagnosis and typing of amyloidosis, and for the diagnosis of glomerulonephritis resulting from immune complex deposition and/or complement deposition [139].

Cell cultures also are of interest for mechanistic and molecular marker qualification studies. It is assumed that a cell on average expresses 10,000 proteins. If a cell culture contains multiple types of cells, then this number is higher. The preparation of cell cultures is simpler than that of tissue and often involves direct lysis of the cells in dish, after removal of the cell culture medium as a first step. After solubilization, the sample is transferred and sonicated. The following steps involve those as previously described for tissue sample preparation. Tissue samples, cells, and purified samples should be stored for long term at −80°C [125–127].

Techniques used in proteome analysis

The human genome consists of more than 20,000 protein-coding genes, but there could be more than 1 million different protein products, when taking into account splice variants and posttranslational modifications [141]. Therefore, all proteomics analyses start with some kind of protein separation [142].

Many different strategies are used to study the proteome of cells or tissues, all of which involve the use of MS technology. The choice of the technique is strongly dependent on the biological question to be addressed as well as on the availability and amount of fresh cells and/or tissues to be tested [89]. Each method is used to determine different types of information and has its own set of strengths and limitations [5].

Proteomics can be classified into three main fields: expression proteomics, bioinformatics analysis, and functional proteomics [103, 143]. *"Expression proteomics"* may be a necessary initial step to prepare the samples for proteomic analysis. This process includes extraction or isolation of proteins from cells and tissues, protein separation, protein identification, and protein quantification. *"Bioinformatic analysis"* expands the initial protein information obtained from expression proteomics and guides the future directions of functional proteomics. During this process, important information such as the primary, secondary, and tertiary structures can be obtained. However, the most important part of proteomic analysis is "*functional proteomics,*" which plays a key role in understanding the multiple functions of targeted proteins [103, 144, 145].

In other words, a proteomic analysis begins with protein extraction [146], followed by protein separation using either a gel-based (two-dimensional polyacrylamide gel electrophoresis; 2-D PAGE) or a gel-free method (liquid chromatography; LC). Separated proteins are then identified by various types of MS such as: matrix-assisted laser desorption/ionization time-of-flight mass spectrometry (MALDI-TOF MS), electrospray ionization (ESI)-TOF MS, and protein microarrays [105, 147, 148]. Bioinformatics is utilized to obtain additional protein information about candidate proteins to guide further functional proteomic study generating a new hypothesis [149]. The final results of a complete proteomic analysis are new hypotheses that can be addressed by functional proteomics and/or conventional molecular biology methods, better understanding of normal physiology and disease mechanisms, biomarker discovery, identification of new therapeutic targets, and drug discovery [145].

In depth, semiquantitative discovery proteomics are usually performed in small sample numbers due to the complexity and cost of analysis. Therefore, targets identified with these techniques are strictly hypothesis-generating and require additional quantitative validation in larger cohorts [7]. Hence, the ideal sequence for biomarker discovery would be MS-based discovery followed by enzyme-linked immunosorbent assay (ELISA)-based validation and clinical application [105]. Over the last years, profiling approaches that allow the use of MS-based techniques in the discovery/validation/clinical phase for the analysis of the urinary proteome have emerged: surface-enhanced laser desorption/ionization (SELDI)-TOF and capillary electrophoresis (CE)-MS [150].

In general, proteomics can be divided into two broad areas based on the detection methods used: (1) approaches using MS to detect and identify proteins and (2) approaches using arrays or ensembles of binding molecules to detect and identify proteins. The latter approach most commonly utilizes antibodies as the binding molecules [94].

Proteomics methods based on mass spectrometry

MS-based proteomic technologies have played an important role in each area of clinical diagnosis as well as the development of a more comprehensive understanding of the underlying disease process using a myriad of diverse sample types and techniques [151].

MS-based proteomics is an approach to studying protein expression, posttranslational modifications, and interactions that provides a wide range of possibilities for analyzing protein functions on a global level [152]. Technological platforms are indispensable in the proteome-wide research, which involve an initial protein separation step (i.e., gel-based or gel-free based) to reduce sample complexity. This is necessary to decrease the "noise" generated by high-abundance proteins, so peptide/protein identification with fingerprinting or tandem mass spectrometry (MS-MS) can be performed [7, 94].

First step: Protein separation

Gel-based proteome analysis

a. ***Two-dimensional polyacrylamide gel electrophoresis (2-D PAGE)*** is a powerful and widely used method for the analysis of complex protein mixtures extracted from cells, tissues, or biological fluids [98]. It was first described in 1975 by O'Farrell and Klose who, independently, developed the procedure [153, 154]. This technique permits separation and characterization of proteins according to their charge/ion strength and molecular weight, in two consecutive gel electrophoresis steps: proteins are first separated by isoelectric focusing (IEF), according to their isoelectric points, and then distinguished according to their molecular mass in sodium dodecyl sulfate (SDS) polyacrylamide gel electrophoresis (SDS-PAGE) [98]. Proteins are visualized and quantified by staining with silver, Coomassie blue, or fluorescent dyes [89, 92, 94].
b. ***Two-dimensional fluorescence difference in gel electrophoresis (2-D DIGE)*** represents a major improvement in this 45-year-old technique that has also been used to identify novel biomarkers in the renal field. The 2-D PAGE technique has been improved by staining of two samples and a pooled internal standard with different fluorescents dyes before electrophoresis. The two samples are simultaneously separated in the same gel. The individual proteins are visualized separately because of their discrete excitation and emission wavelength [155, 156]. This method provides an improvement of reproducibility and better quantification of proteins compared to the original technique [157, 158].

Several studies in AKI have used the 2-D DIGE technique. For example, in a case-control study of adult cardiac surgery patients, 2-D DIGE was used to identify four differentially regulated proteins in AKI versus no-AKI patients, including zinc-α2-glycoprotein, which was then quantified by western blot and ELISA to demonstrate an AUC of 0.68 [159]. Also, urinary exosomal fetuin-A was identified using 2-D DIGE in rat models of AKI, including ischemia reperfusion injury, cisplatin nephrotoxicity, and prerenal azotemia. Urinary exosomal fetuin-A is significantly increased in AKI, and these findings were evaluated in a small group of ICU patients [160]. 2-D DIGE has also been applied in other renal diseases to define specific biomarkers in the urine such as IgA nephropathy (IgAN) [161] as well as in membranous nephropathy (MN) [162].

Gel-free proteome analysis

Liquid chromatography (LC)

It is used increasingly in studies of urinary proteome. LC is particularly suitable when there is a requirement for high resolution and the need for low-abundance protein identification [89].

a. ***High-performance liquid chromatography (HPLC)*** involves dissolving the protein mixture in buffer and pumping it through a series of columns. The columns are composed of materials with various physical, chemical, and immunological properties, which bind different proteins with varying degrees of affinity depending on the complementary protein properties. The proteins can then be eluted from the columns. The properties on which separation can be based are numerous. The elements most frequently applied to urine are: size exclusion (proteins progress at rates based on their size), reverse phase (elution is based on hydrophobicity, with the more hydrophobic proteins separating first), strong and weak cation binding, and affinity binding (i.e., immunoglobulin adsorbing to protein of interest) [100].
b. ***Reverse-phase liquid chromatography (RP-HPLC)*** is increasingly used in proteomic studies and resolves peptides extremely efficiently. The equipment is routinely configured for direct and continuous spraying of column effluents into a mass spectrometer, which enables online detection of peptide ions and their fragmentation products. Elution of positively charged peptide ions by LC occurs in the order of increasing peptide hydrophobicity, in response to an escalating gradient of organic solvent (acetonitrile) under acidic conditions [100].
   Although RP-HPLC is the standard for separating peptides in proteomic experiments, prefractionation of peptides can markedly increase the number of proteins that can be identified. Prefractionation can be performed using different approaches based on: (1) charge, using cation exchange chromatography, (2) any of the principles used by LC such as reverse phase, anion exchange, hydrophobic interaction, and gel filtration media, which requires trypsinolysis of every collected protein fraction, and (3) chemical-tagging and separation technologies, which have been used to enrich protein or peptide subproteomes, particularly those with specific posttranslational modifications [163].
c. ***Nanoflow liquid chromatography (nLC)*** has the highest sensitivity and resolution, often allowing sequencing of peptides at low femtomol (fmol) levels (i.e., from proteins that are insufficiently abundant to be detected by PAGE and silver staining). Although sample volumes are very low with nanobore columns (low μL), this limitation is offset by using sample enrichment columns prior to nLC. Agilent Technologies has recently advanced nLC separation of peptides using a reverse-phase column scribed

with a laser onto a silicon wafer for direct spray into a mass spectrometer. This allows unprecedented resolution, sub-fmol peptide sensitivity, and sequencing of thousands of peptides in a complex sample [100].

Capillary electrophoresis. *Capillary electrophoresis* can be applied to intact proteins as well as digested proteins, similarly, as previously described for HPLC method. In a silica capillary, proteins or peptides are separated as a function of charge at a desired pH by an electric field in with the capillary is housed [118]. Although CE is a powerful separation technique, it does not yield reliable quantitative information [100].

Second step: Protein identification

Two-dimensional gel electrophoresis coupled with mass spectrometry (2DE-MS). Two-dimensional gel electrophoresis (2DE) coupled with mass spectrometry (2DE-MS) is still the most accessible technique that allows the study of large molecules; it has been used on numerous occasions for the description of urinary proteome [123].

In both forms of 2DE, the intact proteins are separated before digestion and identification. Trypsin digestion breaks a given protein into a unique series of peptide fragments. These fragments are identified by MS [94]. However, 2DE-MS has the disadvantage that the reproducibility is low, time of analysis is long, and the technique is difficult to automate [98].

A major strength of two-dimensional electrophoresis (2DE) for proteomic studies is its capacity for quantitation [152]. However, there are several limitations of 2DE as a separation method for proteomics studies. For example, owing to the incompatibility of the IEF step with ionic detergents, such as SDS, which are required to fully solubilize proteins and break up protein complexes, hydrophobic proteins (especially integral membrane proteins) often do not enter the gel [92, 94, 98]. Furthermore, the range of molecular weights that is resolvable on the second dimension of the gel is limited, eliminating very large and very small proteins from detection [92, 94, 98]. Highly acidic or highly basic proteins may also be lost. In addition, there is strong bias toward high-abundance proteins rather than low-abundance proteins, even though the latter may play critical regulatory roles in a given tissue [92, 94, 98]. Finally, the technique is labor intensive and requires extensive training to achieve reasonable gel-to-gel reproducibility [94, 141]. Although still in use because of its simplicity, the technique is not applicable in clinical settings [98].

Liquid chromatography coupled to mass spectrometry (LC-MS/MS). Tandem-MS can be coupled with HPLC to form a powerful technique for protein and peptide analysis. Complex protein mixtures can be digested without initial separation, and the resultant peptides can be analyzed by micro-HPLC-MS/MS. In micro-HPLC, a capillary column with a flow rate <1 μL/min is used, which provides a very high sensitivity and separation capability. The signal strength of the electrospray/ionization (ESI) is independent of flow rate, making the application an ideal interface for tandem-MS and micro-HPLC. The system can be set to find peptide peaks of interest and switched to MS/MS mode to obtain their MS/MS spectra (representing part of the amino acid sequence) as the peptides are eluted from the column. Hence, hundreds of proteins can be identified by a single analysis [89].

The separation capability can be further increased by 2DE-HPLC-MS/MS, where the total protein sample is digested, acidified, and subjected to cation-exchange chromatography, followed by gradient reverse-phase capillary electrophoresis (RP-HPLC) that enables the separation of peptides based on their hydrophobic properties. The eluted peptides are subsequently analyzed in a mass spectrometer [89].

Some of the limitations of HPLC include that the analysis can be very time consuming, potentially taking an entire day to complete a single sample despite providing an optimal protein coverage in complex samples. Hence, LC-MS/MS is not recommended as profiling proteomics technique for large series of prognostic or diagnostic samples [89, 105].

Several studies in urine proteome using LC-MS/MS have led to a better understanding of the pathogenesis of certain diseases. For example, a study performed in patients with Fanconi syndrome more than 100 different peptides were detected using nLC-MS/MS in urine samples, which includes carrier proteins transported by megalin/cubulin, cytokines, and complement components. In another study involving patients with Dent's disease, carrier proteins transported by megalin/cubulin were also detected at higher concentration than normal in the urinary proteome [165, 166].

Capillary electrophoresis coupled with mass spectrometry (CE-MS). This method provides a relatively fast analysis, within one hour, and high resolution in small volumes [96, 113, 167]. It is rather robust and compatible with most of the buffer and analytes, and it provides a stable constant flow avoiding elution gradients that may interfere with MS detection [92].

A disadvantage of CE is that the analysis is restricted to low-molecular-weight proteins as the larger proteins tend to precipitate at the low pH generally used in the running buffer. This might be seen as a limitation, but it is of little consideration for the urine analysis as the urinary proteome contains a high percentage of low-molecular-weight proteins [168]. Another potential disadvantage is that only a relatively small amount of volume can be loaded onto the capillary, leading to a potentially lower sensitivity of detection. However, improvement of both coupling and the detection limits of mass spectrometers enables detection within the low amol range, making this issue less relevant [169]. Moreover, despite the low amount of sample volume that can be loaded, sequencing of CE-MS-defined biomarkers can be performed by direct interfacing of CE with MS/MS instruments [170] or by subsequent targeted sequencing [171].

Capillary electrophoresis mass spectrometry (CE-MS) has been used to identify unbiased biomarkers in AKI and CKD [172–174]. For instance, CE-MS has been used to identify 20 urine peptides that were the degradation of six proteins from a critical care model of AKI. In this study, adult AKI patients were found to have increased albumin, α1-antytripsin, and β2-microglobulin peptides as well as decreased fibrinogen-α, collagen 1α (I), and collagen 1α (III) peptides [172]. In another study performed by Rossing et al., they analyzed the urine of 305 individuals to find urinary biomarkers in diabetes using high-resolution CE-ESI/MS of urine. A panel of 40 biomarkers distinguished patients with diabetic nephropathy from healthy individuals with 89% sensitivity and 91% specificity [173]. Moreover, Devarajan suggested that some candidate plasma proteins associated with CKD, such as asymmetric dimethylarginine, adiponectin, apolipoprotein A-IV, fibroblast growth factor 23, NGAL, and the natriuretic peptides as well as urinary N-acetyl-β-D-glucosaminidase could serve as discriminatory biomarkers for CKD progression [174].

Matrix-assisted laser desorption/ionization-time of flight coupled with mass spectrometry (MALDI-TOF/MS). Another type of MS that is employed is matrix-assisted laser desorption/ionization (MALDI)-time of flight (TOF) [164]. With this type of approach, up to 96 individual protein samples are spotted onto a stationary target for analysis. Although the sensitivity of MALDI-TOF is limited to characterizing the 10–15 most abundant proteins in each sample, it has the benefit of being a higher-throughput platform for the lower complexity, prefractioned protein mixtures (after gel separation) because each sample is analyzed in minutes; whereas a typical one-dimensional LC run requires an hour and 2-dimensional LC-MS/MS takes 10–12 h [151].

Solid-phase chromatography coupled with mass spectrometry (SELDI-MS). Surface-enhanced laser desorption/ionization (SELDI) coupled with MS (SELDI-MS) is a useful high-throughput screening technique that facilitates relative abundance profiling of individual proteins from different samples such as serum, plasma, urine, and cellular lysates from freshly frozen tissue specimens, including LMD-cells. The relative speed and simplicity of SELDI-MS gives its potential for use in hospital diagnostic laboratories [89, 100, 175].

This approach is a modification of MALDI-TOF, where this technique incorporates chromatographic and MS principles in a single platform [176]. It measures protein ions after the proteins are selectively bound to a plate coated with an affinity surface (binds proteins on the basis of their chemical and physical properties); unbound proteins are washed off. The bound proteins are directly analyzed by MALDI-TOF-MS (SELDI-TOF-MS) [98]. A subset of the proteome is thus selected and the chip plugs directly into the mass spectrometer for analysis. Various chromatographic surfaces (protein CHIP surfaces) are available [89, 100]. Peak height differences between samples correlate with relative abundance in the sample. However, identification of the protein represented by the peak can be difficult [92].

Although the approach can be extremely useful for screening peptide/protein samples for recognition of biomarker ions, it does not enable the protein origin of these ions to be reliable discerned. Another limitation is that a very small fraction of all proteins in a sample binds to the chip surface; therefore, only a fraction of the information contained in a biological sample can be exploited for the presence of biomarkers, even if there are a number of different chip surfaces available. In addition, binding to the different chip surfaces varies depending on sample concentration, pH, salt content, and the presence of interfering compounds [89, 105]. Furthermore, SELDI TOF-MS is optimized to the low-molecular-weight range and has poor resolution for high-molecular-weight proteins [144]. Moreover, this technology is prone to generating artifacts [177, 178], possibly due, in part, to difficulties with calibration and lack of precision of the determined molecular masses of the analytes [105], and to variation of performance between different machines (as does the performance of a single machine over time) [100].

SELDI-TOF results differ from LC-MS/MS and MALDI-TOF-TOF in that the results are given in mass to charge ratios (m/z) rather than peptide sequence, so positive protein identification is not possible. It is, however, useful for rapid analysis of the protein m/z profiles of semicomplex samples by reducing upfront separation while preserving the fast analysis time of a MALDI platform. Although not as desirable for discovery phase, the attributes of relative ease of sample preparation, speed of analysis and data output as well as lower startup and operation costs makes SELDI-TOF a more suitable MS platform for a clinical test [151].

SELDI TOF-MS has been one of the technical approaches used to detect novel biomarkers in AKI, CKD as well as in renal transplantation [83]. In a prospective study, SELDI TOF-MS demonstrated a 2.78-kDa protein peak (hepcidin-25), which was differentially upregulated at postoperatory day 1 in non-AKI vs. AKI patients. This protein was isolated with ion exchange chromatography, purified by RP-HPLC and identified with MS-MS [179]. Other prospective studies validated this result in urine, demonstrating that urinary hepcidin-25:creatinine ratio is strongly and independently associated with avoidance of AKI and preservation of renal function [180–182].

Differential protein profiling. Differential protein profiling is a powerful method for identifying differentially expressed proteins. This affinity tag adds a predictable mass/charge separation and therefore allows for the determination of relative abundance at the same time as protein identification [7]. Some examples of differential labeling include isotope-coded affinity tags (ICAT), O-methylisourea, and isobaric tag for relative and absolute quantitation (iTRAQ). This technique can be limited by the extent to which the label is incorporated in a sample. For example, ICAT labels are incorporated at cysteines, *O*-methylisourea is incorporated at trypsin digestion sites, and iTRAQ labels are incorporated at N-terminal or lysine side chains. Therefore, relative abundances will not be determined for peptides that do not have these specific labeling sites [7].

Differential protein profiling has been used in several studies to identify proteins. For example, chitinase 3-like protein 1 was identified in a rat model of sepsis [183]. In this study, rat septic AKI and non-AKI urine samples were compared by labeling with light versus heavy isotopes using propionylation. The light and heavy isotope-labeled samples were combined and subjected to MS-MS for identification of differentially regulated proteins [183].

a. ***Isotope-coded affinity tags (ICAT)***. It is the most common isotopic labeling of peptides used in LC/MS. This contain three functional regions: an affinity purification region, a peptide-binding region, and an isotopically distinct linker region [184]. A biotin tag is used for affinity purification. A thiol-specific binding moiety is used to covalently link the reagent to cysteine residues in a target peptide. The intervening linker region is isotopically labeled with either ^{12}C or ^{13}C so that peptides labeled with the reagent are chemically identical but can be distinguished in a mass spectrometer based on their mass differences. This advance allows the mass spectrometer to quantify protein abundance differences in two samples [185], but it is limited to labeling only peptides containing the amino acid cysteine [92].
b. ***Isobaric tag for relative and absolute quantitation (iTRAQ)*** is a similar technique that has been introduced. This method labels all peptides allowing increased confidence in the identification of proteins because multiple peptides for the protein are identified, and permits simultaneous quantification of four samples [186].

Mass spectrometry. Over the last few years, MS has seen growing use in the identification of endogenous antigens that are the targets of autoantibodies. For example, a major breakthrough was achieved when two podocyte cell surface proteins, phospholipase A2 receptor (PLA2r) and thrombospondin type 1 domain-containing 7A (THSD7A), were identified in membranous glomerulonephritis [187, 188]. More recently, a novel proteomic biomarker was found for fibrillary glomerulonephritis called DnaJ homolog subfamily B member 9 (DNAJB9), which is a member of the chaperone gene family [189, 190]. Similarly, LDL receptor-related protein 2 (LRP2), also known as megalin, has been identified as a target antigen in antibrush border antibody (ABBA) disease, a newly described form of AKI [191].

A mass spectrometer is an instrument that measures mass-to-charge ratio (m/z) of ions in a gas phase, which is the base of various strategies for identification and characterization of proteins in complex mixtures [94, 152]. Modern mass spectrometers allow identification of peptides and proteins at sub-fmol ($<10^{-15}$ mol) levels, thus, potentially enabling the study of low-abundant proteins, such as most proteins involved in signal-transduction process [152].

Two fundamental mass spectrometric approaches can be used to identify the protein using the trypsin digest: (1) peptide mass fingerprinting, usually employing MALDI-TOF mass spectrometers; applicable to solid-phase samples spotted on a laser target plate and (2) ESI, which is applicable to solution-phase samples, commonly sprayed in a continuous stream from an LC or CE column [100, 152].

(a) Peptide mass fingerprinting. Peptide mass fingerprinting is usually accomplished using MALDI-TOF technology. In this type of ionization technique, a laser beam excites the solid matrix containing the analyte molecules. The ionization reaction takes place in the desorbed matrix-analyte cloud, just above the surface. The ions are then extracted into the mass spectrometer for analysis [152]. The mass analyzer portion of the instrument works on the TOF principle, in which TOF of an ion from the source to the detector is proportional to the mass-to-charge ratio. The resulting spectrum of signal intensity versus m/z represents a fingerprint of the original protein [94].

Peptide mass fingerprinting identifies a protein by measuring the molecular masses of all major trypsin products and matching these molecular masses with databases of theoretical sizes of trypsin fragments from unknown protein sequences [94]. Trypsin cleavage sites are predictable (cleavage after lysines and arginines); therefore, a known protein sequence can be readily converted to a unique set of peptide masses by computer analysis [94].

(b) Electrospray ionization. ESI is the most common soft ionization technique employed for proteomics [192]. A solvent containing ionized analyte is pumped through a very small, charged capillary. The generated aerosol contains droplets of solvent and analyte. Following evaporation of the solvent, the ions move to the mass analyzer of the mass spectrometer [152].

Nano ESI (nESI) produces droplets up to 1000 times smaller than generated by high flow ESI, which leads to improved evaporation, more efficient sample transfer, and potentially more ions detected [193]. The best-performing nESI sources use a smaller internal diameter for higher ionization efficiency. Utilization of nontapered emitters with an internal diameter in the range of 10–30 μm provides an adequate compromise between sensitivity, system robustness, and sample consumption [194–198]. Wider internal diameters reduce potential clogging and facilitate more stable ionization spray, resulting in more reproducible mass spectra and ion chromatograms [194, 195, 197, 199].

(c) Peptide sequencing using tandem mass spectrometers. These instruments combine two mass analyzers in series making them a more powerful proteomic technique [94, 100]. The first stage is conceptually like the instruments used for peptide mass fingerprinting as described earlier. The second mass analyzer is most commonly a TOF device. The differences in molecular weight between successive fragments identify the specific amino acid species by comparison with the known residue masses of the various amino acids. In this manner, the amino acid sequence of an individual trypsin fragment can be read from the product ion spectrum, and the protein can be identified by comparing this sequence with databases of protein sequences using algorithms such as BLAST and FASTA [94]. "TOF-TOF" provides additional information on fragmentation of a parent peptide ion, enabling elucidation of the amino acid sequence for definitive protein identification [94, 100].

Applications of mass spectrometry-based proteomics. MS-based proteomics approaches can be used in a conceptually simple way to survey a given tissue or organelle to identify its proteome, i.e., generate a list of expressed proteins. However, the rapidly developing methodology of this approach has enabled the quantitative and qualitative characterization of proteins in a large scale as well as the regulatory processes manifested at a protein level, which include posttranslational modifications and interaction partners [94, 152].

(a) Abundance profiling. Currently, there is a great interest in learning to detect changes in protein abundance in a given tissue in response to a given physiologic or pathophysiologic change. Difference between healthy and diseased tissues or between drug-treated and drug-untreated cells can be measured using different proteomic approaches. In the most widespread and simplest method for quantitation of protein, abundances are measured by densitometry of stained 2D gel, before MS identification of the protein [152]. An early success of this type of approach utilized comparative analysis of two-dimensional gels to identify proteins whose expressions are increased or decreased in cyclosporine A nephrotoxicity [200]. Such comparisons are often difficult because of gel-to-gel variability, requiring complex algorithms to manipulate gel images so that they are in precise alignment. Moreover, because identification of gel-based proteins is limited to proteins of high or intermediate abundance, alternative MS-based methods have been developed. Overall, they can be classified as methods that are either label-free or that are based on incorporating stable isotopes [201].

In the label-free approach, the intensities of mass spectral peaks are typically compared directly between samples, based on their mass and retention times [202, 203]. This type of approach is well suited for multiple replicate analysis;

however, they require that the sample-preparation and analytical systems be robust and reproducible and that sophisticated data-processing capabilities exist to support the analysis [152].

Another quantification method that is observed in increasing use is the ***isotope-coded affinity tag (ICAT)*** labeling of proteins [184]. With this method, two protein samples (experimental and control) are derivatized at cysteinyl residues using avidin-containing reagents that are chemically identical but differ in molecular mass owing to the presence of eight deuteriums replacing eight hydrogens in one reagent. The derivatized samples are mixed together and subjected to trypsinization, and the labeled peptides are separated by avidin affinity chromatography. The purified, derivatized peptides are analyzed by microcapillary liquid chromatography linked to a tandem mass spectrometry (iLC-MS/MS) in dual mode. One mode analyzes the relative abundance of the same peptide from the two samples (discriminated on the basis of the deuterium-labeling of one), and the other mode sequences the peptide to determine its protein of origin. As noted earlier, this technique has been exploited already in large-scale analyses of protein expression in yeast [204].

Multiplexed isobaric tagging (iTRAQ) represents another group of methods that uses a chemical tagging reagent that allows multiplexing of samples. In single MS mode, the differentially labeled versions of a peptide are indistinguishable. However, in MS-dual mode (in which peptides are isolated and fragmented), each tag generates a unique reporter ion. Protein quantitation is then achieved by comparing the intensities of the four reporter ions in MS- dual mode spectra [152].

(b) Posttranslational modifications. In addition to studies of protein abundance, MS can be used for large-scale investigations of protein modifications. Of considerable interest are large-scale detections of posttranslational modifications, including phosphorylation, acylation, ubiquitination, glycosylation, among others [94, 152]. The greatest progress has been made in development of techniques to investigate targets for protein phosphorylation. The most fundamental approach is to label proteins from experimental and control samples with ^{32}P and then to compare autoradiograms of 2D gels to identify proteins spots in response to the experimental manipulation. This approach has been successful but can misidentify phosphorylated proteins if they overlie more abundant related proteins. Therefore, success with this method requires confirmation of the phosphorylation by tandem MS or other approaches. An alternative approach is to carry out an initial step that enriches the phosphorylated proteins in the sample [94, 152]. This can be done, for example, by immunoprecipitation with phosphorylation-specific antibodies before mass spectrometric identification. Although this approach works well for identification of proteins phosphorylated on tyrosines, it has not been successful for isolation of proteins phosphorylated on serines and threonines. Consequently, alternative techniques have been developed for affinity isolation of phosphorylated proteins by covalent attachment of affinity tags at sites of phosphorylation [205, 206].

*(c) **Protein-protein interactions***. Cell function is dependent on a myriad of protein-protein interactions. The yeast two-hybrid method is an approach for the identification of protein-protein interactions that has been widely employed [207]. On this technique, two yeast strains containing two different cDNA libraries are mated. One has its open reading frames fused with the DNA-binding domain of GAL4, and the other has its open reading frames fused with the activation domain of GAL4. A GAL4-dependent reporter gene is expressed only when the DNA-binding and activation domains of GAL4 are brought together by binding of the proteins to which they are fused. The DNA inserts from the interacting plasmids are sequenced to identify the interacting proteins. An example of application of this technique was to identify Nedd4 as a binding partner for the aldosterone-regulated epithelial sodium channel (ENaC) [208]. Nedd4 is believed to decrease the protein half-life of ENaC in the kidney by ubiquitinating one or more ENaC subunits, a process that has been demonstrated to be down-regulated by the aldosterone-activated kinase sgk [209, 210]. However, the limitation on this approach is that it only identifies binary interactions. Many or most protein complexes consist of more than two proteins; therefore, alternative approaches have been under development to study multiprotein complexes. For example, this can be done by immunoprecipitation or by affinity isolating proteins-of-interest after expressing them as fusion proteins that include an affinity tag such as glutathione S-transferase (GST). Proteins present in affinity-isolated complexes can then be identified by MS. MS-based identification of protein-protein interactions has begun to be carried out at large scale in yeast, identifying thousands of potential protein-protein interactions [211, 212]. A disadvantage of this approach is that protein overexpression in cells may alter binding patterns relative to those present at the native level of expression [94].

Proteomics methods based on ensembles or arrays of binding proteins

A widely heralded new approach to proteomic analysis is the use of ensembles or arrays of binding proteins (or other macromolecules) to identify individual proteins in complex protein mixtures [213].

Ensembles of antibody can be chosen to identify known members of certain populations of proteins in a "targeted proteomics" approach. Examples include antibodies to proteins involved in cancer-related signaling [214], antibodies to proteins involved in cell-cycle regulation and apoptosis [215], and antibodies to sodium transport proteins expressed along the renal tubule [216–218]. Alternatively, large numbers of affinity matched antibodies can be obtained from recombinant sources such as phage display libraries. Such antibody ensembles can provide broad coverage of epitopes found in large populations of proteins, allowing individual proteins to be identified by matching patterns of epitope recognition [94].

Targeted proteomics approaches. Cloning of cDNAs for specific proteins involved in physiologic processes has made possible the production of high-quality polyclonal or monoclonal antibodies to them, using synthetic peptides as immunogens, obviating the need for protein purification. This technological development has allowed investigators to accrue comprehensive sets, or "ensembles," or antibodies against proteins relevant to a given physiologic process. The technical approaches used to detect changes in abundance of proteins have included ***multiplexed immunoblots*** and ***antibody microarrays*** [94].

(a) Multiplexed immunoblots. If the number of proteins in a targeted protein population is relatively low (<20), it is advantageous to assess abundance changes through quantitative or semiquantitative immunoblotting [217].

Semiquantitative immunoblotting (with proper loading controls) allows precise simultaneous comparison of multiple samples, whereas antibody microarrays generally permit only binary comparisons. In addition, in this approach, the throughput can be increased through multiple probing of the same immunoblot with several antibodies, applied either sequentially or simultaneously. In general, when probing with several antibodies simultaneously, it is necessary to choose antibodies that target proteins with different molecular weights to avoid overlap. In addition, blots prepared from two-dimensional gels have been successfully probed with as many as nine antibodies simultaneously [219]. An alternative approach that can detect different proteins simultaneously employs antibodies conjugated with different fluorophores and quantification of fluorescence by using a molecular imager [220]. With this approach, several proteins can be detected even when bands overlap [94].

An example of this approach has been applied to carry out a systemic identification of renal tubule transporters that are targets for adaptive regulation by hormonal and nonhormonal regulatory factors involved in the control of BP and extracellular fluid volume [221, 222]. Additional studies using the same ensemble of antibodies have identified Na transporters dysregulated in animal models of various disorders of NaCl and water balance, including chronic renal failure [223], cirrhosis [224], lithium-induced nephrogenic diabetes insipidus [225], syndrome of inappropriate antidiuretic hormone secretion [226], primary aldosteronism [218], and vitamin D-induced hypercalcemia [227].

(b) Antibody microarrays. Protein-binding arrays use antibodies or synthetic protein-binding small molecules like peptoids [228, 229], or aptamer [230] to visualize protein binding. Using array technologies, hundreds of antibodies can be robotically positioned on a glass slide or nitrocellulose; the latter offers a longer storage life. The sample of interest is then incubated with the slide, unbound proteins washed off, and bound proteins can be detected using either an ELISA or a colorimetric reaction. Problems with this method are the inaccuracy of quantification, the fact that only a limited subset of the proteome is studied, and that the technique is dependent on the antibodies available [100, 231].

An approach identification of differentially expressed proteins in two complex mixtures of proteins using the antibody microarray concept was reported by Haab et al. where lysine moieties in two protein samples were derivatized with two different fluorescence dyes, Cy3 and Cy5, using *N*-hydroxysuccinimide-e-ster-activated dyes. The samples were mixed and exposed to a microarray containing 115 antibodies spotted onto poly-L-lysine-coated glass slides with a robotic microarrayer. The fluorescence signal associated with the bound proteins was quantified using a standard microarray reader, allowing signal normalization and calculation of relative protein abundances for the two samples for each of the proteins on the array [232].

Another study demonstrated the application of the antibody array wherein tissue lysates were biotinylated and exposed to an antibody microarray consisting of 368 antibodies spotted on a glass slide. The antibody ensemble was made up of commercially obtained antibodies to proteins involved in cancer cell growth, including many extracellular and intracellular matrix proteins. After washing the exposed arrays, the antibody-immobilized biotinylated proteins were bound to a streptavidin-horseradish peroxidase conjugate and were detected by chemiluminescence using a standard flatbed scanner to generate array images. With this technique, the two samples to be compared were run on separate arrays, and the images were compared electronically to determine which proteins are differentially expressed [214].

Targeted aptamer-based array chips (DNA-tagged antibodies) is another promising development. Aptamers are highly structures oligonucleotides, which are selected from combinatorial libraries of synthetic nucleic acid by an iterative process referred to as systematic evolution of ligands by exponential enrichment [233, 234]. They can specifically bind to a wide variety of targets ranging from small organic molecules, proteins to supramolecular structures. In general, they show less crossreactivity, have a wider linear range than antibodies, and can be reproducibly manufactured in compliance with the rules and regulations of good manufacturing practices [83].

A recent study by Carlsson et al. showed the associations between the decline in estimated glomerular filtration rate (eGFR) and 92 plasma proteins from an established cardiovascular panel. They used two well-characterized Swedish cohorts of elderly community-based subjects followed for 5 years. The eGFR was calculated from cystatin C, and plasma proteins were measured by a proximity assay that utilized DNA-tagged antibodies (aptamers) binding in close proximity to the same protein, followed by polymerase chain reaction (PCR)-based amplification and quantification. In the discovery cohort, a total of 28 proteins were significantly associated with eGFR decline after adjusting for demographics, lipids, blood pressure (BP), glucose, and treatment of these risk factors. Twenty of these proteins were significant in the replication cohort, and 11 of them also consistently predicted incident CKD. These proteins shared several common pathways that give important information on pathophysiologic mechanisms in CKD progression: phosphate homeostasis, inflammation, apoptosis, proteolysis and extracellular matrix remodeling, angiogenesis, endothelial function, and thrombosis. However, additional adjustments of eGFR at baseline rendered all proteins nonsignificant. The authors, therefore, conclude that their proteomic profiling did not identify a promising biomarker with clinical utility for risk prediction. The study poses some general limitations on generalizability, survival and selection bias, and single assessments of eGFR and proteomics; however, the strengths of the study should also be mentioned, including the use of separate discovery and replication cohorts and a highly specific analytical technique with very good intra- and interassay coefficients of variation [233, 234].

The major limitation of antibody microarrays is the technical approach to antibody production. The conventional means of antibody production, through immunization of animals, is expensive and time consuming. The antibodies produced are highly variable in affinity and selectivity, making it difficult to choose a single condition for antibody microarray incubations that will optimize antibody-binding reactions for all the antibodies spotted on an array [94].

Proteomics raw data processing and feature identification

Identifying features (cognate proteins/peptides and metabolites) and abundance (relative intensities) from raw MS data files are fundamental steps in the analysis pipeline. The process of converting MS precursor ion scans and associated MS/MS fragmentation spectra into identified proteins and metabolites is different for each data type and is addressed using different software tools. Therefore, bioinformatics becomes an important step for data processing and identification [100].

Bioinformatics is the computational branch of molecular biology and it can be envisaged as the application of computerized technology to mapping genome, proteome, and metabolome. All this -omic technologies require bioinformatic mining of data from two basic types of experiments: shotgun proteomic studies, and attempts to identify biomarkers (e.g., from SELDI and CE-MS experiments). The large amounts of information generated by mass spectrometers require sophisticated bioinformatic tools for analysis. List of peak m/z values, ion intensities, fragment ion m/z values and chromatographic elution times must be catalogued and statistically compared—both with previously acquired data and with databases of theoretical protein digests and their theoretical fragmentation patterns [100].

Analyses of shotgun proteomic data are supported by an array of broadly applicable approaches and commercially available software packages that have been reviewed elsewhere [235]. The second more specialized application of bioinformatics to urinary proteomics is classification of data obtained by SELDI or CE-MS [118, 236, 237].

The primary aim of these computational efforts has been to identify spectral patterns that are diagnostic of a disease or physiological condition, without necessarily identifying the protein masses or elucidating the pathophysiological process. Data for this application share two complicating characteristics: the number of variables (proteins) analyzed far outweighs the number of patients in the studies, and, unfortunately, diseases are very unlikely to be defined by a single novel protein with a unique mass charge [100].

The classification methodologies that have become available as a result of computerization have been adapted and applied to abundance profiling, which is semiquantitative at best. Several independent analytical techniques are available to be applied to the data [100].

For proteomics, the peptide/protein search space is inherently limited by biology (genome of the particular species) and chemistry (peptide bond constraints). Peptide database files for many species are available for download from sources such as Uniprot and have well-defined rules for polymeric structure and fragmentation patterns [238]. MaxQuant is a popular and relatively easy to use full-featured tool, which runs the Andromeda search engine as well as associated quantification tools. Other powerful software tools for searching MS/MS spectra data to identify peptides include the popular search engines MS-GF+, Comet, SpectraST, Mascot, XTandem, MyriMatch, Sequest, and Tide [239–245].

Validation of peptide identification is often estimated through empirical false discovery rate (FDR) using a reverse "decoy" database search [246, 247]. Many additional tools are used to interface with these search engines to assign intensity values and to format data for subsequent statistical analysis [248–252].

Several community tools such as the Trans-Proteomic Pipeline, Peptide Shaker, and OpenMS, offer extra functionality that improve overall performance, including permitting running multiple search engines to compare and compile results [245]. Open-source software, such as OpenMS, also provide powerful configurable features, while retaining a simplified user interface even when running complicated analysis pipelines [251].

Proteomics in renal research and nephrology

In recent years, there has been significant progress in the identification of candidate protein markers for the diagnosis of acute and CKDs [253–255]. Many of these protein markers are of potential interest not only for diagnosis but also for monitoring patients with certain renal diseases [83].

Research on new biomarkers have been supported by health agencies like the National Institutes of Health (NIH) and the Food and Drug Administration (FDA), and these have also unveiled a framework of qualification of biomarkers. For example, antiphospholipase A2 receptor (PLA2r) is currently the only FDA-approved CKD biomarker discovered by a proteomic approach. Beck et al. combined serum from patients with membranous GN with healthy glomeruli from kidney donors, modified the Western blotting method slightly, excised the "suspicious" band from electrophoresis gel, digested the proteins with trypsin, and analyzed the resulting peptides with LC-MS [187]. PLA2r, located on the podocytes, has been recognized as the major antigen and has later been used in clinical trials and clinical practice to differentiate primary from secondary forms of MN, risk stratification, and to monitor response to treatment. Moreover, the test has good diagnostic accuracy (sensitivity of 0.78; specificity of 0.82; positive likelihood ratio of 4.38; negative likelihood ratio of 0.26; and AUC of 0.82 for patients with nephrotic syndrome before immunosuppressive treatment) [256]. Furthermore, immunohistochemical staining for detection of PLA2r has been developed and is now used in kidney biopsies in order to differentiate primary from secondary forms of membranous GN. This stain is an easy and quick test that can be performed in any pathology laboratory. Moreover, the PLA2r stain is even helpful when patients have a false-negative serology test for PLA2r, which can happen in about 15% of patients at clinical presentation or at early stages of the disease [257, 258].

First approaches of using urinary proteomics as diagnostic tool were reported more than 10 years ago [118]. Although these first studies were underpowered, they already highlighted the potential of urinary proteome analysis to support diagnosis of CKD and etiology. Some of these studies demonstrated specific urinary biomarkers for IgAN that could distinguish them from other patients with MN and healthy controls [118, 259, 260].

The need for detection of biomarkers by urinary proteomics has also been applied in renal transplant patients to detect early acute and chronic rejection [261–264], but it will require further study in the appropriate patient populations. Although most of the studies have been done in rat models, proteomics has been used to study the toxicity of immunosuppressants [265] as well as acute rejection [266] and chronic allograft injury [267].

In addition, the use of urine proteomics as a tool in CKD has been considered for early detection as well as for prognosis of progression. Several studies, although some in small cohorts, have demonstrated the benefit of urinary proteomics for this purpose [173, 268–273]. For example, a new proteomic classifier based on 273 urinary peptides, termed CKD273, has been used to assess early detection of CKD in diabetic patients and for prognosis of progression [268, 269, 272, 274].

Moreover, many of the biomarkers that are employed in (early) detection can also inform the response to therapy. Some examples of the studies that have demonstrated this include the response to angiotensin receptor blocker (ARB) in diabetic nephropathy [275, 276], and response to immunosuppressive treatment in antineutrophil cytoplasmic antibody (ANCA)-associated vasculitis [259].

More recently, the FDA issued a biomarker letter of support encouraging further development of the CKD273 urinary peptides developed by Mischak and Schanstra on the basis of capillary electrophoresis MS technology [277]. The panel has been tested, over the past 8 years, on various CKD diagnoses, but most extensively in diabetic kidney disease (DKD), and has shown promising diagnostic and prognostic value. The best performance of CKD273 has been found in early CKD (stages 1-3a) [278]. CKD273 has outperformed urinary albumin excretion rate in predicting overt albuminuria, kidney function decline, and future ESRD [279]. Important panel components are diverse collagen fragments found in reduced amounts and, most importantly, in early CKD, indicating increased collagen deposition [173]. Later CKD stages also include increased levels of blood-derived proteins, indicating inflammation and filtration barrier damage (fibrinogen, α-1-antitrypsin, apolipoproteins, etc.) [141]. CKD273 is a very promising panel for kidney disease because these small peptides originate from the kidney and are stable in earlier stages of CKD, making them more reproducible by MS. The panel is currently available for patients with diabetes in Germany [141].

Applications of proteomics methodologies are growing importance in drug discovery, largely, because drugs may act by mechanisms manifested at the level of protein function rather than gene transcription changes. Thus, adverse effects of drugs in many cases result from direct interaction with proteins, without direct manifestations at a nucleic acid level [280, 281].

The integration of proteomics early in the drug development process could improve the rate of clinical approval success in clinical/pharmacological trials. First, validation of the drug targets in human physiology and disease (i.e., comparing proteomic signatures identified preclinically with human phase 1–2 studies) is crucial as differences between animal models and clinical medicine are more often the rule rather than the exception. Proteomics can also be of help in pharmacokinetics and pharmacodynamics studies. Second, proteomics could distinguish responders from nonresponders, identify subjects with toxic reactions, and potentially serve as surrogate end points. In total, it is hoped that proteomics, accompanied by other omics, could contribute to more streamlined inclusion and exclusion criteria, smaller sample size requirements, and improved efficiency of clinical trials. This is, however, a challenging process and only a few pharmacoproteomic studies have been conducted, most in the field of oncology [280].

In summary, proteomics is expected to give important contributions to clinical nephrology in the coming years. Admittedly, similar statements have repeatedly been put forward over the last decade, but the technology is now further refined, simplified, and more available, and so hopefully we will begin to see proteomics-based diagnostic and therapeutic products incorporated into clinical care [141].

Metabolomics and the kidney

Metabolome is a quantitative descriptor of all endogenous low-molecular-weight components (small molecules defined as <1500 Da) contained in a biological sample such as urine or plasma [83, 282]. They participate in general metabolic reactions and are required for the maintenance, growth, and normal function of a cell. The "metabolome" refers to the complete set of metabolites, such as metabolic substrates and products, lipids, small peptides, vitamins, amino and fatty acids, nucleic acids, carbohydrates, organic acids, polyphenols, intermediates of many biochemical pathways, other protein cofactors, and inorganic and elemental species, which are present in an organism [83, 144, 283].

Metabolomics, the latest of the "omics" sciences, refers to the systematic study of metabolites and their changes in biological samples due to physiological stimuli and/or genetic modifications [284]. In contrast to the genome and proteome that are restricted to an individual organism, the metabolome is an open system. Because the metabolites represent the downstream expression of genome, transcriptome, and proteome, they can closely reflect the phenotype of an organism at a specific time [285] and they can measure the interactions with food, medications, environment, and gut microbiome [88, 286].

Metabolomics is a rapidly expanding area of scientific research. It can reflect various phenotypes at a functional level and is in close relation to the disease pathogenesis [287]. However, bioanalytical metabolomics assays are more challenging as the metabolome covers a wide range of compounds with very different physicochemical properties, and wide coverage requires the combination of multiple assays [288, 289] (68,83 from renal 8413). Moreover, the metabolome is open to the environment and the microbiolome, therefore, the metabolome is constantly in flux and more complex than the genome and proteome [83].

The potential of metabolomics for disease diagnosis, prognosis and, in clinical trial setting, for monitoring drug therapy relies on its ability to extract a disease signature from the multivariate analysis of the metabolic profiles of statistically relevant ensembles of samples derived from different donors. Hence, the reliability of the approach requires that the chemical nature and the relative concentration of all the metabolites present in the specimens are neither affected by the preanalytical treatment used to store the samples nor by the analytical methodology. And, this is why metabolomics itself is assuming a growing importance in the definition of operating procedures aimed at collecting and preserving biological samples and for standardizing protocols [63].

Metabolomic samples and preparation

The analysis of human biological samples provides characteristics number of metabolites, depending on the biofluid or tissue under examination and the analytical instrumentation used, and they can reach values in the thousands. The Human Metabolome Project has the goal to identify, quantify, catalog, and record all metabolites potentially found in human specimens. Over 100,000 metabolites have already been identified and stored in the human metabolome database (HMDB) [290].

Samples

Metabolomic samples generally include biofluids (urine, blood, cerebrospinal fluid, saliva, among others), intact tissues (heart, liver, kidney, brain, among others), tissue extracts (fresh-frozen and FFPE tissues), and exhaled breath [59]. Most metabolomic studies in kidney diseases involve common biofluids such as urine and serum/plasma, with the dialysate fluid also offering potential benefits [291]. During sampling and sample preparation, the state of the biological sample must be preserved as much as possible in order to maintain the information contained within a sample [292]. This could be challenging due to rapid induction of metabolic stress, degradation, and dynamics of protein modification [199].

Urine, in particular, is a very appealing biofluid for analysis because it is abundant, can be collected noninvasively, and is particularly rich in terms of its chemical diversity. Consequently, urine offers significant opportunities for data mining, data modeling, and biomarker discovery, particularly with respect to human health and disease [293]. Furthermore, urine exhibits a strong phenotypic or metabotypic stability [294], which strengthens its potential for biomedical research and clinical utility [59].

In addition, metabolomic profiling of urine gives a time-averaged representation of an individual's recent (typically within 24 h) homeostatic condition. Some metabolites may associate with individual's physiological or pathological state, whereas others may associate with an individual's genotype, environmental exposures, dietary habits or drug intake as well as the time of collection [59]. As such, it is important that in quantitative metabolomics the inter- and intraindividual metabolite variance within the normal/control group be properly identified and minimized as much as possible. This may be facilitated by requiring a 12-h fast prior to urine collection or restricting the consumption of supplemental protein performance enhancing food, such as protein shakes, prior to any sample collection [295].

Blood ***derivatives***, such as serum and plasma, are also very common biofluids in metabolomic studies. The disadvantage with respect to urine is that the collection of blood samples is slightly more invasive. Therefore, it may be more difficult to obtain multiple collections from patients, and even more from healthy individuals. On the other hand, blood is less affected by daily variations and daily diet than urine [63].

Sample preparation

For metabolomics sample preparation, there is additional challenge created by the greater diversity of physical and chemical properties of molecules, which are not easily isolated with a single method [292, 296]. Protocols used for metabolite extraction from complex biological matrixes, such as biological fluids, tissue, cell pellets, and cell media, separate metabolites and perform preconcentration prior to analysis. In order to reduce or eliminate matrix effect, metabolites are separated from salts and macromolecules, such as nucleic acids and proteins [296, 297]. Liquid-liquid extraction (LLE), solid-phase extraction (SPE), protein precipitation, dialysis, and ultrafiltration are common methods used for metabolomics sample preparation [298–301]. LLE is the most widespread extraction procedure in the integrated one-pot methods of proteomics and metabolomics analysis [199].

For integrative proteomics and metabolomics from a single sample, a good starting point is protein precipitation and metabolite depletion from the whole sample using organic solvent [302]. After addition of the extraction solvent, metabolites may be isolated in the supernatant, while the remaining cell pellet is used for protein extraction [303]. Sample preparation for bottom-up proteomics must include chemical or enzymatic protein digestion to peptides prior to LC-MS analysis [304].

Other methods are reported in the literature for simultaneous extraction of DNA, RNA, proteins, and metabolites from a single biological sample [302, 305]. The advantage of this method is that it can be used to separate RNA and DNA from proteins to reduce complexity of protein mass spectra allowing integration with transcriptomics and genomics studies [199].

Technology in metabolomics for data acquisition

The two main analytical techniques consist of NMR spectroscopy and MS [306]. These techniques can handle complex biological samples with a high sensitivity, selectivity, and throughput [307]. Both NMR and MS can be used to characterize metabolite data, either in a targeted or nontargeted (pattern recognition) manner [308]. The targeted analysis involves the identification and quantification of specific analytes in a given biofluid or tissue extract by comparing the spectrum of interest with a library of reference spectra of pure compounds.

Nuclear magnetic resonance (NMR) spectroscopy

NMR spectroscopy is a powerful analytical approach for both identification and quantification of analytes with superior advantages, such as being nondestructive, highly reproducible, and most importantly, requiring minimal sample preparation. However, NMR equipment is expensive and relatively insensitive compared with MS-based techniques. It gives detailed simultaneous information on both the structure and molecular mobility of metabolites without the need for the preselection of analytical parameters or sample derivatization procedures (i.e., chromatography) [59]. As it is a nondestructive technology, it can be used for the analysis of tissue biopsies and can be directly compared with histopathological findings using the same tissue samples. However, sensitivity is a limiting factor and often metabolite concentrations in range of 1–10 μmol/L are required for detection and quantification by NMR spectroscopy [309–311].

NMR spectroscopy is a quantitative technique as the intensity of an NMR signal is proportional to the concentration of detectable (usually ^{1}H) nuclei in the receiver coil. The ***1D ^{1}H NMR*** spectra of urine samples are highly complex, with thousands of distinct signals visible in a single spectrum. Consequently, signal overlap and signal distortions from nearby (strong) peaks are often evident even in 1D ^{1}H NMR spectra collected at 800 MHz and above. Over the past two decades, several methods had been proposed or developed to address all these issues (i.e., lack of sensitivity and speed as well as difficulties in reproducible quantitation compared to 1D NMR) [59].

2D ^{1}H INADEQUATE with sparse sampling/nonlinear sampling (NLS) seems to be a promising method able to characterize and quantify low-abundance metabolites [59]. More recently, another approach has been proposed, wherein three different collection and processing techniques were combined, including ***J-compensated 2D heteronuclear single quantum correlation (HSQC), NLS, and forward maximum entropy (FM) reconstruction***. This combination resulted in a 22-fold reduction in NMR recording time (relative to a conventional HSQC spectrum) while at the same time yielding precise metabolite quantitation in both native and lyophilized urine samples [309].

Ultrafast 2D NMR spectroscopy is another promising method; however, this approach requires sophistication in pulsed-field gradient (PFG) performance and specific processing software, currently not available on most commercially available NMR spectrometers. Nevertheless, given its speed, resolution, and sensitivity advantages over fast heteronuclear NMR and even 1D homonuclear NMR methods, this approach may soon become the method of choice for identifying and quantifying low-abundance metabolites in urine samples [59].

PFG versions of NMR pulse sequences such as J-resolved, homonuclear correlated spectroscopy (COSY), ^{1}H-^{1}H total correlated spectroscopy (TOCSY), ^{1}H-^{13}C heteronuclear single quantum correlation (HSQC), nuclear Overhauser effect spectroscopy (1D NOESY), and flip angle adjustable 1D NOESY (FLIPSY) have been used to allow more rapid data acquisition through faster signal recovery along with suitable signal correction methods. However, gradient pulse sequences are not recommended, despite their advantages, due to instability and gradient ring down periods. These limitations will manifest as peak distortions to the internal chemical shift and quantitation references [312–315].

To assist with peaks identification and metabolite identification process, it is possible to use several approaches. One approach is spectral simplification through statistical analysis or chemometric analysis using spectral binning/alignment and statistical total correlation spectroscopy (STOCSY). These methods statistically align and compare the NMR spectra between two groups (diseased and healthy), and the most significantly different peaks are then identified. This leads to a reduction in the number of peaks that need to be analyzed or identified. However, the drawback of this method is that can lead to problems in compound identification and quantification, as the spectra have been extensively averaged as part of the statistical processing [316, 317].

Another approach is known as spectral deconvolution, in which involves identifying and quantifying as many compounds as possible prior to determining any statistical differences between groups. However, the process is time consuming, as dozens of compounds and hundreds of peaks must be identified and quantified through a process known as spectral deconvolution [317]. Several companies though, have developed spectral deconvolution tools and software (e.g., Chenomx NMRSuite from Chenomx, Inc.; AMIX from Bruker, Inc.; MnovacScreen from Mestralab Research; CRAFT or Complete Reduction to Amplitude Frequency Table from Agilent) making this process easier [59].

Mass spectrometry

MS platforms tend to have much greater analytical sensitivity, enabling broader surveys of the metabolome. Prior to analysis, the samples need to be separated using chromatography, commonly either gas or liquid chromatography (GC or LC) followed by ionization in a fluid or matrix; and metabolites are subsequently identified using a mass spectrometer on the basis of their mass-to-charge ratio (m/z) and their representation on a spectrum. The combination of chromatographic separations with MS increases the biological information obtained and enhances the sensitivity and the ability to identify metabolites in complex biological systems [306, 310].

An advantage of MS-based technique is the ability to identify metabolites, either through the measurement of molecular mass (indicative of the molecular formula) to a high mass accuracy (typically parts per million, ppm) or by collection of fragmentation mass spectra (indicative of molecular structure). This allows the identification of novel metabolites not currently described in metabolomic databases as well as previously characterized metabolites across large sample sets. Ion suppression in complex biological samples may be caused by the interaction of multiple analytes that are present in the ionization source at the same time, thus limiting the ability of MS to quantify metabolites [310, 318, 319].

Given nLC-MS is widely and successfully applied in proteomics research labs, there is significant interest in adapting the same technology for metabolomics analyses to enable combined proteomic and metabolomic studies using the same platform [320]. Recently, nLC-MS workflows have been reported for the use in more targeted metabolite and small-molecule analyses, including the detection and quantification of amino acids, fatty acids, lipids, prostaglandins, di/tripeptides, steroids, vitamins, nucleic acids, xenobiotics, and, less frequently, in global metabolomics surveys of urine, plasma, and cell samples [321, 322]. Nanoflow techniques do pose limitations beyond column blockage; for example, ESI in negative mode is complicated by reduced solvent desolvation and electrical discharge [323].

While HPLC-MS is widely used for high-throughput nontargeted metabolic analysis, matrix effects and poor ionization efficiency due to ion suppression may produce lower sensitivity and restrict detection to highly abundant metabolites [324]. This limitation has motivated the use of nLC-MS for metabolomics, too. Nanoflow HPLC columns with internal diameters between 10 and 75 μm and flow rates of 10–500 nL min^{-1} offer lower chromatographic dilution as well as enhanced peak capacity, resolution, and detection sensitivity [325, 326]. Nanoflow also provides improved analyte preconcentration and more efficient introduction of biomolecules via ESI to the MS system as compared to conventional high-flow LC-MS systems [327, 328]. Compared to conventional HPLC-MS, nLC-MS systems are up to 2000 times more sensitive, while achieving up to a 300-fold lower limit of detection and limit of quantification [329, 330].

Moreover, in cases where specimens are limited, nLC-MS is also beneficial because it requires less sample, though reproducibility and speed may be compromised as compared to HPLC. It is possible to use both hydrophilic interaction chromatography (HILIC) and RP-HPLC for nLC-MS based metabolomics profiling; however, running samples in negative ionization mode is more difficult for metabolomics studies using nanoflow ESI [323, 331]. HILIC is often most effective because it provides consistent separations of polar metabolites [332].

In addition, several methods are also used to produce consistent nanoflow rates, including split flow, direct fusion, and direct flow [199]. ***Split flow*** was the first introduction of nLC, which utilized dual HPLC pumps with a diverter that split the higher flow fluidics to deliver nanoscale flow rates to the mass spectrometer [324]. This results in about 99% of the sample and most of solvent being lost to waste [333]. ***Nanoflow direct infusion*** MS has long been successfully applied for metabolomics sample analysis from many sample types, such as liver extracts, plasma, urine, and embryos [326, 334–338]. These studies take advantage of the high-throughput offered by direct infusion. However, the absence of chromatography before MS enhances the matrix effect and makes the differentiation difficult between isobars or isomers [193, 339]. ***Nanoflow direct flow*** MS analysis methods may use nanoflow reciprocal or syringe pumps and microfluidic flow [340, 341]. These methods improve compound retention and spray characteristics of the nanoflow platform. Application has been reported in metabolomics analysis of complex samples such as plasma, tissue, urine, breath, sweat, cerebrospinal fluids, and cell extracts [342–345]. Moreover, deviation in back pressure and surface tension due to variability in mobile phase viscosities can create limitations in obtaining stable analyte retention times, especially with gradient separations [346, 347]. Direct flow is not as high throughput as direct infusion; however, direct flow does provide increased resolution via pre-MS separations without waste inherent with sample split flow. Nanocolumns, up to 0.01-mm internal diameter, can provide stable low flow gradients, reducing chromatographic dilution while requiring low sample and mobile-phase consumption [348–350].

Two other parameters influencing the quality of the recorded metabolomics data are retention time drift and peak intensity. Methods and techniques should be optimized to improve retention time stability and minimize the coefficient of variation (CV) between replicate analyses. Variation in peak intensity by replicate nLC-MS analyses may be as high as 30% [66], and CVs consistently less than 20% for targeted and nontargeted metabolomics analysis are difficult but not impossible to achieve by nLC-MS [351].

More recently, unimolecular approaches have been reported for the parallel analysis of the proteome and metabolome based on most existing LC-MS strategies [352, 353]. In these studies, molecular extraction and analysis workflows are performed independently for both proteomics and metabolomics measurements on unique replicate samples using different LC-MS techniques. However, utilizing methods to perform simultaneous molecular extractions from a single sample minimizes experimental variation and reduces the amount of sample consumed and the number of handling steps performed during sample preparation [354]. This is especially valuable in cases where the biological samples being studied are patient-derived, extremely limited, or otherwise hard to generate or obtain [199]. Hence, a sample preparation method for extracting proteins and metabolites from the same biological sample must be compatible with both classes of biomolecules and produce sufficient yield to be practical [320].

Both proteomics and metabolomics benefit greatly from continued progress in LC-MS instrument engineering. High-resolution mass spectrometers with higher mass accuracy and faster scan speeds provide more accurate, in-depth, raw data. Improving ion resolution and throughput facilitates better feature identification and makes the analysis feasible in a larger number of sample [355, 356].

Data analysis and interpretation

Metabolomic studies often require the analysis of many samples prepared simultaneously to reduce variability and improve workflow efficiency. Utilization of preservative techniques, such as the addition of protease inhibitors to biospecimens, should be considered in terms of study design and generalizability. While the addition of such preservatives may allow identification of a larger range of metabolites, which may inform detailed mechanistic insights, their use does not necessarily reflect the "real-world" bedside scenarios in the era of bedside point-of-care testing. Therefore, the specific goals and long-term objectives of the study must be considered carefully [306].

In metabolic studies, different approaches are used for the collection, processing, and interpretation of the data depending on the specific problem: metabolic profiling, metabolomics, and metabolic fingerprinting. The aim of ***metabolic profiling*** is to identify and quantify a selected number of predefined metabolites in a given sample, whereas ***metabolomics*** allows the complete analysis of all the metabolites in a sample, which are quantified and identified (targeted analyses). Finally, ***metabolic fingerprinting*** is intended to measure the global profile of the metabolites characterizing the sample, without specific identification and quantification (nontargeted analyses) [283].

In metabolomic studies, many hundreds of samples are routinely analyzed, and a minimum of several hundreds of metabolites are usually detected. All the data obtained from the collection stage should be further analyzed using statistical multivariate methods (chemometrics) to extract biological, physiological, and clinically relevant information. Before performing the statistical analysis, the data are subjected to several processing steps: (1) the spectrum requires different processing depending on the specific analytical techniques employed; (2) a data matrix is produced from the analytical measurements with m rows (i.e., observations, samples) and n columns (i.e., variables, frequencies, integrals); (3) data normalization and scaling are performed; and (4) multivariate statistical modeling of the data is completed [283].

Statistical multivariate analysis

Metabolomic studies yield huge amounts of data, making the application of pattern-recognition methods (also known as chemometric or multivariate statistical analysis) important in interpreting complex data and identifying metabolites as potential biomarkers. The complexity of the spectroscopic data due to the commonly high number of samples and variables (i.e., metabolites) inevitably requires the use of data reduction techniques to extract latent metabolic information and enable sample classification and biomarker identification [283].

After data collection, a statistical analysis method is chosen to suit the study objective. Univariate analysis of variance, Mann-Whitney *U* test, Wilcoxon signed-rank test, and logistic regression are applied to identify metabolites that are capable of differentiating between groups. Multivariate analysis methods are used to develop statistical pattern recognition models. By extracting and displaying the systematic variation in the data based on projection methods, multivariate analysis reduces the variability of the data and combines complex interaction [306].

Multivariate statistical analysis can be further categorized into unsupervised and supervised methods, in which principal component analysis and partial least squares discriminant analysis are the most widely used tools [283].

Among the *unsupervised methods*, the most classic example is principal component analysis (PCA), which is applied by summarizing the data into much fewer variables called scores, which are weighted averages of the original variables. Each principal component (PC) is a linear combination of the original data parameters and each successive PC explains the maximum amount of variance possible, not accounted for by the previous PC [306]. Hence, the PCA is referred to as the scores plot, which provides an initial overview of all of the data for the detection of outliers, trends, patterns, or clusters [283].

In contrast, *the supervised approaches* include the partial least-squares discriminant analysis (PLS-DA) and the orthogonal partial least-squares discriminant analysis (OPLS-DA). PLS-DA enables the identification and characterization of metabolic perturbations signatures through a linear regression model by projecting the predicted variables and the observed variables to a new space. Therefore, PLS-DA is the main tool used in chemometrics for classification and discrimination purposes [357]. The OPLS-DA regression is an extension of PLS-DA, where only the variation in the set of predictor variable that correlates with the response variable is retained [306]. The supervised methods classify samples into different categories or classes based on a priori knowledge. Hence, these methods are also referred to as the loadings plot, which shows the importance of different variables (i.e., the metabolites) in defining the model [283].

Discriminant analysis of principal component (DAPC) has recently been proposed as a way of combining the flexibility and efficiency of PCA and the discriminating power of the discriminant analysis (DA). The use of dimension reduction techniques in a regression context requires the specification of the number of latent variables to be retained in the model. This usually relies on crossvalidation procedures aiming at the identification of the number of components that optimizes both interpretability and prediction error [358].

Metabolomic data sets are intrinsically multidimensional with the number of measured metabolites typically ranging from a few dozen to hundreds. A key step in placing statistically significant findings from chemometric analyses into a meaningful biological context is to identify significantly altered pathways represented by certain spectral peaks. Therefore, the metabolites identified are integrated into metabolic correlation networks in order to better understand the complex relationships among various metabolites. The analysis and interpretation of these data-generated networks reflect the structure of the underlying biochemical pathways. Furthermore, pathways mapping represents a snapshot of the physiological state at a given point in time, which is a useful tool in metabolic fingerprinting [306].

Metabolomics raw data processing and feature identification

Metabolomics feature identification is a more difficult challenge in that the search space for identifications is much greater due to the increased structural diversity. This is further confounded by the detection of multiple alternately adducted ion species by LC-MS, and the lack of constrained MS/MS fragmentation rules, compared to polypeptides, which tend to cleave in a more predictable manner [199]. In order to establish a knowledge base that can serve as reference, metabolomics databases have been actively developed, such as the Human Metabolite Database (HMDB; http://www.hmdb.ca/), METLIN (https://metlin.scripps.edu/), and MetaCyc (http://www.metacyc.org/) [290, 359]. While these publicly available databases are accessible, proprietary databases do exist and may be accessed through contractual/research agreements with the entities that have developed them. Alternatively, the nontargeted (global) analysis serves as a hypothesis-generating tool whose results often require follow-up with more targeted approaches. The global pattern-recognition method can also screen for a multitude of key compounds in specific metabolic pathways such as carbohydrates, amino acids, fatty acids, phospholipids, and the nitric oxide (NO) synthesis pathway, which provides valuable information for metabolic fingerprinting. Such pathway analysis can provide insights into real-time disease processes, developmental processes, and potentially reveal biological therapeutic targets [306].

METLIN has an extensive library of experimentally defined MS2 spectra, while HMBD features extensive compound annotation information and, as of the most recent release, has made all curated data, including spectral libraries, which are downloadable in open-source formats. Unfortunately, the community still lacks a common identification nomenclature for conversion between resources, limiting their utility [199].

Metabolomics data can be noisy, leading to spurious compound assignments, which heightens the need for confident metabolite identifications from features of interest [360, 361]. In order to achieve reliable metabolite feature matching, planning the overall experimental design, data analysis pipeline, and validation strategy in a stringent manner early on in the project lifecycle is crucial [199].

Many open-source and commercial software tools are available for performing metabolomics spectral filtering, peak detection, retention time alignment, and normalization. These include MZmine, MS-DIAL, and R packages and associated with the mzR parser, and other tools like MAIT provide additional functionality for routine detection of differential metabolites from LC/MS-based metabolomics data [362–364]. XCMS and MetaboAnalyst are also popular, user-friendly tools associated with the METLIN and HMDB databases, respectively. Although available as standalone R packages, powerful features like raw data file processing and pathway enrichment analysis are only or most easily accessed via online web-based versions of these tools, providing a hurdle for integration into more advanced computational workflows [236, 357, 365]. OpenMS likewise offers plugins for metabolite feature identification that use advanced signal processing techniques to identify features in raw data and permit the querying of public databases to find matches based on exact mass and structural fragmentation patterns [366].

Despite these powerful resources, metabolomics feature identification and quantification remain a challenge. Compared to the thriving ecosystem of robust open-source sequencing analysis tools, metabolomics and proteomics search tools tend to be more closed and different algorithms often do not converge to similar results with the same degree of confidence. Hopefully, more robust, user-friendly, and open-source solutions will continue to advance this area of research [199].

Metabolomics in renal research and nephrology

The recent rapid development of a variety of analytical platforms, including MS in conjunction with radioactive and enzymatic methods, methyl chloroformate based on the principle of the alkylation reaction, and NMR spectroscopy [59, 144, 306], have enabled separation, characterization, detection, and quantification of these chemically diverse structures. Although these techniques are still in development process and have not been widely applied in clinical use, a growing number of reports have described the use of metabolomics in diabetic nephropathy, MN, IgAN, autosomal dominant polycystic kidney disease (ADPKD), urinary tract infections (UTIs), interstitial nephritis [367–372] as well as in the detection of AKI and CKD and prognosis of progression of renal function [373]. In addition, metabolomic profiling approaches have been used in kidney transplantation for the assessment of donor organ quality, storage, ischemia/reperfusion injury, early-stage organ rejection as well as evaluation of biochemical effects of immunosuppressants and their combinations in the kidney (i.e., drug-induced kidney injury) [265, 374].

Bell et al. performed NMR studies of blood plasma and serum and found that during the process of renal damage, amines, particularly trimethylamine-N-oxide, might buffer and stabilize proteins and protect them from biochemical denaturation secondary to urea, guanidine, and guanidine derivatives, among others [375].

A metabolomic profile study performed in patients with UTI compared with healthy individuals and patients who had recovered showed molecular discriminators that characterize different patient groups. Some of the discriminators (e.g., acetate, trimethylamine) showed association with the degree of bacterial contamination of urine, whereas others (e.g., para-aminohippuric acid [PAH], scyllo-inositol) were identified as more likely to be markers of morbidity [370].

Gronwald et al. were able to differentiate subjects with ADPKD from those with other kidney diseases and from subjects with normal kidney function, using NMR metabolic fingerprinting of the urine. They grouped a total 178 patients with CKD into 5 cohorts. From the total of patients 54 had ADPKD with slightly reduced eGFR, other 10 patients had ADPKD on hemodialysis with residual renal function, 16 kidney transplanted patients, 52 with type 2 diabetes, and 46 healthy volunteers with normal function. In this study, ADPKD patients showed increased excretion of proteins and methanol compared with the control and other patient groups, suggesting the existence of set of urinary compounds specific to the early stage of ADPKD [369].

In another study performed by Gao et al., they applied parallel metabolomics in urine and serum from patients with MN, in which patients were divided into two groups: one with low level of urinary proteins <3.5 g/24 h (low urinary protein MN group; LUPMN) and the other group with high levels of urinary proteins >3.5 g/24 h (high urinary protein MN group; HUPMN). Citric acid and four amino acids were markedly elevated only in the serum samples of HUPMN patients, implying more impaired filtration function of kidneys in this group than in LUPMN patients. Additionally, dicarboxylic acids, phenolic acids, and cholesterol were significantly increased only in the urine of HUPMN patients, suggesting more severe oxidative stress [368].

Several experimental studies have utilized metabolomics to study AKI induced by toxins and antibiotics in animal models [376–379]. Xu et al. showed that the excretion of amino acids which included glucose, lactate and ketone was significantly increased upon treatment with certain nephrotoxins such as gentamicin and cisplatin [379]. Similarly, Boudonck et al. performed a global, nontargeted metabolomics analysis on Sprague-Dawley rats treated with gentamicin, cisplatin, or tobramycin. Using a combination of GC/MS and LC/MS, they found a significant increase in polyamines and amino acids in urine from drug-treated rats after a single dose and prior to observable histological kidney damage and conventional clinical chemistry indications of nephrotoxicity [377].

More recently, Zhang et al. demonstrated that early-stage obstructive nephropathy can be diagnosed in an animal model base on plasma metabolomics, wherein they performed unilateral ureteral occlusion surgery on rats to induce renal interstitial fibrosis. At the first postoperative day, they found 13 differential plasma metabolites and 14 differential urine metabolites using ultra performance liquid chromatography (UPLC)-MS-based metabolic approach. The altered metabolic pathways included glycerophospholipid metabolism, tryptophan metabolism, glutamate metabolism, and purine metabolism. Additionally, they identified a panel of five plasma biomarkers, which offered a good diagnostic performance for early diagnosis of obstructive nephropathy [247].

Other small studies have also demonstrated the relevant changes in metabolomic profile associated to therapeutic interventions. For instance, metabolomic analysis in healthy individuals showed changes in urine metabolite patterns within the first 4 h after a single 5 mg/kg dose of cyclosporine (oral formulation) [380].

Metabolomics is emerging as a powerful tool for AKI and CKD research as well as for therapeutic intervention and its effect in kidney function. However, it must be coupled with thoughtful study design, careful consideration of potential confounders, and rigorous quality control to yield meaningful results. Metabolomics studies can be seen as an important starting point for discovery and hypothesis generation, with subsequent efforts required for independent replication, individualized assay development, and mechanistic investigation [286].

Limitations of proteomics and metabolomic technologies

Although nontargeted proteomics and metabolomics technologies are valuable discovery tools and hypothesis-generating technologies, there are methodological limitations. In an ideal world, nontargeted proteomics and metabolomics technologies would be truly unbiased, would capture the complete proteome and metabolome (as this is nowadays possible in the field of genomics), and would allow for at least a semiquantitative comparison of the proteomes and metabolomes of interest. However, current technologies capture only a part of the metabolome and proteome and, therefore, produce inherently biased results [288].

Important factors that introduce bias into metabolomics analyses may include, but are not limited to, timing of sample collection, sample collection procedure, sample processing, stabilization, stability and storage, extraction procedures, dilution of sample, type and number of analytical methods used, preferences of analytical assays for metabolites with certain physicochemical properties, ion suppression (LC-MS), derivatization (gas-chromatography-MS), sensitivity of the assay, and range of reliable response [288]. In term of proteomics, the situation is similar [83].

At present, even the best metabolomics platforms provide incomplete coverage of the human metabolome, detecting less than one-quarter of the currently known endogenous and exogenous metabolites in a given biologic specimen. Different platforms provide different coverage of the metabolome, such that a metabolite detected on one platform may not be identified on another, complicating comparisons across studies. Within a platform, not all detected metabolites are measured equally well [381]. Thus, it may be prudent to corroborate the assignments of promising metabolites with experiments that analyze samples spiked with the authentic standard [382].

The sensitivity and throughput of MS-based omics studies does lag considerably behind genomic and transcriptomic sequencing technologies. This challenge makes studies with large sample sizes difficult. While some large-scale efforts, such as the Clinical Proteomic Tumor Analysis Consortium (CPTAC), are addressing this shortcoming, proteomics and metabolomics is still catching up to the massively parallel nucleic acid profiling approaches in terms of ease of data generation, standardization, and accessibility [383, 384]. Further improvements in multiplexing, more rapid and standardized separations, and improved MS scan speed and automation will be important criteria for continuing to advance the field [199].

The many computational, chemometric, and biostatistical steps required to link changes in metabolite and protein patterns to biochemical and signal transduction pathways have to address multiple challenges such as the very large number of variables (metabolite and/or proteins) that often greatly exceed the number of observations. Most of the hundreds and thousands of data points generated are not relevant to the disease or drug effect and do not convey additional information. This increases the risk of false-positive results and may mask valid information. In such settings, classical statistical approaches are inadequate. There is often a severe lack of degrees of freedom generally because of the relatively small sample sizes resulting in lack of statistical power and false positives [385–387]. Moreover, the large sparsity of informative variables tends to make metabolomics and proteomics data extremely noisy. Spurious correlations and colinearities often exist between variables. This may be due to the nature of the data as well as due to dimensionality artifacts [388].

In this context, the complexity and heterogeneity of metabolites are considerably greater than those of genes and proteins. These problems have been addressed by the development of algorithms for data reduction and filtering, for FDR control, and high-dimensional statistical modeling strategies [389]. However, it is impossible to completely exclude that the selection of the chemometric/bioinformatics strategy may affect the results and conclusion. An overly conservative strategy may reject valid molecular markers (high false-negative rate), while a liberal strategy may result in false-positive results [83].

Tightly controlling and analyzing potential confounders such as age, gender, comedications, and diet are critical and can be very challenging in complex patient populations. For example, it has been shown that age has a significant effect on the urinary proteome [390, 391].

Also, it is important to take into account the time dependency of protein as well as metabolite pattern changes. Studies comparing sets of single samples per individual may be difficult to interpret with confidence. Especially during earlier biochemical stages of a disease process, protein and metabolite pattern changes may vary quickly as the injury progresses. This may include compensatory mechanisms, secondary mechanisms such as oxygen radical formation and damage, changes in cell function and regulation, and the triggering of additional systemic responses, such as immune reactions and inflammation [83]. In addition, one of the most important confounders in metabolomics and proteomics studies of CKD is kidney function. Typically assessed as GFR, kidney function is related to approximately one-third of detected metabolites in both general and CKD populations [392, 393]. Hence, the number of statistically significant associations between metabolites and subsequent mortality is greatly reduced after adjustment for GFR [286]. Therefore, careful measurement and consideration of baseline GFR are critical.

Proteomics and metabolomics technologies can create molecular fingerprints of body fluids such as plasma and urine that are characteristic for diseases and drug effects. It is an attractive idea to build expert systems based on nontargeted/nonbiased analytical strategies that will generate a holistic view of a patient's plasma and urine metabolome/proteome and that in combination with genomics will allow for a systems biology-based approach to medicine. Today, using truly nontargeted screening technologies in decision making is not yet feasible, mostly because of the complexity of the data generated, the difficulties of validating such assays, the lack of verification, and the lack of algorithms to convert this information into robust clinically relevant information [83]. Continuing to build out and improve the quality and scope of integrative omics analysis tools that exploit user-defined protein-protein interactions, protein modification sites, and protein-metabolite interaction information will be an important endeavor to enhance the informativeness of integrative studies [199].

Details known to be important for driving biological processes, such as macromolecular structure, interaction binding kinetics, equilibrium states, and spatiotemporal control, are not incorporated into most analyses. Borrowing from the fields of Bioengineering and Physics, there is hope for more advanced models of dynamical systems and their application to omics data analysis to drive better understanding of the molecular mechanisms driving cell biology [394].

Developing innovative computational tools to make the application of such integrative models more accessible to more biological researchers and robust enough for a variety of use cases will remain a challenge. However, the potential exists to be able to more accurately describe changes in biological systems and to better predict cellular responses to genetic and interventional perturbations, for example in the case of combinatorial therapeutics [199].

Transcriptomics in kidney disease

Transcriptomics is the study of the complete set of RNA transcripts produced by the genome, in a specific cell or under certain circumstances. Over the last two decades, transcriptomics studies on renal disease have been carried out using different approaches, either studying the entire transcriptome or focusing on individual biomarkers [395].

Transcriptomics builds on genomic knowledge [144]. Initially, it was thought that the human genome was mostly composed of protein-coding genes, but it was not until 2001 when the sequencing of the human genome was completed revealing that only a small proportion of human genes were indeed able to encode proteins (~2%). Meanwhile, about 80% of our genes are transcribed leading to the synthesis of RNA molecules. These findings by the ENCODE Project Consortium have suggested that the majority of human genes are transcribed into nonprotein-coding RNAs or noncoding RNAs (ncRNAs) [396–398].

The mammalian transcriptome is composed of protein-coding and ncRNAs. Only 2% of RNAs are protein-coding (i.e., messenger RNA; mRNA), while the remainder are ncRNAs, which include (1) those involved in protein synthesis: ribosomal RNA (rRNA) and transfer RNA (tRNA) and (2) those involved in regulation of gene expression: small noncoding RNA and long noncoding RNA [395].

The noncoding RNAs (ncRNAs) family is a large and diverse class of RNAs with multiple functions. A simple classification of these RNA molecules relies on their size. An arbitrary cutoff of 200 nucleotides separates small ncRNAs from long ncRNAs (lncRNAs). The most investigated small ncRNAs are microRNAs (miRNA), which are 20- to 25-nucleotide-long RNA molecules able to downregulate the expression of target protein-coding genes by translational repression, mRNA cleavage, and deadenylation. These latter groups have been proposed as both disease biomarkers and therapeutic targets for personalized medicine [399]. Also, the lncRNA groups are involved in regulation of gene expression and constitute a reservoir of potential novel disease markers and drug targets. This group comprises both linear lncRNAs (named by default as lncRNAs) and circular RNA (cicRNAs) [400].

miRNAs are secreted into the circulation by many different mechanisms such as exosomes, microvesicles, and apoptotic bodies. Also, they can be protected from degradation by binding to RNA-binding proteins [401]. Due to their stability, miRNAs are readily quantified in serum, plasma, and other body fluids. Hence, researchers have focused their interest into noncoding RNAs (ncRNAs) to improve the precision of kidney diseases diagnosis and find novel treatment targets, as they might have a potential in diagnosis as well as treatment in renal diseases [400, 402].

Transcriptomics and samples

Most transcriptomic studies in kidney diseases involve common biofluids such as urine and serum/plasma as well as tissue extracts (i.e., fresh-frozen and FFPE tissues). During collection and sample preparation, the state of the biological sample must be preserved as much as possible in order to maintain the information contained within a sample [60, 62].

More recently, urine has caught the attention of many researchers. Urine samples have the potential to provide a wealth of information both for clinical applications and translational investigations as well as to provide a perfect medium for biomarker injury [402]. Urine in general contains very little RNA [403]. Efforts to improve the quality of RNA to be extracted from urine specimen may decrease the interplatform variability observed in some studies [402]. Recently, as more commercially available kits are designed for this purpose, more investigators are evaluating the quality and quantity of miRNAs extracted from urine [404]. These investigations along with improved methodology to enrich for exosomal RNAs to be used in profiling platforms hold promise for improved analysis and concordance between profiling platforms [402, 404].

Transcriptomics and technology

Over the years, research in the field has progressed from analysis of a few candidate genes to high-throughput expression profiling driven by the advent of transcriptomic techniques, such as: microarrays [405], serial analysis of gene expression (SAGE), genome-wide association studies (GWAS) [406], expressed sequence tag (EST) sequencing, cDNA-amplified fragment length polymorphism (cDNA-AFLP) [407], massive parallel signature sequencing (MPSS), NGS, and more [408].

The most commonly used platform for the analysis and profiling of miRNAs is the quantitative real-time PCR (qRT-PCR). This method of analysis may detect some isomeric forms of the miRNA with extra bases on either ends but it will miss a wealth of information, most importantly the vast majority of miRNAs in the isomeric forms with deletions/substitutions that are not detected by the predesigned qRT-PCR primers available using conventional qRT-PCR technology such as TaqMan Low Density Array (TLDA-A, Life Technologies) [402].

Newer technologies are becoming more available and have promised an increased sensitivity and specificity. One of the technical approaches that has gained popularity and successfully been used to characterize miRNA profiles in various tissues as well as biofluids is the NGS. NGS platform has shown to generate highly reproducible, accurate data with high correlation to other platforms in high-quality RNA samples [409–415].

Moreover, NGS is specific and can detect the vast majority of miRNA in their isomeric form; however, it may not detect certain low abundant miRNAs in low-quality RNA specimens like those obtained from urine samples. NGS platform seems to be highly sensitive to the presence of contaminants, including PCR primers to bacterial and fungal genomes that could negatively affect the detection of low expressing miRNAs [409–415].

Each platform has its strengths and weakness; therefore, careful considerations should be made when selecting a platform for miRNA profiling.

Transcriptomics in renal research and nephrology

Since 1994, over 80 study reports have been published in scientific journals where the authors have tried to identify key regulators of different pathways involved in several renal conditions, such as CKD, ESRD, DKD, primary glomerulonephritis, and systemic lupus erythematosus (SLE), among others. Most of these studies have been done in small cohort of patients, with only a few studies analyzing results from over 100 patients [416–420].

For instance, a recent integrative bioinformatics analysis of 250 gene expression datasets of healthy renal tissues and those with various types of established CKD (i.e., DKD, hypertensive nephropathy, and glomerulonephritis) has identified a total of nine genes with significant and different expression in both diseased glomeruli and tubules. These genes were retrieved from the Gene Expression Omnibus repository (https://www.ncbi.nlm.nih.gov/geo/) and include *IFI16*, *COL3A1*, *ZFP36*, *NR4A3*, *DUSP1*, *FOSB*, *HBB*, *FN1*, and *PTPRC* [420].

Similarly, Perco et al. performed three extensive microarray studies on CKD using kidney tissue biopsy material to identify abnormalities at the mRNA level.

They found several genes that were differentially expressed across the transcriptomic data set, including collagen type XV alpha 1 (*COL15A1*), uromodulin (*UMOD*), prostaglandin D2 synthase (*PTGDS*), and apolipoprotein A-I (*APOA1*) [421].

Furthermore, Rudnicki et al. isolated renal proximal tubular epithelial cells (RPTEC) from 19 CKD patients and five controls using LCM and performed transcriptome analysis using cDNA microarray technology. This study found that 12 transcription factors (*MADP-1*, *PWP1*, *SP3*, *ZNF3*, *SNAPC3*, *MORF4L2*, *YEATS4*, *TFCP2*, *HR*, *RARB*, *FOXK2*, and *GLIS1*) were upregulated and other 16 (*GTF2F2*, *POU2AF1*, *EGR2*, *NR2C2*, *ZNFN1A5*, *ELK3*, *RBM15*, *CHD2*, *MLL5*, *VDR*, *FOXJ3*, *ZFHX4*, *SPI1*, *ZNF292*, *EEF1A2*, and *KIAA0863*) were downregulated in CKD patients [422].

More recently, using real-time PCR (RT-PCR) it was discovered that levels of mRNAs for urinary podocyte markers such as synaptopodin, podocalyxin, CD2-AP, α-actin 4, and podocin were significantly increased in DKD patients [423]. Also, another study found that urinary mRNA level of α-SMA, fibronectin, and MMP-9 was significantly higher in DKD group compared with controls ($P < .05$), and those mRNA levels showed a positive correlation with DKD progression [424].

Several other studies have attempted to find transcriptomic biomarkers associated with a higher risk of CKD or biomarkers that predict changes in eGFR or proteinuria. However, most of the studies are limited in sample size and power and the biomarkers are not replicated in other studies. For instance, expression of some cytokines such as transforming growth factor alpha (TGF-α), transforming growth factor beta 1 (TGF-β1) [425–427], and the matricellular protein connective tissue growth factor (CTGF) have been proposed in different studies as potential biomarkers of renal function in CKD [428]. A study that involved 39 patients with lupus nephritis showed that *CTGF* mRNA expression was significantly correlated with TGF-β1, but inversely correlated with eGFR, being higher in patients with CKD stages 3–5 compared to patients with CKD stages 1–2 [428]. However, more studies need to be done in larger cohorts to confirm the utility of such biomarkers.

Recent studies have also reported significant levels of miRNAs, lncRNAs, and cicRNAs in body fluids that could serve as useful biomarkers in kidney function [400, 402]. miRNAs can potentially impact multiple signaling pathways and their dysregulation may lead to a variety of different renal disorders, such as diabetic nephropathy, hypertensive nephropathy, glomerulonephritis, renal cancer, and polycystic kidney disease [429–436]. For example, in a small study, Nassirpour et al. demonstrated that let-7d, miR-203, and miR-320 may potentially serve as promising novel urinary biomarkers for drug-induced renal tubular epithelial injury [402].

Another miRNA biomarker associated with a variety of kidney diseases is miR-155, which have been found to be inversely correlated with eGFR as well as differentially expressed in CKD [236, 437–441], IgAN [442], focal segmental glomerulosclerosis (FSGS) [443], SLE patients [444, 445], and nephrolithiasis [441]. Meanwhile, the upregulation of members of miR-29 has been found to be associated with CKD [446] and overt albuminuria in DKD [447]. On the other hand, other studies have found that miR-29 is downregulated in CKD conflicting with the results of those studies previously described in this paragraph [448].

It is clear that transcriptomic analysis is a profound tool aimed to understand biological and pathological events such as AKI and CKD. However, this method alone cannot provide sufficient information because translation and posttranslation modifications cause functional changes in proteins, which are considered the real executors of biological events [405]. Hence, future progress in this area and its integration with all the "-omics" technologies (i.e., genomics, epigenomics, proteomics, and metabolomics) will allow more precision in the diagnosis and prognosis of renal diseases.

Extracellular vesicles and potential clinical use as biomarkers for renal disease

Extracellular vesicles are constituted by microvesicles, exosomes, apoptotic bodies, and argosomes, which are classified by size, content, synthesis, and function. Currently, the best characterized are exosomes and microvesicles [449]. In the last decade, urinary extracellular vesicles, including the so-called exosomes, have become of great interest among the nephrology and research community as promising markers for kidney disease [450–452].

Exosomes

In 1981, Trams et al. proposed to use the term "exosomes" for those exfoliated membrane vesicles, appearing as large (500–1000 nm) and small (approximately 40 nm) vesicles, which were identified to be secreted by a variety of cell types [453]. Later, Johnston et al. discovered specific proteins (i.e., transferrin receptor) that are shed via secretion of <100 nm vesicles during reticulocyte maturation; these vesicles were labeled as "exosomes" [454]. Currently, exosomes are defined as one type of extracellular vesicle measuring 40–100 nm diameter that are secreted upon fusion of multivesicular bodies (late endosomes) with the plasma membrane [449, 455–457]. The multivesicular bodies are formed from early endosomes and are known to be involved in numerous endocytic and trafficking functions, including protein sorting, recycling, transport, storage, and release [458].

Exosomes are released by any kind of cells including T-cells, B-cells, dendritic cells, fibroblasts, and endothelial cells. They are found in a wide range of body fluids like blood, saliva, bronchial, urine, seminal fluid, amniotic fluid, pleural fluid, and breast milk [459–468].

Exosomes contain various proteins, specific and nonspecific ones, and various nucleic acids including mRNAs, miRNA, lncRNAs as well as nuclear and mitochondrial DNA. Each of them, separately, could be used as potential exosomal biomarker [469].

Exosomes as a source of protein biomarkers

Proteomic analysis has shown that many of the proteins detectable in exosomes, from all cell types, are common among themselves. These include ribosomal components, cytoskeletal proteins, small and heterotrimeric GTPases, tetraspanin proteins, and the components of the endosomal sorting complex required for transport (ESCRT), which is also involved in the multivesicular bodies formation [456, 470, 471].

Furthermore, exosomes contain many cell-specific proteins making them a potential biomarker tool for diagnosis and therapy that can be used in renal diseases. Proteome analysis of urinary exosomes has identified proteins from all segments of the nephron, including glomerular podocytes (e.g., podocin and podocalyxin), proximal tubules (e.g., megalin, cubilin, aquaporin-1, and type IV carbonic anhydrase), thick ascending limb of Henle (e.g., type 2 Na-K-2Cl cotransporter), distal convoluted tubule (e.g., thiazide-sensitive Na-Cl co-transporter), and the collecting duct (e.g., aquaporin-2) [456, 472]. Also, they have shown to reflect AKI and are considered as potential diagnostic biomarkers [160].

Besides the probable role of exosomes to eliminate the excess or senescent proteins and lipids, there is considerable evidence that they could be involved in intercellular signaling in a cell-selective manner [473–480]. Exosomes may elicit a response on target cells by at least three possible mechanisms: (1) they can adhere with high specificity to the target cell surface (without membrane fusion) through adhesion molecules and receptors present on their surfaces, leading to receptor activation and downstream signaling in the target cell [474, 475]; (2) they could fuse directly with target cells, and consequently transferring the contents of exosomes (mRNA, miRNAs, proteins, and signaling molecules); and (3) via endocytosis, the exosomes may incorporate their content into target cells and subsequently being processed through endosomal pathway; this possibility is better supported in the literature [476, 478–480].

In addition, it is possible that the exosomes secreted into the blood and extracellular fluid may be involved in renal physiology and pathophysiology, especially among cell types with their plasma membranes in direct contact with the vascular compartment such as cells of the immune system and endothelial cells. A great example is the role played by exosomes and microvesicles in the cell-cell communication during immune and stem cell signaling [477, 481–486]. Additionally, exosomes transported by blood may be involved in tumor angiogenesis since tumors promote their vascularization not only through the secretion of known angiogenic cytokines and growth factors but also via exosomes [478, 487, 488].

Urinary proteomics studies have identified potential urinary biomarkers for several pathological entities, such as ADPKD [489], acute kidney transplant rejection [452], and diabetic nephropathy [490]. Despite these and other successes, the number of kidney-derived proteins and peptides detectable in whole urine or "minimally processed" urine by MS has been limited in part due to the presence of filtered plasma proteins and very abundant kidney-derived proteins, especially Tamm-Horsfall protein (i.e., uromodulin). Abundant proteins compete with less abundant proteins for identification in the mass spectrometer. Therefore, we may be missing the biomarker candidates that would provide the best sensitivity and specificity for diagnosis of a given disease. Hence, enrichment of kidney-derived proteins has been done through exosomal isolation from urine to improve the detection of specific biomarker candidates [456]. Since about 3% of the total urinary protein is derived from exosomes, their isolation provides more than 30-fold enrichment of exosomal proteins allowing easier detection of those minor components from the whole urine by immunochemical methods or MS-based systems [121].

The identification and quantification of thousands of proteins from a single sample has become possible due to the use of shotgun proteomics technique based on MS systems, which combine liquid chromatography and tandem mass spectrometry (MS/MS). The combination of new and advanced technology developments in MS and the advent of detailed protein sequence data from human genome have led to the discovery of more protein biomarkers [491, 492]. Moreover, this advancement has enabled to carry out large-scale profiling of proteins present in urinary exosomes from normal humans providing more data, which is available on public database (http://dir.nhlbi.nih.gov/papers/lkem/exosome/). This database provides a listing of 1160 proteins present in urinary exosomes and contains potential biomarker proteins that can be the basis of hypotheses regarding the mechanism of the disease [492].

The study analysis of exosome as a source of protein biomarkers could be a good fit on those diseases that require clinical decision making based on current diagnostic methodologies, which may not be as optimal or time efficient for patient care. On the basis of these considerations, one example of a good target for biomarker discovery in urinary exosomes may be in renal allograft patients who experience an increase in serum creatinine levels. The discrimination between rejection and kidney injury as well as the discrimination between different mechanism of rejection is generally addressed through renal biopsy and in some places may take longer than expected to get the results. Hence, a rapid immunological test could speed the initiation of appropriate therapy and it may allow a much earlier detection of disease than a renal biopsy. However, this is a point of controversy and great debates on opinions [493].

Another target that exosomes could be useful would be in patients with early diagnosis of AKI in surgical and intensive care settings. In several studies, potential markers have been identified, including KIM-1 [494], heat shock protein-72 (HSP72) [495], Klotho (KL) [496], IL-6 [497], NGAL [498], L-FABP [499], netrin-1 [500], IL-18, cystatin C, and fetuin-A [160]. Among these, the study of fetuin-A, IL-18, cystatin C, and NGAL has been conducted using exosomes [160, 491, 501, 502]. In 2013, Alvarez et al. reported that urinary exosomal NGAL was significantly higher in patients with recent kidney transplant that would develop delayed graft function (DGF) than in those without DGF, being considered as a need for dialysis during 1 week after transplantation. However, the results were not significant for urinary cellular NGAL [501]. Moreover, other kidney injury markers such as IL-18 and cystatin C could also be detectable in urinary exosomes and provide important information about kidney or transplant outcome [502]. Furthermore, CD3+ T-cells exosomes have been described to be higher in the graft of patients with acute cellular rejection than in those without rejection [503].

Urinary exosome analysis has been increasingly used to identify noninvasive markers of disease, including ADPKD [489, 504, 505], lysosomal storage disease (e.g., Nieman-Pick disease and cystinosis), and transporter mutations (e.g., Gitelman and Bartter syndromes) [492]. For instance, a recent quantitative proteomic study on urinary extracellular vesicles in patients with ADPKD has identified higher abundances of urinary complement-related proteins (i.e., C3 and C9) as well as cytoskeletal proteins (i.e., villin-1 and plakins, such as: envoplakin and periplakin), suggesting the use of these proteins as disease markers [489]. Moreover, it has been suggested that urinary exosomes analysis could be performed in patients with hepatorenal syndrome [506], and in those with hypertension [507] in order to find biomarkers to predict the most effective drug as well as for the purpose of being used as a noninvasive diagnostic and monitoring test in various glomerular diseases, including FSGS [508].

Exosomes as a source of RNA biomarkers

The RNAs contained in exosomes are potentially useful as disease biomarkers. Most studies of urinary RNAs have bypassed exosome isolation, opting for direct analysis of mRNA levels using RT-PCR in sediments from whole urine, which contains RNA from both exosomes and whole cells [509–511]. For example, a recent study has shown an increased glycoprotein B7-1 to nephrin mRNA ratio in urinary sediments from patients with minimal change disease compared with FSGS [512]. Also, another study has found that urinary granzyme A mRNA levels might potentially distinguish patients with cellular rejection from those with AKI [513].

miRNA profiling could also be used to identify potential biomarkers. For example, Lv et al. have studied urinary exosomal miR29 and miR200 families to explain their role in renal fibrosis. They have shown that members of both of these families, including miR-29a, miR-29c, miR-200b, and miR-200c, are significantly downregulated in CKD patients with mild fibrosis in contrast to those patients with moderate and severe fibrosis [448]. Two other studies have confirmed these results [514, 515]. On the other hand, other studies have demonstrated that decreased levels of miR-200b isolated from urine and plasma are accompanied by kidney graft fibrosis [516, 517]. Additionally, a study with urine exosomes without enrichment have found several miRNAs (miR-200a, miR-200b, and miR-429) to be decreased in urinary sediments from patients with immunoglobulin A nephropathy, and the degree of reduction correlates with the severity of the disease [518].

It may be possible to increase both the sensitivity and the specificity of RNA biomarker approaches through flow cytometry by enrichment of exosomes specific to the given cell type [519, 520].

Exosomes as potential therapeutic agents

The evidence learned about exosomes and their contents, including mRNAs and miRNAs, along with their involvement in mechanisms of cell-cell communication, may suggest a new direction for the use of exosomes as delivery vehicles for targeted therapy. RNA-bearing exosomes can potentially deliver their contents to specific target cells in order to transiently correct dysregulated processes [452]. Some of the studies have demonstrated the application of exosomes in melanoma and lung cancer immunotherapy; however, patients included in those clinical trials have shown a minor effect on reduction of disease progression [521–526].

Another potential use of exosomes is as vehicles for the delivery of specific antigens, for example in vaccinations [527, 528]. Additionally, exosomes may play an important role in the mechanism of paracrine effect between cells which have been observed in experimental stem cell therapy; for instance, in an experimental stem cell therapy study for AKI the mesenchymal stem cells have been shown to improve recovery in part due to paracrine factors derived from secreted exosomes [529, 530].

Although the use of exosomes as diagnostic biomarker and therapeutic agents in kidney diseases is highly promising, more studies are required to confirm the previously described findings.

Epigenomics in kidney disease

Epigenetics is defined as "the study of changes in gene function that are mitotically and/or meiotically heritable and that do not entail a change in DNA sequence" [531]. Epigenetics acts as a bridge between genotype and phenotype, helping to understand that some genetic alterations may not necessarily result in an altered phenotype. Therefore, it is important to consider epigenomic biomarkers alongside genomic and transcriptomic analyses to discover the role of different pathways in disease, as changes in gene expression may not be solely associated to the genetic aberrations and it may not be detected using purely genetic analysis [395].

Epigenetics is subdivided by mechanisms, in which covalent modifications of DNA and histone proteins (i.e., DNA methylation) as well as the action of noncoding RNAs (i.e., miRNAs) have the ability to interfere with gene expression without changing the genetic sequence of the gene in question, including any of its promoter or enhancer regions [395].

Epigenetics regulation of gene expression results in altered levels of target gene mRNA available for translation, which subsequently affects protein synthesis. Moreover, epigenetic regulation is required to maintain the regulation of gene activity, transcription, and genomic stability [532]. Some examples of epigenetic dysregulation that have been described include: (1) Prader-Willi syndrome, which is an imprinting disorder, (2) Fragile X syndrome, which is caused by loss of genomic stability, and (3) multiple different cancers [533]. Also, there is evidence suggesting that loss of epigenetic control might be associated with CKD due to differential expression of microRNAs [534] and differential DNA methylation detected in kidney fibrinogenesis; the latter being a hallmark of CKD [535].

DNA methylation is the transfer of a methyl group ($—CH^3$) from S-adenosylmethionine to the fifth carbon of a cytosine nucleotide in the DNA sequence and is catalyzed by DNA methyltransferases (DNMTs) [536–538]. The presence of these cytosine-bound methyl groups can either prevent the binding of transcription factors in the promoter region of genes or attract and bind repressor proteins; both leading to a reduction in gene transcription. The majority of the methylated cytosines are followed by a guanine nucleotide and these sites have been denoted as CpG sites [539].

DNA methylation has been associated with renal, colon, and breast cancer, with all of them showing high levels of epigenetic dysregulation [540]. In CKD, DNA methylation has also been described in several genes. In some studies, hypermethylation of methylenetetrahydrofolate (*MTHFR*), uromodulin like 1 (*UMODL1*), and klotho (*KL*) has been associated with rapid CKD progression;

however, due to conflicting results in other studies showing hypomethylation in some of these genes (i.e., *MTHFR*), the usefulness of any panel of noninvasive epigenetic biomarkers remains controversial [236, 437, 541–546]. Hence, further investigations using larger number of CpG sites, phenotypes, and ethnicities in larger cohorts are required to allow accurate, phenotype-specific epigenetic biomarkers to be established [395].

It is important to have a better understanding of DNA methylation and its involvement in disease development that subsequently will lead to the discovery of potential points of therapeutic intervention as well as the discovery of diagnostic and prognostic biomarkers [395].

Genomic medicine in kidney disease

Genomics refers to the complete study of the genome of an organism. It gained popularity in the 1990s with the initiation of the HGP, which was completed in 2003 [547]. The human genome contains approximately 3 billion DNA nucleotides [548] and about 25,000 human protein-coding genes [549], of which nearly 4000 have been implicated in human disease [550]. The human genome as well as genomes of other species have been delineated and placed into public databases, which can be easily searched [5, 549]. Genomics consists of three interacting parts: (1) structural genomics, which handles the mapping and sequencing of genomes; (2) functional genomics, dealing with the analysis of mRNA expression in biological development; and (3) computational genomics, which focuses on new strategies to analyze the vast and complicated genomic databases [551].

The ultimate aim of genetic testing is to provide a molecular explanation for a patient's disease or certain aspects of it such as better understanding of the pathophysiology of the specific disease as well as confirmation of diagnosis to provide precise genetic counseling and appropriate clinical management [552]. However, identification of the mutation for a disease can be a difficult task due to the wealth of variation within the human genome.

Though individual inherited kidney diseases are rare, together they account for approximately 10% of adult ESRD [553–555] and at least 70% of pediatric nephropathy [556, 557]. Besides, 10%–29% of adult patients with ESRD report a positive family history across different ethnicities and etiologies [558–560].

CKDs with Mendelian causes often differ considerably from acquired forms of disease in their clinical prognosis, course, and management [553, 561], and they can be difficult to detect with the use of traditional diagnostic approaches such as renal biopsy [562–564]. Moreover, patients may not present symptoms until they reach ESRD, since early-stage of CKD (stages 1, 2, and 3) is often clinically silent. Hence, in about 15% of adult and 19% of pediatric patients with newly diagnosed ESRD, the clinical diagnosis is ambiguous and classified as "other" or "unknown," making patient's clinical management even more difficult and impeding tailored therapeutic interventions [348, 565].

Technologies such as chromosomal microarray (CMA) and NGS were first introduced as research tools in the past decade and have empowered investigation of genetic contributions to disease through assessing variation across the genome [566, 567]. These technologies are being increasingly deployed in the clinic, reflecting a shift in the work-up of suspected hereditary disorders in clinical medicine. Traditionally, diagnosis of individuals suspected to have an inherited form of nephropathy involved multiple clinical visits and complex and/or invasive studies (such as renal biopsy or biochemical testing) to identify the most likely etiology, in which the clinician would select one or a few associated genes for genetic testing. Now, a shift toward genomic medicine is occurring, where genome-wide testing approaches are used to identify the causal etiology of disease in a patient and used to guide their clinical care, once the genetic variant or variants has/have been identified [568].

As sequencing costs are rapidly declining, genome-wide testing has been advocated as a sensitive and ultimately cost-effective first-line diagnostic test [569–571] in adults and pediatric patients with early-onset CKD [557, 561]. The use of such testing could help to surmount the diagnostic challenges posed by the substantial genetic and phenotypic heterogeneity of many hereditary nephropathies [572].However, many complex questions must be answered, which includes how to identify patients for whom genetic testing is indicated, choose the appropriate test, identify causal variants, and translate genetic findings into personalized care. In order to address these questions, it would require a thorough investigation in large patient cohorts of diverse ages, ethnicities, and disease etiologies. In addition, the ethical, legal, and social implications (ELSIs) that accompany genetic testing on an increasingly broad scale must be a priority [562].

It is imperative to remain mindful of the limitations of our knowledge. Genetic testing does not give absolute answers, but rather provides a probabilistic biomarker, the meaning of which must be interpreted in the overall genomic and clinical context [573]. We need to be aware that in many cases, the "one gene, one disease" model does not apply due to the presence of genetic and environmental modifiers. Therefore, physician and geneticist must incorporate diagnostic sequence interpretation with traditional tools such as clinical history and renal biopsy as well as with other sources of "omic" data, all of which can provide crucial insight into the genetic findings. As we consider each patient individually based on his/her clinical context, genomic medicine will be able to deliver truly personalized care for patients with kidney disease [562].

When is recommended to do genetic testing?

Currently, genetic testing is recommended as part of the diagnostic work-up for pediatric patients with nephropathy of unknown etiology [557, 561]. In adult patients, genetic testing is suggested only for those who are strongly suspected to have a known hereditary form of nephropathy, particularly in the setting of a compelling family history of early-onset renal failure, and patients with clinical features (phenotypes) that have a strong hereditary basis such as congenital anomalies of the kidney and urinary tract (CAKUT) [562, 574–578].

Certain clinical situations might also merit genetic testing, such as those in which diagnostic findings would enable patients to avoid undergoing unnecessary invasive procedures or prevent them from receiving ineffective and costly treatment with substantial adverse effects (e.g., steroid therapy in patients with hereditary etiologies of steroid-resistant nephrotic syndrome [SRNS]) [576].

In females with clinical features and/or history suggestive of monogenic X-linked nephropathy, such as X-linked Alport syndrome and Fabry disease, genetic testing is also advised because these patients can develop a severe form of the disease despite being female carriers who display a milder (often subclinical) phenotype than seen in male patients [579–581].

Genetic testing has also been recommended in the evaluation of potential living kidney donors, with donation contraindicated among those found to have autosomal dominant forms of inherited kidney disease such as ADPKD or those sharing genetic susceptibility factors for atypical hemolytic uremic syndrome [582]. Furthermore, the 2017 Kidney Disease: Improving Global Outcomes (KDIGO) guidelines have recommended apolipoprotein L1 (*APOL1*) risk alleles (G1 and G2 alleles) testing for candidate kidney donors but has acknowledged that the evidence based is moderate as well as insufficient evidence exists regarding the impact of donation on lifetime risk of nephropathy [582]. Nevertheless, *APOL1*-mediated disease has been reported to follow a recessive pattern of inheritance, with individuals with these two risk variants (either G1/G1, G2/G2 or G1/G2) displaying a 7- to 10-fold higher risk for hypertension-associated ESRD; 10- to 17-fold increased risk for FSGS-associated ESRD [583, 584], and a 29-fold increased risk for HIV-associated nephropathy [562]. Consequently, the *APOL1* risk genotypes have also been associated with increased risk of other forms of kidney disease, progression to ESRD, and allograft failure [585, 586].

Type of genetic defects

Copy number variation

Copy number variation or copy number variants (CNV) refers to the structural variant that results in gain or loss of DNA at relevant locus (i.e., a deletion or a duplication). The variation can affect only a single gene, larger regions involving multiple genes, whole chromosomes (e.g., Down's or Turner's syndromes) or even complete haploid genomes (triploidy). While all genetic variations can occur de novo and thus are not inherited, this scenario is especially common with CNVs [552, 562].

Though the ability to distinguish benign CNV from pathogenic CNV could represent a major challenge, typically large deletions and duplications are considered to be pathogenic resulting in gene dosage changes likely to have deleterious effects. Meanwhile, gene interruptions, gene fusions, and position effects are increasingly recognized as mechanisms that can mediate the downstream effects of CNVs [552]. An example is the region at 17q12, which harbors the hepatocyte nuclear factor 1 homeobox B (*HNF1B*) gene known to cause renal cysts and diabetes syndrome (RCAD), in which the whole-gene deletion of *HNF1B* constitutes about half of all mutations identified in this syndrome [587, 588], and when the deletion was present often extended beyond the region harboring *HNF1B* to encompass other genes [589]. In a recent study involving >15,000 subjects, some patients with deletions at 17q12 had an additional neurodevelopmental phenotype; the locus was also found to confer a high risk of autism and schizophrenia [590]. Up to date, there have been no reported neurocognitive or psychiatric phenotypes in patients with RCAD resulting from point mutations in *HNF1B*. The co-occurrence of multiple, apparently unrelated clinical findings in patients with 17q12 deletion are indicative of a contiguous gene syndrome that extends beyond RCAD [552].

Single-gene mutation

Most of our understanding of genetic disease is based on mutations that cause single-gene disorders. Sequence changes might occur at any of the nucleotide sites and include single-nucleotide variants (SNVs), small insertions or deletions involving <5–10 bp (indels), and structural variants involving large (≥1 kb) DNA variants that can present as balanced (e.g., inversions or reciprocal translocations) and/or imbalanced alterations (e.g., copy number variants). These can lead to amino-acid substitutions (missense mutation), frameshift mutations, the introduction of premature stop codons (nonsense mutations), exon skipping and exon deletion (splice site mutations), and regulatory region (promoter and enhancer) mutations. Owing to the abundance of rare, predicted damaging variants in a typical human genome, the risk of falsely attributing causality is high. Therefore, a major challenge in genetic diagnostics is to identify which variants are disease-causing mutations [552, 591].

The majority of these mutations are routinely detected by dideoxy-DNA sequencing (Sanger) techniques. Individual amplicons covering the regions of interest (i.e., 5′ untranslated region/promoter region and all exons including exon/intron splice sites) are sequencing using standard methodologies [552].

Diagnostic genomics technologies

The improvement in genomics technologies now allows the identification and characterization of disease genes in a matter of a few months with relatively minimal resources. From a diagnostic service point of view, mutation identification is either performed at a chromosomal (i.e., cytogenetics) level in order to identify CNVs, or at a DNA sequence (i.e., molecular genetics) level to identify single-gene mutations [552].

Clinicians are generally advised to start with a disease-specific genetic panel using targeted forms of genetic testing such as Sanger sequencing or targeted NGS panels. If the results are negative, then they can proceed to a: (1) Mendeliome panel, which can detect variants in all known disease-causing genes; (2) whole-exome sequencing (WES), which can detect variants in all coding regions; and/or (3) whole-genome sequencing (WGS), which can detect variants in all coding and noncoding regions. The clinical sequence interpretation should be performed according to consensus guidelines [592–595].

The most common modalities used for genetic testing are: Sanger sequencing, chromosomal microarray (CMA), and NGS approaches. The latter includes targeted NGS, WES, and WGS.

Chromosomal karyotyping

Chromosomal karyotyping has been the longest technique used to identify genetic defects where chromosomal disorders, translocations, and other large genomic imbalances can be detected. Cells are arrested in metaphase and chromosome spreads are prepared. Subsequently, the chromosomes are stained to produce the characteristic G-banding pattern and analyzed using light microscopy. Then, the chromosomes are assessed for changes in copy number (e.g., trisomy, triploidy), rearrangements (balanced and unbalanced translocations), large deletions or duplications and structural abnormalities (i.e., ring chromosomes) [552].

Chromosomal microarray (CMA)

Karyotyping has been the technique used for genomic disorders. However, many genomic disorders are caused by CNVs that fall below 1–2 Mb resolution of karyotyping [596]. Hence, CMA is a technique to detect CNVs of 200–400 kb [8, 597].

The two major types of CMA used in the clinical setting include: (1) array comparative genomic hybridization (aCGH) and (2) single-nucleotide polymorphism (SNP) arrays. Both of these techniques offer excellent genome-wide coverage and use enrichment of probes in clinically relevant regions to enable resolution at the single-exon level [567, 598, 599].

a. **Microarray-based comparative genomic hybridization (aCGH)**

This technique was first used in 1990s to investigate CNV in tumor cells. Nowadays, it is used as a standard diagnostic procedure to detect structural anomalies [552, 600].

aCGH is a type of CMA in which patient's and control's DNA are labeled with different fluorescent dyes and cohybridized to an array containing immobilized oligonucleotide probes (i.e., a single DNA probe) in order to directly compare genomic copy number at that genomic region. Therefore, aCGH identifies net gain (duplication) or loss (deletion) of DNA sequences. Because the DNA sequences of the oligonucleotide probes are known, it is possible to accurately map and define the size of any CNV [552, 562]. This technology is already having a significant impact on clinical practice, enabling the identification of unknown deletions and duplication syndromes [601].

This method can detect only unbalanced chromosomal abnormalities because balanced chromosomal abnormalities such as reciprocal translocations, inversions, or ring chromosomes do not affect the copy number [600]. This method can also detect genomic gains or deletions leading to loss of heterozygosity (LOH). LOH occurs when one allele of a gene is mutated in a deleterious way and the normally functioning allele is lost. LOH occurs commonly in oncogenesis. For example, tumor suppressor genes help keep cancer from developing. If a person has one mutated and dysfunctional copy of a tumor suppressor gene, and his/her second (functional) copy of the gene gets damaged, then, that person may become more likely to develop cancer [602].

b. **Single-nucleotide polymorphism array (SNP)**

SNP array is another type of CMA, which is used to detect polymorphisms within a population. This can be divided into: (1) phenotype-specific SNP panels, which offer a low-cost test containing alleles that are known to drive specific phenotypes [603, 604] and (2) genome-wide SNP panels, which is a large-scale and cost-efficient platform to assess risk of multiple common genetic disorders with variably documented associations in one test [597, 605, 606].

A SNP is the most frequent type of variation in the genome. Around 339 million SNPs have been identified in the human genome [607], 84.7 million of which are present at frequencies of 1% or higher across different populations worldwide [548].

The basic principles of the SNP array are the same as the aCGH, in which a patient's DNA is hybridized to an array containing immobilized allele-specific oligonucleotides (ASO) probes (i.e., DNA probes corresponding to SNPs) and the hybridization pattern is compared with previously analyzed controls. The

ASO probes are often chosen based on sequencing of a representative panel of individuals; positions found to vary in the panel at a specified frequency are used as the basis for probes. Two probes must be used for each SNP position to detect both alleles; if only one probe were used, experimental failure would be indistinguishable from homozygosity of the nonprobed allele [608].

An SNP array is a useful tool for studying slight variations between whole genomes. The most important clinical applications of SNP arrays are for determining diseases susceptibility and measuring the efficacy of drug therapies designed specifically for individuals [597].

In research, SNP arrays are most frequently used for GWAS [609]. Since each individual has many SNPs, SNP-based genetic linkage analysis can be used to map disease loci and determine disease susceptibility genes in individuals. The combination of SNP maps and high-density SNP arrays allows SNPs to be used as markers for genetic diseases that have complex traits. For example, GWAS have identified SNPs associated with diseases such as rheumatoid arthritis [610], prostate cancer [611], and type 2 diabetes mellitus [612].

An SNP array can also be used to generate a virtual karyotype using a software to determine the copy number of each SNP on the array and then align the SNPs in chromosomal order [613]. SNP arrays, however, have an additional advantage of being able to detect copy-neutral LOH (also called uniparental disomy or gene conversion). Copy-neutral LOH is a form of allelic imbalance. In copy-neutral LOH, one allele or whole chromosome from a parent is missing. This problem leads to duplication of the other parental allele. Copy-neutral LOH may be pathological. For example, say that the mother's allele is wild type and fully functional, and the father's allele is mutated. If the mother's allele is missing and the child has two copies of the father's mutant allele, disease can occur [613].

Indications for the use of chromosomal microarrays

- It is useful in patients with phenotypes commonly resulting from genomic imbalances, such as multiple congenital anomalies [562].
- It is used to study genetic abnormalities in cancer [602].

Advantages of chromosomal microarrays

- It has an increased diagnostic yield, in part, due to the improvement in microarray design and fabrication. This allows to produce slides with an excess of 2 million individual features (probes), which means that is possible to achieve a resolution as high as 1–2 kb [552].
- CMAs have as much as 10-fold increased diagnostic yield versus karyotyping for intellectual disability, autism, and multiple congenital anomalies [567, 614, 615]. The higher resolution enables detection of

CNVs missed by karyotyping. Hence, CMAs are recommended as first-line diagnostic testing for the hitherto indications [567, 616].
- The use of genome-wide CNV detection increases diagnostic sensitivity [562].

Disadvantages of chromosomal microarrays
- It cannot detect SNVs, indels, and small CNVs [562].
- It has limited ability to detect balanced chromosomal rearrangements, positional information for unbalanced rearrangements, low-grade somatic mosaicism, and CNVs in certain regions such as pseudogenes and repetitive elements [562].

Examples of chromosomal microarrays applications
- CMA has been shown to be an effective first-line diagnostic tool in the diagnostic evaluation of pediatric nephropathy, such as CAKUT, including those patients who have the syndromic form and those who does not have it. Some of the deletions detected are at the 17q12, 22q11.2, and 16p11.2 loci [617–623].
- For the detection of whole-gene deletion of *HNF1B* in patients with renal hypodysplasia and autism [590].
- To detect 22q11.2 deletion (DiGeorge syndrome) in patients with renal agenesis and neonatal hypocalcemia [562].
- Several personal genome companies now provide versions of commercial clinical genotyping services (genome-wide SNP tests) to consumers, such as the Personal Genome Service from 23andMe, Pathway Genomics, and Navigenics, to name a few [624].

Multiplex ligation-dependent probe amplification

Intragenic rearrangements (i.e., whole/multiple exon deletions/insertions), which may be below the resolution of aCGH, can be detected using a technique called multiplex ligation-dependent probe amplification (MLPA) [625].

This is a high-throughput method developed to determine the copy number of up to 45 different DNA or RNA sequences that can be targeted in a single, semi-quantitative PCR-based approach. The sequences detected can be small (~60 nucleotides in length), enabling the analysis of fragmented DNA [625].

The MLPA reaction is fast, relatively inexpensive, and easy to perform. It requires only 20 ng of sample DNA and can distinguish sequences differing by only a single nucleotide. The MLPA reaction results in a mixture of amplification fragments ranging between 100 and 500 nt in length, which can be separated and quantified by capillary electrophoresis. The equipment necessary for MLPA is identical to that used for standard sequencing reactions: a thermocycler and a fluorescent capillary electrophoresis system. Comparison of the peak pattern obtained on a DNA sample to that of a reference sample indicates which sequence show aberrant copy numbers [626].

It relies on probes to hybridize to the gene region of interest before amplification. The probes are designed to lie immediately adjacent to each other, so that they can be ligated together in the subsequent step before amplification by PCR. The primer is fluorescently labeled, thereby enabling the PCR product to be quantified. If parts of the gene are deleted and a probe could not hybridize, no amplification can occur. Conversely, if there is duplication, more probes will hybridize leading to an increased amplification [552].

The fundamental for the MLPA technique is that it is not the sample DNA that is amplified during the PCR reaction, but MLPA probes that hybridize to the sample DNA. Each MLPA probe consists of two probe oligonucleotides, which should hybridize adjacent to the target DNA for a successful ligation. Only ligated probes can be exponentially amplified by PCR. In contrast to standard multiplex PCR, only one pair of PCR primers is used for the MLPA, resulting in a more robust system. This way, the relative number of fragments present after the PCR reaction depends on the relative amount of the target sequenced present in a DNA sample. Although frozen tissue is preferred over FFPE tissue, both can be used for MLPA. However, is necessary to have an adequate amount of material for reliable detection of chromosomal aberrations. For instance, in tumor cases, it is important to avoid contaminants from necrotic cells; therefore, the percentage of tumor cells in tumor samples should be at least 50% and the DNA used for MLPA analysis should be of high quality and contain about 250–450 ng of DNA per tumor to enable reliable identification of aberrations [626].

Sanger (dideoxy) DNA sequencing

The Sanger-dideoxy method of DNA sequencing refers to the traditional sequencing approach, developed by Frederick Sanger and colleagues [627]. This is a DNA sequencing method that uses labeled chain-terminating dideoxy nucleotides in the DNA strand being sequenced. This method generates a sequence chromatogram that can be analyzed to detect genetic variants [562].

Sanger sequencing has high analytical validity in detecting causal SNVs and indels (small insertions or deletions involving <5–10 bp) within a DNA segment of <1000 bp. Hence, this modality remains the gold standard for molecular diagnosis when a single-gene disorder is suspected and for confirmation of NGS findings [593, 600, 628].

The sequence of interest is amplified by PCR and sequenced using a single primer and a nucleotide mix containing normal and fluorophore-labeled dideoxy nucleotides. Since, these dideoxy nucleotides are chain terminators, the reaction results in a number of fragments of varying lengths, all ending in a fluorophore-marked nucleotide. Capillary electrophoresis is used to fractionate products according to length; thus, enabling the sequence of the gene to be determined. Sanger sequencing can reliably resolve the sequence of DNA fragments of 50–800 bp [552].

Indications for the use of Sanger DNA sequencing

- To confirm the findings from NGS analysis [562, 600].
- To identify genomic regions refractory to NGS, such as those segments that are highly repetitive, homologous, or guanine-cytosine (GC)-rich [562, 600].
- It can be used in patients whose phenotype is indicative of a disorder caused by mutations in one specific gene [562, 600].

Advantages of Sanger DNA sequencing

- It offers higher analytical accuracy [562, 600].
- The sequence interpretation is easier and faster compared with multigene testing leading to faster turnaround time [562, 600].
- There is no risk of secondary findings (genetic findings that are not related to the primary indication for testing; also called incidental findings) [562, 600].

Disadvantages of Sanger DNA sequencing

- The resolution is limited to single DNA fragments of <1 kb (<1000 bp) [552, 628]; in other words, it cannot detect larger structural variants.
- It could become time- and cost-inefficient with increasing gene length and/or number of genes tested, limiting its utility for genetically heterogeneous conditions [594, 600].

Examples of Sanger DNA sequencing applications

- To confirm frameshift *COL4A3* variant detected by NGS [593, 594, 600, 628].
- It can be used as a diagnostic testing for Fabry disease [593, 594, 600, 628].
- To detect *CTNS* mutation (nephropathic cystinosis) in a patient with corneal cysteine crystals and Fanconi syndrome [593, 594, 600, 628].

Next-generation sequencing (NGS)

NGS applies targeted capture and massively parallel sequencing in order to simultaneously assess variation in selected regions of the genome, enabling rapid and cost-effective large-scale genetic investigation. NGS can detect SNVs and small indels (<1000 bp) within selected regions of the genome based on the type of approach used: multiple genes of interest (using *targeted NGS panels*), coding regions of the genome (using *WES*), or both coding and non-coding regions of the genome (using *WGS*) [600, 628–631].

Newly emerging NGS technologies and instruments have further contributed to a significant decrease in the cost of sequencing nearing the mark of $1000 per genome sequencing [547, 632]. Among the main instrument providers for the current generation of sequencers include Illumina (San Diego, CA; HiSeq analyzer, MiSeq analyzer, and Genome Analyzer IIX), Life Technologies (Carlsbad, CA; ABI SOLiD4 and Ion Proton), Roche (454 Life Sciences, Branford, CT), GS FLX Titanium, Complete Genomics, Helicos Biosciences (Heliscope), and Pacific Biosciences (PacBio single-molecule real-time "SMRT" and Sequel System) sequencing [547, 552, 632].

DNA sequencing with commercially available NGS platforms is generally conducted through the following steps. The first step involves the selective amplification or capture of DNA sequences of interest by PCR in vitro, such as a comprehensive collection of candidate genes or a defined linkage region, which may even extend to the entire exome. Second, the amplified or captured DNA is then fragmented, and sequencing libraries are prepared. At this stage, samples from multiple, different individuals can be indexed by the ligation of unique sequence oligonucleotide adapters (multiplexing). Third, the enriched fragments are sequenced multiple times ("re-sequencing"), with the number of times each fragment is sequenced being variable (referred to as the "depth of coverage"). Typically, a mean coverage of 30x-50x is recommended for reliable sequence variant identification [552].

Targeted NGS panels

NGS gene panels use targeted enrichment of selected genes to provide rapid and inexpensive sequencing at higher coverage than that achieved with WES or WGS. In this approach, patients are tested for a set of genes that are commonly associated with their clinically suspected phenotype [593, 594].

As NGS panels are quickly becoming a first-line diagnostic test, it is critical that they can accurately detect whether a particular genetic variant is present in the region of interest [633, 634]. Organizations such as the American College of Medical Genetics and Genomics (ACMG), British Society for Genetic Medicine (BSGM), Association for Clinical Genomic Science (ACGS), and European Society of Human Genetics (ESHG) have published technical guidelines for clinical NGS regarding *sequencing coverage* and *depth* as well as other quality metrics [592, 593, 595]. ***Sequencing coverage*** denotes the percentage of bases in the DNA region targeted by sequencing that is sequenced at a given number of times; while the ***sequencing depth*** refers to the average number of times that a given nucleotide is read in a set of DNA sequence reads. ***High coverage and depth*** means that more of the targeted genomic region has been sampled at a greater number of times, which increases the accuracy of the resulting data [562].

Despite the high coverage offered by targeted NGS analysis, certain regions are poorly covered by NGS alone [594]. Some of these regions include those with GC-rich content, as seen for example with the first exon of *COL4A3* associated with Alport syndrome and thin basement membrane disease, or in those cases with causal variants of *MUC1* gene associated with autosomal dominant tubulointerstitial kidney disease (ADTKD-*MUC1*) [562]. Other undercovered regions include those with high sequence homology (e.g., *PKD1* gene associated with ADPKD). Consequently, laboratories often incorporate other methods such as Sanger sequencing and long-range PCR to ensure that all targeted regions are comprehensively covered at sufficient depth [593].

Indications for the use of targeted NGS panel

- It is useful for those patients with phenotypes that are fairly specific for a particular disorder [562, 593, 600].
- It can be used in those disorders with low genetic and/or phenotypic heterogeneity [562, 593, 600].

Advantages of targeted NGS panel

- The sequencing is selective for the genes that are related to the clinical indication facilitating interpretation and minimizing the risk of secondary findings [593, 594, 635].
- It can be optimized to ensure sufficient coverage of variants in targeted regions. If the targeted panel testing is negative, the clinician can select another panel with broader content or proceed directly to WES or WGS [593, 594, 635].

Disadvantages of targeted NGS panel

- Testing a limited number of genes decreases diagnostic sensitivity, especially for genetically and/or phenotypically heterogeneous disorders [562].
- It can encounter with challenges due to panel design and gene selection; consequently, it needs frequent updates to include newly discovered disease-associated genes [636–638].
- It has minimal capacity for sequence analysis. Hence, this approach needs supplementation with other techniques such as CMA, Sanger sequencing, and long-range PCR in order to identify large genetic rearrangements and causal variants in regions that are not well captured by NGS alone [562].
- The incorporation of this technology into a routine diagnostic laboratory workflow presents with a number of key challenges involving the bioinformatic field, which includes data handling, storage, and interpretation [552].

Examples of targeted NGS panel applications

- Testing *AGXT*, *HOGA1*, and *GRHPR* for primary hyperoxaluria in a patient with childhood-onset calcium oxalate urolithiasis [639, 640].
- Testing for *COL4A3*, *COL4A4*, and *COL4A5* mutations in patients with suspected Alport syndrome and thin basement membrane disease [641, 642].
- Testing patients with nephrotic syndrome using a panel containing genes that are commonly implicated in hereditary forms of this disorder [557, 643, 644].
- Testing for mutations in patients with nephronophtisis-related cliliopathies (NPHP-RC) [645, 646].
- Testing for mutations in genes encoding complement and complement-regulatory proteins that have been reported in patients with C3 glomerulopathy/Dense deposit disease such as *C3*, C8A, *CFB*, *CFH*, *CFHR2*, *CFHR3*, *CFHR4*, *CFHR5*, *CFI*, and *MCP* [647–649].
- Testing for mutations in genes encoding complement and noncomplement regulatory proteins in patients with renal biopsy showing morphology and/or clinical features of thrombotic microangiopathy (TMA), atypical hemolytic uremic syndrome (aHUS), thrombotic thrombocytopenic purpura (TTP), and in transplant-associated TMA following allogeneic stem cell transplantation [650–654].

Whole-exome sequencing (WES)

WES refers to NGS-based analysis of the exome, which is the protein coding regions of the genome that contain the majority of known causal variants for Mendelian disorders [655, 656]. This type of approach is considered one of the best techniques for clinical diagnosis and genetic discovery aside from its value due to time- and cost efficiency [657–659].

Targeted capture and sequencing of the protein-coding regions of the genome through exome sequencing is increasingly applied as a first-line diagnostic tool in clinical medicine, and has been successfully used for a variety of clinical indications, particularly for the diagnosis of metabolic and neurodevelopmental disorders in children [569, 660] as well as for the detection of causal mutations in cancer [661–663]. In those contexts, exome sequencing can provide information for appropriate and individualized medical management, including targeted therapy [570, 660, 664, 665].

Several smaller studies have supported the usefulness of exome sequencing for the diagnosis of early-onset or familial nephropathy [666–669]. More recently, Groopman et al. detected diagnostic variants by WES in about 10% of patients from a diverse adult cohort of 3315 patients with CKD [576], confirming the previous results from those smaller studies.

Indications for the use of whole-exome sequencing

- In patients with high degree of genetic and phenotypic heterogeneity of hereditary nephropathies [576].
- In patients with CKD of unknown etiology [576].
- In those patients that have left undiagnosed by targeted NGS panels [576].

Advantages of whole-exome sequencing

- It offers an unbiased approach that increases diagnostic sensitivity of disorders with high genetic and/or phenotypic heterogeneity [600].
- It has the ability to achieve a specific diagnosis when traditional clinical methods are unsuccessful [576].
- The interrogation of the coding regions that are enriched for known disease-causing mutation is a cost-effective approach to genome-wide testing [576].
- The genetic findings can reclassify disease or provide a cause for undiagnosed nephropathy, emphasizing the usefulness of exome sequencing, which assess genes that otherwise may have gone unevaluated with the use of single-gene or phenotype-driven panel testing [576].
- It enables sequence reanalysis, which may include recalling variants from raw data, reannotating called variants by use of novel bioinformatics tools, and/or re-examining annotated variants in light of newly discovered gene-disease associations [591, 657, 670–674].
- The genetic diagnosis resulting from WES may impact clinical care, including guidance regarding adequate choice of therapy as well as surveillance of associated extrarenal comorbidities, and transplant evaluation for prognosis and choice of donor [576].

Disadvantages of whole-exome sequencing

- It has lower analytical sensitivity and specificity than WGS due to limited coverage of some clinically relevant regions, such as mitochondrial genome [675]. For instance, this limitation can be seen in about 50% of reported pathogenic variants in the *WT1* gene, which is associated with hereditary nephrotic syndrome and Deny-Drash syndrome [676]. Another example includes the duplicated regions of *PKD1* [594].
- It has the limitation and inability to detect indels and copy-number variants [669, 677].
- This approach can lead to multiple candidate variants, increasing time required for interpretation, and need for follow-up testing [576, 669].
- It could generate a burden of secondary findings in genes unrelated to the primary indication for testing [576, 669].

Example of whole-exome sequencing applications

- Diagnosis of nephronophthisis-related ciliopathies (NPHP-RC) with a detection rate in about 60%–70% of patients [667, 678–680].
- Identification of causal variants in pediatric patients with SRNS, expanding the phenotype spectrum associated with *COL4A*-mediated nephropathy [666, 669, 681, 682].
- WES has helped to broaden the genetic spectrum of PKD, in which mutation-negative families have been found to have *GANAB* in ADPKD [683] and *DZIP1L* in ARPKD [684].
- It has a diagnostic value for patients who present with nonspecific renal phenotypes or kidney disease of unknown etiology [685–688] (e.g., congenital chloride diarrhea in an unresolved case of suspected Bartter syndrome where WES detected a pathogenic variant in *SLC26A3*) [685].
- Diagnosis of LIM homeobox transcription factor 1 β (*LMX1B*) glomerulopathy in familial ESRD of unknown origin, which is associated with the Nail-patella syndrome [574, 669].
- Diagnosis of genetic disorders in patients with CKD of unknown origin and in those with early-onset or familial nephropathy [567, 568, 576, 669, 689].

Whole-genome sequencing (WGS)

WGS refers to NGS-based analysis of the whole genome, including protein-coding and noncoding regions [562].

Up to date, the majority of diagnostic variants identified in clinical WGS analysis have been found in exonic regions [690–693]. However, noncoding variants have been involved in various kidney disorders [694–698]. For example, WGS detected a deep intronic mutation in *DGKE* gene in two unrelated families with infantile-onset atypical hemolytic uremic syndrome (aHUS) who had been left undiagnosed by WES use [698]. Subsequent analysis of patient's RNA showed that the variant created a novel splice site that abrogated normal protein function [698].

Genome-wide sequencing increases diagnostic sensitivity relative to targeted panels but holds the challenges of identifying causal variants in the vast amount of data generated and interpreting secondary findings [562].

Indications for the use of whole-genome sequencing

- In those patients with highly genetically heterogeneous phenotypes [562].
- In those patients with nonspecific phenotypes [562].
- In patients with CKD of unknown etiology [562].
- In patients that have left undiagnosed by all other genetic testing modalities [562].

Advantages of whole-genome sequencing

- WGS has superior diagnostic and analytical sensitivity to WES owing to its ability to assess for SNVs, indels, and CNVs in coding and noncoding regions and more complete per-base coverage. WGS has been reported to detect causal variants in about 20%–40% of patients left undiagnosed by WES and/or CMA [699–701].
- It is considered an unbiased approach that has an increased sensitivity for diagnosis of disorders with high genetic and/or phenotypic heterogeneity [562].
- It has the ability to achieve a specific diagnosis when traditional clinical methods are unsuccessful [562].
- WGS enables sequence reanalysis, which may include recalling variants from raw data, reannotating called variants by use of novel bioinformatics tools, and/or re-examining annotated variants in light of newly discovered gene-disease associations [591, 657, 670–674].
- It has the ability to avoid capture bias and provide more complete per-base coverage of coding and noncoding regions [702, 703], facilitating accurate detection of variant genes with high homologous regions such as *PKD1* gene [704], and detection of structural variants, such as those found in several patients with Joubert syndrome [705].

Disadvantages of whole-genome sequencing

- WGS has the difficulty of interpreting noncoding variants [562].
- Some types of variants such as those that contains a highly repetitive, GC-rich sequence, remain wholly refractory to detection using current NGS technology and have been missed by NGS-based regional capture, WES, and WGS [562].
- The large amount of data generated results in a substantial time and monetary costs, hindering return of results [562].
- It can provide secondary findings in genes unrelated to the primary indication for testing [562].

Examples of whole-genome sequencing applications

- Detection of causal intronic variants that have been observed in genetically unresolved case of Alport syndrome [696], Schimke immune-osseous dysplasia [695], and Gitelman syndrome [697].
- Genetic diagnosis of ADPKD (owing to high sequence homology of *PKD1* gene) [704].
- Detection of causal balanced translocations for congenital anomalies [706, 707].

Genetic linkage mapping and association approaches

Recent progress in characterization human genetic variation and its organization in several ethnically diverse groups coupled with advances in computational algorithms and computing power have yielded unprecedented opportunities to interrogate the human genome. It is now possible and commonplace to conduct genomic scans using large panels of markers to identify locations that likely harbor disease genes and to investigate interactions with other genetic or environmental factors. The primary analytic approaches for gene mapping, discovery, and statistical characterization of the genotype-phenotype relationship are: (1) linkage and (2) association studies [144].

Linkage analysis (rare disease: Mendelian inheritance)

The chromosomal region (locus) harboring the new mutation can be located by linkage analysis when the disease segregates through the patient's family following an inheritance model (i.e., autosomal/sex-linked, recessive/dominant). Linkage analysis is the process by which past meiotic events are inferred within a pedigree and compare which loci segregate in a manner consistent with the trait under the assumption that the mutation-harboring gene is in linkage with some of the genetic markers. In order to do this, it is necessary to know the kindred's familial relationships, disease status, and the genotypes of as many individuals as possible (both affected and unaffected). The accuracy and likelihood of identifying the disease-causing genetic locus varies with each family. For instance, large consanguineous families with many individuals available for genotyping are most informative, while several smaller families affected by the same disease might provide information if these families are under the assumption of genetic homogeneity [708–711].

Results from linkage analysis are reported as logarithm^{10} of odds (LOD) scores plotted along the length of the chromosome to produce LOD curve. These represent the likelihood ratio of the disease-causing trait being linked to each genetic marker versus the observation being mere chance. A LOD score of ≥ 3 (odds of 1000:1) is considered the gold standard for demonstrating linkage to a trait [552].

In the case of having a single locus with an interesting candidate gene, Sanger DNA sequencing can be employed to identify whether any mutations are present. In the case of the peaks being wider (i.e., large linkage intervals) or more numerous (i.e., evidence of multiple linkage regions), NGS might be more appropriate [552].

Homozygosity mapping

Homozygosity mapping refers to a simplified version of linkage analysis, which is employed for mapping recessive traits in consanguineous families. This method is used under the assumption that affected individuals will share large regions of DNA that are identical by descent and which are homozygous for the chromosomal region carrying the disease-causing mutation. However, if the assumption of homozygosity is incorrect (there are several examples of compound heterozygous mutations in consanguineous patients), this approach will fail [552].

Genome-wide association studies (GWAS; common diseases: complex inheritance)

GWAS is an observational study of a genome-wide set of genetic variants in different individuals to see if any of those variants are associated with a trait. With this method, a large number of patients with identical diagnosis are genotyped, following by statistical calculations to test for an association between each individual SNP, or a series of contiguous SNPs (haplotype) and the disease [552].

The vast majority of diseases are not caused by a single-gene defect, but are multifactorial (e.g., diabetes mellitus) and influenced by complex interactions between multiple genes and environmental factors. Because GWAS examine SNPs across the genome, they represent a promising way to study complex, common diseases in which many genetic variations contribute to a person's risk. Those SNPs with a statistically significant association are therefore likely to identify genetic contributors to the disease [552].

The general rule is that the more common the disease and the more diverse the underlying etiology, the more subjects are needed to find significant associations. It is not uncommon for a large consortium projects to enroll thousands of cases and controls. In an early landmark study, seven SNPs were identified that showed a statistically significant association with type 1 diabetes, based on 2000 cases and 3000 control [712]. In contrast, in another study with fewer than 600 cases of MN, the authors were able to identify significant association with alleles at two genomic loci (SNPs within the *HLA-DQA1* and *PLA2R1* genes) by GWAS [713]. Hence, the likelihood of success is dependent upon the effect size and allele frequency of the disease-associated marker.

Clinical implications of genetic diagnosis

Genetic findings are increasingly used to inform the clinical management of many nephropathies, enabling targeted disease surveillance, choice of therapy, and family counseling. Genetic analysis has excellent diagnostic utility in pediatric nephrology, as illustrated by sequencing studies of patient with CAKUT and SRNS. Although additional investigation is needed, pilot studies suggest that genetic testing can also provide similar diagnostic insight among adult patients [576].

In order to realize the promise of genomic medicine for kidney disease, many technical, logistical, and ethical questions that accompany the implementation of genetic testing in nephrology must be addressed. The creation of evidence-based guidelines for the utilization and implementation of genetic testing in nephrology will help to translate genetic knowledge into improved clinical outcomes for patient with kidney disease [562].

Utility in patient care

Confirmation of diagnosis

The potential use of genetic data is to improve diagnosis accuracy as well as to allow disease stratification for risk assessment and treatment selection. A definitive diagnosis has psychological implications for the patient and their families because it provides the patient and their family with accurate information about the condition and enable them to have a better plan in medical, social, and educational care. Also, the confirmed diagnosis would allow to make recurrence risk calculations for future pregnancies and family planning [552, 714].

Newborn and prenatal screening

Parents can also be offered prenatal testing for selected disorders with appropriate counseling [600]. The identification of a specific mutation in a proband enables the determination of carrier or affected status in other family members who consent to being tested (so-called cascade screening). In some conditions, early diagnosis is critical for an optimal outcome before the development of overt symptoms. For example, treatment with cysteamine (Cystagon) dramatically delays the onset of renal failure and other complications in cystinosis [715, 716]. As another example, a new and promising drug has been developed for patients with ADPKD known as Tolvaptan, which is a competitive vasopressin V2 receptor antagonist [717]. Torres et al., in a recent study, estimated that Tolvaptan treatment when started in a patient with ADPKD and eGFR of 41 would delay in about 3 years (from 6.2 to 9 years) the time to reach CKD stage 5 [718–721].

Cascade screening, however, could be controversial if the screened individual does not have symptoms and there is no established therapy to prevent or delay the onset of the disease [552]. Nevertheless, with the increasing availability of comprehensive genetic testing (WES and WGS), our knowledge of the relative contributions and effects of different genes will dramatically expand, providing new targets for drug development and novel treatment regimens [552, 576, 600].

Better understanding of pathophysiology

A molecular diagnosis allows the cohorting of patients and review of their clinical features that, subsequently, reveals surprising aspects of gene function and better understanding of the pathophysiology. Mutational analysis also allows correlation of genotype and phenotype, which may reveal functional properties of the encoded protein [552]. An example of this is the *HNF1B* gene, which was originally identified as the cause of a form of autosomal dominant maturity-onset diabetes in the young (MODY5) [722]. Later, it was noted that many mutation carriers also had renal malformations (notably cysts), resulting in the diagnostic label of RCAD syndrome [723]. Subsequently, genital abnormalities and defects in renal urate and magnesium handling were also noted [587, 724–726]. Hence, genetic testing for *HNF1B* mutations has tremendously enhanced our understanding of the gene's function.

Supporting clinical management

This is the area with probably the greatest potential to improve patient care. Genetic information will provide not only valuable information on a patient's prognosis but also improved direct clinical management. Through clinical observation of molecularly defined patient cohorts, it is possible to gain experience with the progression of the disease and response to treatment. For example, patients with mutations in *NPHS2* (podocin) do not respond to standard steroid treatment of nephrotic syndrome [727]. Some may improve with cyclosporine treatment; however, available evidence showed no benefit from the commonly used cytotoxic treatment regimens or high-dose corticosteroids [728]. Therefore, the ability to stratify patients according to their podocin mutation status will obviate the need to prescribe ineffective treatments, which carry the risk of unwanted side effects.

Interpretation challenges and reporting

Return of results

Clinical genetic testing is rapidly moving toward GWAS [593, 600, 729]. The aim of clinical sequence interpretation is to identify the genetic variant that is responsible for the phenotype of an individual patient. Although this expanded genetic scope increases diagnostic sensitivity, it also has the potential to identify variants that are unrelated to the primary indication for testing [562]. Such secondary findings must be considered with respect to their clinical validity and actionability [730]. Clinically valid findings include those that can be used to accurately predict that a patient will have the associated condition [731]; these encompass variants in genes for highly penetrant Mendelian diseases, pharmacogenomic variants that are informative regarding drug metabolism, and risk variants that affects susceptibility for a given condition. This category also includes clinically actionable variants, the detection of which would enable a physician to implement interventions that prevent or lessen the clinical consequences of the disease for which the variant confers increased risk. Clinically actionable variants have been recommended for return by both the ACMG [732, 733] and the ESHG [734].

It is critical that clinical sequence interpretation be highly accurate and reproducible because the identified variant is used to guide subsequent care. Hence, the diagnostic interpretation is based on established guidelines [592, 595] in which geneticists have arrived at an overall variant level of classification, including the following types: (1) pathogenic, (2) likely pathogenic, (3) likely benign, (4) benign, and (5) variant of uncertain significance (VUS) [595].

Currently, the ACMG advises returning known and expected pathogenic variants in 59 genes to patients regardless of their age or indication [733]. Approximately 1%–3% of the general population has a pathogenic mutation in one of these 59 genes encompassing conditions deemed to be highly penetrant and actionable, and predominantly associated with various hereditary forms of cancer and cardiovascular disease [735, 736].

The genetic diagnosis could initiate referral and evaluation for previously unrecognized extrarenal features of the associated diseases. In other patients, the genetic diagnosis could inform adequate therapy; for example, by disfavoring immunosuppression among patients who were found to have monogenic forms of FSGS, prompting the referral to clinical trials targeted to the genetic disorder identified. Another example is the institution of tailored therapies in patients with Dent's disease who could benefit from thiazide diuretics and a high citrate diet [576].

Secondary findings of pathogenic variants would lead to targeted subspecialty referral and work up, such as oncologic evaluation and mammography for patients with *BRCA2* mutations. For each patient, these secondary genetic findings also had implications for nephrology care. For instance, these secondary findings could inform the physician to use immunosuppression medication in patients with findings for hereditary cancers. Meanwhile, knowing other pathogenic variants may influence the doctor's decision to indicate dialysis or the use of diuretics for patients who were found to have a genetic predisposition to cardiac arrhythmias such as *KCNQ1* mutation associated with long QT syndrome [576].

Continuous review

New genetic knowledge acquired from experimental data may reclassify a mutation that was previously deemed diagnostic as a VUS or benign variant as pathogenic [562]. Currently, there are no explicit standards for reviewing clinical genetic testing data, and the practice is rare with few laboratories routinely engaging in this practice [737]. Furthermore, continuous review of sequence data not only involves questions regarding the optimal frequency and analytical methodology but also involves a multitude of ELSIs, including who should be responsible for requesting reanalysis, physician liability and duty to inform versus the right of patient not to know, and the psychosocial impact of recontact patients and their families [738, 739].

Genetic education and counseling

The clinician ordering the genetic test is expected to ensure informed consent by providing patients with comprehensive pretest counseling, including discussion of the test and its limitations, the potential for secondary findings, and the complexity of the genetic interpretation [592, 740]. Nephrologists are expected to have a basic understanding in genetics, a detailed knowledge of genetic forms of renal disease as well as to know the general best practices for genetic testing. In the absence of a geneticist in the clinical team, patients should be referred to genetic counselors for counseling before and after testing [562].

In order to facilitate implementation of clinical care based on genetic findings, it is important to provide genetics education to physicians, promote interactions between referring physicians and clinical testing laboratories, and create clinical decision support tools [739, 741].

In the future, nephrogenetics may emerge as a superspecialty, similar to transplant or interventional nephrology. However, genetic counselors specialized in nephrology should become part of multidisciplinary teams fully integrated into the clinical setting, as similarly seen in oncology [742] and cardiology [743].

Ethical, legal, and social implications (ELSIs) of genomic research

The ELSI research program was founded in 1990 as an integral part of the HGP to foster basic and applied research on the ELSI of genetic and genomic research for individuals, families, and communities [744]. Here, we highlight some of the key ELSIs that must be considered when implementing genomic medicine in nephrology.

Participation in genetic research

In genomic medicine, the boundaries between bench research and bedside care have become increasingly blurred. Participating in research can directly impact the clinical care and disease course of patients with hereditary nephropathies [745, 746]. The results generated through taking part in genetic studies differ markedly from those that are returned from clinical diagnostic testing. A clinical genetic test provides a genetic diagnosis for a patient within a defined time period, whereas sequencing performed in the context of genetic research produces generalized knowledge useful for future patients [747]. Although the latter approach might lead to diagnostic results, the time frame for their return is generally indefinite [748, 749]. Some have argued for the unification of research and clinical genetic testing, including requiring all research-level sequencing to be conducted in a Clinical Laboratory Improvement Amendments (CLIA) laboratory environment to enable genetic discoveries to be rapidly implemented in patient care [750, 751].

Minority populations

Evaluation of sequence data in the context of a patient's ancestry is critical to avert variant misinterpretation. In addition, diverse population control databases are needed to address such disparities in genetic diagnostic testing among minority populations to avoid false-positive results that subsequently may lead to inaccurate assessment of risk status and medical management and substantial psychological distress for patients and their families [673].

Knowledge of ancestry-specific alleles can have clinical utility for Mendelian nephropathies and for more common, complex forms of CKD. For example, founder mutations for autosomal recessive Alport syndrome [752], Fanconi anemia type C [753], and Zellweger syndrome [754] have been documented among Ashkenazi Jews. Another example is observed among patients of sub-Saharan African descent [583, 585] and those with sickle cell trait (HBB variant) who are affected by *APOL1* risk genotypes conferring an increased risk for CKD [755, 756].

To ensure that the benefits of genomic nephrology are available to all patients with kidney disease, it is imperative that minority populations have equal access to participation in genetic research and to clinical genetic assessment [757].

Pediatric genetic testing

As early-onset CKD is enriched for genetic etiologies, genetic testing has been advocated as a first-line diagnostic for pediatric nephropathy [557, 561, 564]. However, the ethical implications and psychosocial impact of genetic testing in minors should include important issues that need to be addressed, such as: the age at which consent for testing should be provided by the child rather than by the parents, how to balance the potential for early clinical intervention versus the right of the patient to an open future [758], how to explain the primary findings to the patient, and the potential impact of knowledge of a genetic condition on the psychoemotional development and health of the child. Furthermore, with the advent of genome-wide testing, the question of "how secondary findings should be treated among pediatric patients" has also arisen. Although this knowledge could resolve parental anxiety regarding risk status and inform family planning, it threatens the right of the patient to autonomy and confidentiality and could result in considerable psychological harm [759]. Moreover, at this moment there are no clear consensus guidelines regarding this topic. The American Academy of Pediatrics recommends deferring predictive testing for later-onset conditions until adulthood, whereas the ACMG advises that the decision be made on a per-family basis [760, 761]. This topic becomes more complex with adolescents, as they may want to decide independently of their parents or other family members and/or may not want to have their genetic findings shared with their families [762].

Longitudinal study of the medical and psychosocial impact of genetic testing on pediatric patients and their families will help to guide creation of best practice guidelines regarding these topics [562].

Legal protection

In 1997, the United Nations Educational, Scientific and Cultural Organization (UNESCO) Declaration on the Human Genome and Human Rights ruled that genetic information should not be used to infringe upon "human rights, fundamental freedoms, and human dignity," establishing and international posture against genetic discrimination (GD). In Europe, the 1997 Convention on Human Rights and Biomedicine and the 2012 Charter of Fundamental Rights of the European Union protect individuals against GD; under the primacy of European law these protections take precedence over any conflicting national legislation, and many European nations have also created their own anti-GD status [763, 764].

Other nations, such as Australia, the US, and Canada, have also passed anti-GD legislation. On the other hand, in Latin America, Africa, and the Middle East anti-GD protection remains scant, with the exception of Mexico, Chile, Malawi, and Israel [763]. Similarly, anti-GD legislation is sparse across Asia, despite the strong presence of genetic research and personalized medicine in many Asian nations; to date, only South Korea and Taiwan have created specific prohibitions [763, 765].

Legal protection for individuals undergoing genetic testing has focused on regulating the use of genetic findings by insurers and employers, reflecting concerns that these parties will use genetic information to deny individuals employment and/or associated insurance benefits [763, 764]. However, the degree of legal protection varies substantially between nations. The 2016 Council of Europe recommendation requires insurers to justify use of all "health-related personal data" and forbids insurers from requiring individuals to undergo genetic testing for insurance purposes [766]. Similarly, in Israel, the Genetic Information Law of 2000 prohibits the use of genetic information for employment and insurance purposes [767]. In the UK, insurers may use genetic test results if approved by the government [768]. In Canada, insurers have pledge to not "use genetic test results for life insurance coverage of $250,000 or less" [769].

In the USA, the lack of national health-care system has magnified concerns, as individuals usually obtain employment-based insurance. In an effort to protect individuals against GD, in 2008, the US government enacted the Genetic

Information Nondiscrimination Act (GINA), which forbids employers and health insurers to request genetic information from individuals and/or discriminate against them on the basis of any available genetic information. These protections do not, however, extend to other types of insurance, including life and disability coverage, and have been threatened by subsequent legislative initiatives [770].

Data storage

Currently, bioinformatics, analytical approaches for data mining, are a major limitation in maximizing the knowledge to be gained from expression libraries. Data analysis of expression profiles ultimately will depend on integrating kidney mRNA libraries with external information resources and will require software development, such as the VectorArray application used to analyze the gene expression profiles during branching morphogenesis [771].

Some of the necessary bioinformatics tools include links between kidney genes identified in an expression profile and Genbank, links to user-friendly biological pathway databases, and access to databases that can identify functional nucleotides of importance (e.g., regulatory elements) or protein motifs (e.g., kinase domain) [772].

Informatics challenges continuously emerge as the technology in this field evolves. These include new analytic pipelines, image management systems, patient privacy concerns, and laboratory regulations. The data derived from these molecular tests, if well-managed in the clinical laboratory, will redefine the clinical practice in medicine [773, 774].

Costs and settings

In low-income and middle-income countries (LMICs), access to clinical screening and care, including genetic testing, can often be limited [775, 776]. Although WES or WGS are becoming the standard for work-up of undiagnosed disorder, the high cost and limited availability of these technologies are substantial barriers to their use in LMICs. Therefore, some strategies such as targeted genetic testing for specific variants are proposed approaches to implement genetic testing in LMICs to provide clinically useful and cost-effective services tailored to fit the needs of the given population [777]. Mendeliomes have been recommended as a means of providing broad assessment of the disease associated genome at a feasible cost and have shown promising results as diagnostics in this context [778, 779]. Organizations such as Human Hereditary and Health in Africa (H3Africa) and the Mexico National Institute of Genomic Medicine (INMEGEN) have already begun to implement such strategies, which will help to make the benefits of genomic medicine available globally [550, 767, 780].

Open-data sharing resources and the principle of open data itself can help to reduce the cost of conducting genetic research. They can also limit the number of conflict-of-interest problems that occurs as academic medical centers partner with commercial activities. Although this type of infrastructure does not help to solve generalized funding issues, it does set a precedent for sharing data rather than keeping it proprietary. In doing so, it reduces the scope and impact of potential conflicts and helps to ensure that commercial relationships are based on open principles [750].

References

[1] Levin A, Tonelli M, Bonventre J, et al. Global kidney health 2017 and beyond: a roadmap for closing gaps in care, research, and policy. Lancet 2017;390:1888–917.

[2] Garreta E, Gonzalez F, Montserrat N. Studying kidney disease using tissue and genome engineering in human pluripotent stem cells. Nephron 2018;138:48–59.

[3] Katz R. Biomarkers and surrogate markers: an FDA perspective. NeuroRx 2004;1:189–95.

[4] Biomarkers Definitions Working G. Biomarkers and surrogate endpoints: preferred definitions and conceptual framework. Clin Pharmacol Ther 2001;69:89–95.

[5] Nguyen MT, Devarajan P. Biomarkers for the early detection of acute kidney injury. Pediatr Nephrol 2008;23:2151–7.

[6] Chen BS, Wu CC. Systems biology as an integrated platform for bioinformatics, systems synthetic biology, and systems metabolic engineering. Cells 2013;2:635–88.

[7] Ho J, Dart A, Rigatto C. Proteomics in acute kidney injury—current status and future promise. Pediatr Nephrol 2014;29:163–71.

[8] Carter JL, Lamb EJ. Evaluating new biomarkers for acute kidney injury: putting the horse before the cart. Am J Kidney Dis 2014;63:543–6.

[9] Steyerberg EW, Pencina MJ, Lingsma HF, Kattan MW, Vickers AJ, Van Calster B. Assessing the incremental value of diagnostic and prognostic markers: a review and illustration. Eur J Clin Investig 2012;42:216–28.

[10] Van Calster B, Vickers AJ, Pencina MJ, Baker SG, Timmerman D, Steyerberg EW. Evaluation of markers and risk prediction models: overview of relationships between NRI and decision-analytic measures. Med Decis Mak 2013;33:490–501.

[11] Ioannidis JP. Expectations, validity, and reality in omics. J Clin Epidemiol 2010;63:945–9.

[12] Murray PT, Mehta RL, Shaw A, et al. Potential use of biomarkers in acute kidney injury: report and summary of recommendations from the 10th Acute Dialysis Quality Initiative consensus conference. Kidney Int 2014;85:513–21.

[13] Pepe MS, Etzioni R, Feng Z, et al. Phases of biomarker development for early detection of cancer. J Natl Cancer Inst 2001;93:1054–61.

[14] Parikh CR, Thiessen-Philbrook H. Key concepts and limitations of statistical methods for evaluating biomarkers of kidney disease. J Am Soc Nephrol 2014;25:1621–9.

[15] Pickering JW, Endre ZH. New metrics for assessing diagnostic potential of candidate biomarkers. Clin J Am Soc Nephrol 2012;7:1355–64.

[16] Pencina MJ, D'Agostino Sr. RB, D'Agostino Jr RB, Vasan RS. Evaluating the added predictive ability of a new marker: from area under the ROC curve to reclassification and beyond. Stat Med 2008;27:157–72. discussion 207-12.

[17] Pepe MS, Janes H, Longton G, Leisenring W, Newcomb P. Limitations of the odds ratio in gauging the performance of a diagnostic, prognostic, or screening marker. Am J Epidemiol 2004;159:882–90.

[18] Zweig MH, Campbell G. Receiver-operating characteristic (ROC) plots: a fundamental evaluation tool in clinical medicine. Clin Chem 1993;39:561–77.

[19] Baker SG. The central role of receiver operating characteristic (ROC) curves in evaluating tests for the early detection of cancer. J Natl Cancer Inst 2003;95:511–5.

[20] Warnock DG, Peck CC. A roadmap for biomarker qualification. Nat Biotechnol 2010;28:444–5.

[21] Pepe MS, Cai T, Longton G. Combining predictors for classification using the area under the receiver operating characteristic curve. Biometrics 2006;62:221–9.

[22] Hanley JA, McNeil BJ. The meaning and use of the area under a receiver operating characteristic (ROC) curve. Radiology 1982;143:29–36.

[23] Ransohoff DF. How to improve reliability and efficiency of research about molecular markers: roles of phases, guidelines, and study design. J Clin Epidemiol 2007;60:1205–19.

[24] DeLong ER, DeLong DM, Clarke-Pearson DL. Comparing the areas under two or more correlated receiver operating characteristic curves: a nonparametric approach. Biometrics 1988;44:837–45.

[25] Chambless LE, Diao G. Estimation of time-dependent area under the ROC curve for long-term risk prediction. Stat Med 2006;25:3474–86.

[26] Heagerty PJ, Zheng Y. Survival model predictive accuracy and ROC curves. Biometrics 2005;61:92–105.

[27] Parikh CR, Han G. Variation in performance of kidney injury biomarkers due to cause of acute kidney injury. Am J Kidney Dis 2013;62:1023–6.

[28] Waikar SS, Betensky RA, Emerson SC, Bonventre JV. Imperfect gold standards for kidney injury biomarker evaluation. J Am Soc Nephrol 2012;23:13–21.

[29] Vanmassenhove J, Vanholder R, Nagler E, Van Biesen W. Urinary and serum biomarkers for the diagnosis of acute kidney injury: an in-depth review of the literature. Nephrol Dial Transplant 2013;28:254–73.

[30] Siew ED, Ware LB, Ikizler TA. Biological markers of acute kidney injury. J Am Soc Nephrol 2011;22:810–20.

[31] Cook NR. Use and misuse of the receiver operating characteristic curve in risk prediction. Circulation 2007;115:928–35.

[32] Kerr KF, Meisner A, Thiessen-Philbrook H, Coca SG, Parikh CR. Developing risk prediction models for kidney injury and assessing incremental value for novel biomarkers. Clin J Am Soc Nephrol 2014;9:1488–96.

[33] Parikh CR, Coca SG, Thiessen-Philbrook H, et al. Postoperative biomarkers predict acute kidney injury and poor outcomes after adult cardiac surgery. J Am Soc Nephrol 2011;22:1748–57.

[34] McGough JJ, Faraone SV. Estimating the size of treatment effects: moving beyond p values. Psychiatry (Edgmont) 2009;6:21–9.

[35] Janes H, Pepe MS. Adjusting for covariates in studies of diagnostic, screening, or prognostic markers: an old concept in a new setting. Am J Epidemiol 2008;168:89–97.

[36] Huang Y, Pepe MS. Biomarker evaluation and comparison using the controls as a reference population. Biostatistics 2009;10:228–44.

[37] Kerr KF, Pepe MS. Joint modeling, covariate adjustment, and interaction: contrasting notions in risk prediction models and risk prediction performance. Epidemiology 2011;22:805–12.

[38] Zou KH, Resnic FS, Talos IF, et al. A global goodness-of-fit test for receiver operating characteristic curve analysis via the bootstrap method. J Biomed Inform 2005;38:395–403.

[39] Begg CB, Gonen M, Seshan VE. Testing the incremental predictive accuracy of new markers. Clin Trials 2013;10:690–2.

[40] Demler OV, Pencina MJ, D'Agostino Sr. RB. Misuse of DeLong test to compare AUCs for nested models. Stat Med 2012;31:2577–87.

[41] Pepe MS, Kerr KF, Longton G, Wang Z. Testing for improvement in prediction model performance. Stat Med 2013;32:1467–82.

[42] Chen HC, Kodell RL, Cheng KF, Chen JJ. Assessment of performance of survival prediction models for cancer prognosis. BMC Med Res Methodol 2012;12.

[43] Pencina MJ, D'Agostino Sr. RB, Demler OV. Novel metrics for evaluating improvement in discrimination: net reclassification and integrated discrimination improvement for normal variables and nested models. Stat Med 2012;31:101–13.

[44] Pencina MJ, D'Agostino RB, Pencina KM, Janssens AC, Greenland P. Interpreting incremental value of markers added to risk prediction models. Am J Epidemiol 2012;176:473–81.

[45] Pencina MJ, D'Agostino Sr. RB, Steyerberg EW. Extensions of net reclassification improvement calculations to measure usefulness of new biomarkers. Stat Med 2011;30:11–21.

[46] Kivimaki M, Batty GD, Hamer M, et al. Using additional information on working hours to predict coronary heart disease: a cohort study. Ann Intern Med 2011;154:457–63.

[47] Leening MJ, Vedder MM, Witteman JC, Pencina MJ, Steyerberg EW. Net reclassification improvement: computation, interpretation, and controversies: a literature review and clinician's guide. Ann Intern Med 2014;160.

[48] Kavousi M, Elias-Smale S, Rutten JH, et al. Evaluation of newer risk markers for coronary heart disease risk classification: a cohort study. Ann Intern Med 2012;156:438–44.

[49] Pepe MS. Problems with risk reclassification methods for evaluating prediction models. Am J Epidemiol 2011;173:1327–35.

[50] Pepe MS, Janes H. Commentary: reporting standards are needed for evaluations of risk reclassification. Int J Epidemiol 2011;40:1106–8.

[51] Leening MJ, Cook NR. Net reclassification improvement: a link between statistics and clinical practice. Eur J Epidemiol 2013;28:21–3.

[52] Cook NR, Paynter NP. Comments on 'Extensions of net reclassification improvement calculations to measure usefulness of new biomarkers' by M. J. Pencina, R. B. D'Agostino, Sr. and E. W. Steyerberg. Stat Med 2012;31:93–5. author reply 6-7.

[53] Hilden J, Gerds TA. A note on the evaluation of novel biomarkers: do not rely on integrated discrimination improvement and net reclassification index. Stat Med 2014;33:3405–14.

[54] Kerr KF, Wang Z, Janes H, McClelland RL, Psaty BM, Pepe MS. Net reclassification indices for evaluating risk prediction instruments: a critical review. Epidemiology 2014;25:114–21.

[55] Kerr KF, Bansal A, Pepe MS. Further insight into the incremental value of new markers: the interpretation of performance measures and the importance of clinical context. Am J Epidemiol 2012;176:482–7.

[56] Sackett DL, Haynes RB. The architecture of diagnostic research. BMJ 2002;324:539–41.

[57] Hlatky MA, Greenland P, Arnett DK, et al. Criteria for evaluation of novel markers of cardiovascular risk: a scientific statement from the American Heart Association. Circulation 2009;119:2408–16.

[58] Vickers AJ, Pepe M. Does the net reclassification improvement help us evaluate models and markers? Ann Intern Med 2014;160:136–7.

[59] Emwas AH, Roy R, McKay RT, et al. Recommendations and standardization of biomarker quantification using NMR-based metabolomics with particular focus on urinary analysis. J Proteome Res 2016;15:360–73.

[60] Kirwan JA, Brennan L, Broadhurst D, et al. Preanalytical processing and biobanking procedures of biological samples for metabolomics research: a white paper, community perspective (for "Precision Medicine and Pharmacometabolomics Task Group"—The Metabolomics Society Initiative). Clin Chem 2018;64:1158–82.

[61] Yuille M, Illig T, Hveem K, et al. Laboratory management of samples in biobanks: European consensus expert group report. Biopreserv Biobank 2010;8:65–9.

[62] Paskal W, Paskal AM, Debski T, Gryziak M, Jaworowski J. Aspects of modern biobank activity—comprehensive review. Pathol Oncol Res 2018;24:771–85.

[63] Bernini P, Bertini I, Luchinat C, Nincheri P, Staderini S, Turano P. Standard operating procedures for pre-analytical handling of blood and urine for metabolomic studies and biobanks. J Biomol NMR 2011;49:231–43.

[64] Schaub J, Schiesling C, Reuss M, Dauner M. Integrated sampling procedure for metabolome analysis. Biotechnol Prog 2006;22:1434–42.

[65] Yuan M, Breitkopf SB, Yang X, Asara JM. A positive/negative ion-switching, targeted mass spectrometry-based metabolomics platform for bodily fluids, cells, and fresh and fixed tissue. Nat Protoc 2012;7:872–81.

[66] Want EJ, Masson P, Michopoulos F, et al. Global metabolic profiling of animal and human tissues via UPLC-MS. Nat Protoc 2013;8:17–32.

[67] Fan TW, Lane AN, Higashi RM. Stable isotope resolved metabolomics studies in ex vivo tissue slices. Bio Protoc 2016;6.

[68] Fan TW, Lane AN, Higashi RM, et al. Altered regulation of metabolic pathways in human lung cancer discerned by (13)C stable isotope-resolved metabolomics (SIRM). Mol Cancer 2009;8:41.

[69] Ly-Verdu S, Schaefer A, Kahle M, et al. The impact of blood on liver metabolite profiling—a combined metabolomic and proteomic approach. Biomed Chromatogr 2014;28:231–40.

[70] Kelly AD, Breitkopf SB, Yuan M, Goldsmith J, Spentzos D, Asara JM. Metabolomic profiling from formalin-fixed, paraffin-embedded tumor tissue using targeted LC/MS/MS: application in sarcoma. PLoS One 2011;6.

[71] Mutter GL, Zahrieh D, Liu C, et al. Comparison of frozen and RNALater solid tissue storage methods for use in RNA expression microarrays. BMC Genomics 2004;5:88.

[72] Lopez-Bascon MA, Priego-Capote F, Peralbo-Molina A, Calderon-Santiago M, Luque de Castro MD. Influence of the collection tube on metabolomic changes in serum and plasma. Talanta 2016;150:681–9.

[73] Jobard E, Tredan O, Postoly D, et al. A systematic evaluation of blood serum and plasma pre-analytics for metabolomics cohort studies. Int J Mol Sci 2016;17.

[74] Breier M, Wahl S, Prehn C, et al. Targeted metabolomics identifies reliable and stable metabolites in human serum and plasma samples. PLoS One 2014;9.

[75] Mateos J, Carneiro I, Corrales F, et al. Multicentric study of the effect of pre-analytical variables in the quality of plasma samples stored in biobanks using different complementary proteomic methods. J Proteome 2017;150:109–20.

[76] Skogholt AH, Ryeng E, Erlandsen SE, Skorpen F, Schonberg SA, Saetrom P. Gene expression differences between PAXgene and Tempus blood RNA tubes are highly reproducible between independent samples and biobanks. BMC Res Notes 2017;10.

[77] Roux A, Thevenot EA, Seguin F, Olivier MF, Junot C. Impact of collection conditions on the metabolite content of human urine samples as analyzed by liquid chromatography coupled to

mass spectrometry and nuclear magnetic resonance spectroscopy. Metabolomics 2015;11:1095–105.

[78] Walsh MC, Brennan L, Malthouse JP, Roche HM, Gibney MJ. Effect of acute dietary standardization on the urinary, plasma, and salivary metabolomic profiles of healthy humans. Am J Clin Nutr 2006;84:531–9.

[79] Carotenuto D, Luchinat C, Marcon G, Rosato A, Turano P. The Da Vinci European BioBank: a metabolomics-driven infrastructure. J Pers Med 2015;5:107–19.

[80] Laparre J, Kaabia Z, Mooney M, et al. Impact of storage conditions on the urinary metabolomics fingerprint. Anal Chim Acta 2017;951:99–107.

[81] Gika HG, Theodoridis GA, Wilson ID. Liquid chromatography and ultra-performance liquid chromatography-mass spectrometry fingerprinting of human urine: sample stability under different handling and storage conditions for metabonomics studies. J Chromatogr A 2008;1189:314–22.

[82] Rotter M, Brandmaier S, Prehn C, et al. Stability of targeted metabolite profiles of urine samples under different storage conditions. Metabolomics 2017;13.

[83] Bohra R, Klepacki J, Klawitter J, Klawitter J, Thurman JM, Christians U. Proteomics and metabolomics in renal transplantation-quo vadis? Transpl Int 2013;26:225–41.

[84] Koop R. Combinatorial biomarkers: from early toxicology assays to patient population profiling. Drug Discov Today 2005;10:781–8.

[85] Wishart DS. Metabolomics: the principles and potential applications to transplantation. Am J Transplant 2005;5:2814–20.

[86] Dieterle F, Riefke B, Schlotterbeck G, Ross A, Senn H, Amberg ANMR. MS methods for metabonomics. Methods Mol Biol 2011;691:385–415.

[87] Griffiths WJ, Wang Y. Mass spectrometry: from proteomics to metabolomics and lipidomics. Chem Soc Rev 2009;38:1882–96.

[88] Xu EY, Schaefer WH, Xu Q. Metabolomics in pharmaceutical research and development: metabolites, mechanisms and pathways. Curr Opin Drug Discov Devel 2009;12:40–52.

[89] Groenen PJ, van den Heuvel LP. Teaching molecular genetics: Chapter 3—proteomics in nephrology. Pediatr Nephrol 2006;21:611–8.

[90] Lee JW, Figeys D, Vasilescu J. Biomarker assay translation from discovery to clinical studies in cancer drug development: quantification of emerging protein biomarkers. Adv Cancer Res 2007;96:269–98.

[91] Anderson NL, Anderson NG. Proteome and proteomics: new technologies, new concepts, and new words. Electrophoresis 1998;19:1853–61.

[92] Janech MG, Raymond JR, Arthur JM. Proteomics in renal research. Am J Physiol Renal Physiol 2007;292:F501–12.

[93] Peng J, Gygi SP. Proteomics: the move to mixtures. J Mass Spectrom 2001;36:1083–91.

[94] Knepper MA. Proteomics and the kidney. J Am Soc Nephrol 2002;13:1398–408.

[95] Davis MT, Spahr CS, McGinley MD, et al. Towards defining the urinary proteome using liquid chromatography-tandem mass spectrometry. II. Limitations of complex mixture analyses. Proteomics 2001;1:108–17.

[96] Hampel DJ, Sansome C, Sha M, Brodsky S, Lawson WE, Goligorsky MS. Toward proteomics in uroscopy: urinary protein profiles after radiocontrast medium administration. J Am Soc Nephrol 2001;12:1026–35.

[97] Spahr CS, Davis MT, McGinley MD, et al. Towards defining the urinary proteome using liquid chromatography-tandem mass spectrometry. I. Profiling an unfractionated tryptic digest. Proteomics 2001;1:93–107.

[98] Muller GA, Muller CA, Dihazi H. Clinical proteomics—on the long way from bench to bedside? Nephrol Dial Transplant 2007;22:1297–300.

[99] Lawrance IC, Klopcic B, Wasinger VC. Proteomics: an overview. Inflamm Bowel Dis 2005;11:927–36.

[100] O'Riordan E, Gross SS, Goligorsky MS. Technology Insight: renal proteomics—at the crossroads between promise and problems. Nat Clin Pract Nephrol 2006;2:445–58.

[101] Konvalinka A, Scholey JW, Diamandis EP. Searching for new biomarkers of renal diseases through proteomics. Clin Chem 2012;58:353–65.

[102] Prunotto M, Ghiggeri G, Bruschi M, et al. Renal fibrosis and proteomics: current knowledge and still key open questions for proteomic investigation. J Proteome 2011;74:1855–70.

[103] Thongboonkerd V, Malasit P. Renal and urinary proteomics: current applications and challenges. Proteomics 2005;5:1033–42.

[104] Mischak H. Pro: urine proteomics as a liquid kidney biopsy: no more kidney punctures! Nephrol Dial Transplant 2015;30:532–7.

[105] Decramer S, Gonzalez de Peredo A, Breuil B, et al. Urine in clinical proteomics. Mol Cell Proteomics 2008;7:1850–62.

[106] Fliser D, Novak J, Thongboonkerd V, et al. Advances in urinary proteome analysis and biomarker discovery. J Am Soc Nephrol 2007;18:1057–71.

[107] Rodriguez-Suarez E, Siwy J, Zurbig P, Mischak H. Urine as a source for clinical proteome analysis: from discovery to clinical application. Biochim Biophys Acta 2014;1844:884–98.

[108] Vlahou A, Schanstra J, Frokiaer J, et al. Establishment of a European network for urine and kidney proteomics. J Proteome 2008;71:490–2.

[109] Mischak H, Kolch W, Aivaliotis M, et al. Comprehensive human urine standards for comparability and standardization in clinical proteome analysis. Proteomics Clin Appl 2010;4:464–78.

[110] Stalmach A, Albalat A, Mullen W, Mischak H. Recent advances in capillary electrophoresis coupled to mass spectrometry for clinical proteomic applications. Electrophoresis 2013;34:1452–64.

[111] Kushnir MM, Mrozinski P, Rockwood AL, Crockett DK. A depletion strategy for improved detection of human proteins from urine. J Biomol Tech 2009;20:101–8.

[112] Weissinger EM, Schiffer E, Hertenstein B, et al. Proteomic patterns predict acute graft-versus-host disease after allogeneic hematopoietic stem cell transplantation. Blood 2007;109:5511–9.

[113] Kolch W, Neususs C, Pelzing M, Mischak H. Capillary electrophoresis-mass spectrometry as a powerful tool in clinical diagnosis and biomarker discovery. Mass Spectrom Rev 2005;24:959–77.

[114] Omenn GS, States DJ, Adamski M, et al. Overview of the HUPO Plasma Proteome Project: results from the pilot phase with 35 collaborating laboratories and multiple analytical groups, generating a core dataset of 3020 proteins and a publicly-available database. Proteomics 2005;5:3226–45.

[115] Vestergaard P, Leverett R. Constancy of urinary creatinine excretion. J Lab Clin Med 1958;51:211–8.

[116] Schiffer E, Mischak H, Novak J. High resolution proteome/peptidome analysis of body fluids by capillary electrophoresis coupled with MS. Proteomics 2006;6:5615–27.

[117] Fliser D, Wittke S, Mischak H. Capillary electrophoresis coupled to mass spectrometry for clinical diagnostic purposes. Electrophoresis 2005;26:2708–16.

[118] Weissinger EM, Wittke S, Kaiser T, et al. Proteomic patterns established with capillary electrophoresis and mass spectrometry for diagnostic purposes. Kidney Int 2004;65:2426–34.

[119] Schaub S, Wilkins J, Weiler T, Sangster K, Rush D, Nickerson P. Urine protein profiling with surface-enhanced laser-desorption/ionization time-of-flight mass spectrometry. Kidney Int 2004;65:323–32.

[120] Zerefos PG, Vlahou A. Urine sample preparation and protein profiling by two-dimensional electrophoresis and matrix-assisted laser desorption ionization time of flight mass spectroscopy. Methods Mol Biol 2008;428:141–57.

[121] Zhou H, Yuen PS, Pisitkun T, et al. Collection, storage, preservation, and normalization of human urinary exosomes for biomarker discovery. Kidney Int 2006;69:1471–6.

[122] Fiedler GM, Baumann S, Leichtle A, et al. Standardized peptidome profiling of human urine by magnetic bead separation and matrix-assisted laser desorption/ionization time-of-flight mass spectrometry. Clin Chem 2007;53:421–8.

[123] Thongboonkerd V. Practical points in urinary proteomics. J Proteome Res 2007;6:3881–90.

[124] Wu J, Chen YD, Gu W. Urinary proteomics as a novel tool for biomarker discovery in kidney diseases. J Zhejiang Univ Sci B 2010;11:227–37.

[125] Ahmed FE. Sample preparation and fractionation for proteome analysis and cancer biomarker discovery by mass spectrometry. J Sep Sci 2009;32:771–98.

[126] Hu S, Loo JA, Wong DT. Human body fluid proteome analysis. Proteomics 2006;6:6326–53.

[127] Matt P, Fu Z, Fu Q, Van Eyk JE. Biomarker discovery: proteome fractionation and separation in biological samples. Physiol Genomics 2008;33:12–7.

[128] Havanapan PO, Thongboonkerd V. Are protease inhibitors required for gel-based proteomics of kidney and urine? J Proteome Res 2009;8:3109–17.

[129] Charonis A, Luider T, Baumann M, Schanstra JP. Is the time ripe for kidney tissue proteomics? Proteomics Clin Appl 2011;5:215–21.

[130] Guo T, Wang W, Rudnick PA, et al. Proteome analysis of microdissected formalin-fixed and paraffin-embedded tissue specimens. J Histochem Cytochem 2007;55:763–72.

[131] Hood BL, Conrads TP, Veenstra TD. Unravelling the proteome of formalin-fixed paraffin-embedded tissue. Brief Funct Genomic Proteomic 2006;5:169–75.

[132] Hood BL, Conrads TP, Veenstra TD. Mass spectrometric analysis of formalin-fixed paraffin-embedded tissue: unlocking the proteome within. Proteomics 2006;6:4106–14.

[133] Chaurand P, Sanders ME, Jensen RA, Caprioli RM. Proteomics in diagnostic pathology: profiling and imaging proteins directly in tissue sections. Am J Pathol 2004;165:1057–68.

[134] Sedor JR. Tissue proteomics: a new investigative tool for renal biopsy analysis. Kidney Int 2009;75:876–9.

[135] Emmert-Buck MR, Bonner RF, Smith PD, et al. Laser capture microdissection. Science 1996;274:998–1001.

[136] Hobeika L, Barati MT, Caster DJ, McLeish KR, Merchant ML. Characterization of glomerular extracellular matrix by proteomic analysis of laser-captured microdissected glomeruli. Kidney Int 2017;91:501–11.

[137] Hohne M, Frese CK, Grahammer F, et al. Single-nephron proteomes connect morphology and function in proteinuric kidney disease. Kidney Int 2018;93:1308–19.

[138] Waanders LF, Chwalek K, Monetti M, Kumar C, Lammert E, Mann M. Quantitative proteomic analysis of single pancreatic islets. Proc Natl Acad Sci U S A 2009;106:18902–7.

[139] Sethi S, Vrana JA, Theis JD, Dogan A. Mass spectrometry based proteomics in the diagnosis of kidney disease. Curr Opin Nephrol Hypertens 2013;22:273–80.

[140] Sethi S, Vrana JA, Theis JD, et al. Laser microdissection and mass spectrometry-based proteomics aids the diagnosis and typing of renal amyloidosis. Kidney Int 2012;82:226–34.

[141] Hallan SI. The possibilities to improve kidney health with proteomics. Clin J Am Soc Nephrol 2017.

[142] Camerini S, Mauri P. The role of protein and peptide separation before mass spectrometry analysis in clinical proteomics. J Chromatogr A 2015;1381:1–12.

[143] Klein JB, Thongboonkerd V. Overview of proteomics. Contrib Nephrol 2004;141:1–10.

[144] Cao YH, Lu LL, Zhang JD, Liu BC. Application of systems biology to the study of chronic kidney disease. Chin Med J 2012;125:2603–9.

[145] Thongboonkerd V. Proteomics in nephrology: current status and future directions. Am J Nephrol 2004;24:360–78.

[146] Thongboonkerd V, Klein E, Klein JB. Sample preparation for 2-D proteomic analysis. Contrib Nephrol 2004;141:11–24.

[147] Sechi S. Mass spectrometric approaches to quantitative proteomics. Contrib Nephrol 2004;141:59–78.

[148] Thongboonkerd V. Current status of renal and urinary proteomics: ready for routine clinical application? Nephrol Dial Transplant 2010;25:11–6.

[149] Thongboonkerd V, Klein JB. Practical bioinformatics for proteomics. Contrib Nephrol 2004;141:79–92.

[150] Thongboonkerd V. Recent progress in urinary proteomics. Proteomics Clin Appl 2007;1:780–91.

[151] Korte EA, Gaffney PM, Powell DW. Contributions of mass spectrometry-based proteomics to defining cellular mechanisms and diagnostic markers for systemic lupus erythematosus. Arthritis Res Ther 2012;14.

[152] Wisniewski JR. Mass spectrometry-based proteomics: principles, perspectives, and challenges. Arch Pathol Lab Med 2008;132:1566–9.

[153] Klose J. Protein mapping by combined isoelectric focusing and electrophoresis of mouse tissues. A novel approach to testing for induced point mutations in mammals. Humangenetik 1975;26:231–43.

[154] O'Farrell PH. High resolution two-dimensional electrophoresis of proteins. J Biol Chem 1975;250:4007–21.

[155] Tonge R, Shaw J, Middleton B, et al. Validation and development of fluorescence two-dimensional differential gel electrophoresis proteomics technology. Proteomics 2001;1:377–96.

[156] Unlu M, Morgan ME, Minden JS. Difference gel electrophoresis: a single gel method for detecting changes in protein extracts. Electrophoresis 1997;18:2071–7.

[157] Hoorn EJ, Hoffert JD, Knepper MA. Combined proteomics and pathways analysis of collecting duct reveals a protein regulatory network activated in vasopressin escape. J Am Soc Nephrol 2005;16:2852–63.

[158] Hoorn EJ, Hoffert JD, Knepper MA. The application of DIGE-based proteomics to renal physiology. Nephron Physiol 2006;104:p61–72.

[159] Aregger F, Pilop C, Uehlinger DE, et al. Urinary proteomics before and after extracorporeal circulation in patients with and without acute kidney injury. J Thorac Cardiovasc Surg 2010;139:692–700.

[160] Zhou H, Pisitkun T, Aponte A, et al. Exosomal Fetuin-A identified by proteomics: a novel urinary biomarker for detecting acute kidney injury. Kidney Int 2006;70:1847–57.

[161] Liu BC, Lu LL. Novel biomarkers for progression of chronic kidney disease. Chin Med J 2010;123:1789–92.

[162] Ngai HH, Sit WH, Jiang PP, Xu RJ, Wan JM, Thongboonkerd V. Serial changes in urinary proteome profile of membranous nephropathy: implications for pathophysiology and biomarker discovery. J Proteome Res 2006;5:3038–47.

[163] Mann M, Ong SE, Gronborg M, Steen H, Jensen ON, Pandey A. Analysis of protein phosphorylation using mass spectrometry: deciphering the phosphoproteome. Trends Biotechnol 2002;20:261–8.

[164] Powell DW, Rane MJ, Joughin BA, et al. Proteomic identification of 14-3-3zeta as a mitogen-activated protein kinase-activated protein kinase 2 substrate: role in dimer formation and ligand binding. Mol Cell Biol 2003;23:5376–87.

[165] Cutillas PR, Chalkley RJ, Hansen KC, et al. The urinary proteome in Fanconi syndrome implies specificity in the reabsorption of proteins by renal proximal tubule cells. Am J Physiol Renal Physiol 2004;287:F353–64.

[166] Cutillas PR, Norden AG, Cramer R, Burlingame AL, Unwin RJ. Detection and analysis of urinary peptides by on-line liquid chromatography and mass spectrometry: application to patients with renal Fanconi syndrome. Clin Sci (Lond) 2003;104:483–90.

[167] Watzig H, Gunter S. Capillary electrophoresis-a high performance analytical separation technique. Clin Chem Lab Med 2003;41:724–38.

[168] Coon JJ, Zurbig P, Dakna M, et al. CE-MS analysis of the human urinary proteome for biomarker discovery and disease diagnostics. Proteomics Clin Appl 2008;2:964.

[169] Klampfl CW. Recent advances in the application of capillary electrophoresis with mass spectrometric detection. Electrophoresis 2006;27:3–34.

[170] Wittke S, Mischak H, Walden M, Kolch W, Radler T, Wiedemann K. Discovery of biomarkers in human urine and cerebrospinal fluid by capillary electrophoresis coupled to mass spectrometry: towards new diagnostic and therapeutic approaches. Electrophoresis 2005;26:1476–87.

[171] Zurbig P, Renfrow MB, Schiffer E, et al. Biomarker discovery by CE-MS enables sequence analysis via MS/MS with platform-independent separation. Electrophoresis 2006;27:2111–25.

[172] Metzger J, Kirsch T, Schiffer E, et al. Urinary excretion of twenty peptides forms an early and accurate diagnostic pattern of acute kidney injury. Kidney Int 2010;78:1252–62.

[173] Rossing K, Mischak H, Dakna M, et al. Urinary proteomics in diabetes and CKD. J Am Soc Nephrol 2008;19:1283–90.

[174] Devarajan P. The use of targeted biomarkers for chronic kidney disease. Adv Chronic Kidney Dis 2010;17:469–79.

[175] Poon TC. Opportunities and limitations of SELDI-TOF-MS in biomedical research: practical advices. Expert Rev Proteomics 2007;4:51–65.

[176] Clarke CH, Buckley JA, Fung ET. SELDI-TOF-MS proteomics of breast cancer. Clin Chem Lab Med 2005;43:1314–20.

[177] Check E. Proteomics and cancer: running before we can walk? Nature 2004;429:496–7.

[178] Kiehntopf M, Siegmund R, Deufel T. Use of SELDI-TOF mass spectrometry for identification of new biomarkers: potential and limitations. Clin Chem Lab Med 2007;45:1435–49.

[179] Ho J, Lucy M, Krokhin O, et al. Mass spectrometry-based proteomic analysis of urine in acute kidney injury following cardiopulmonary bypass: a nested case-control study. Am J Kidney Dis 2009;53:584–95.

[180] Haase-Fielitz A, Mertens PR, Plass M, et al. Urine hepcidin has additive value in ruling out cardiopulmonary bypass-associated acute kidney injury: an observational cohort study. Crit Care 2011;15:R186.

[181] Ho J, Reslerova M, Gali B, et al. Urinary hepcidin-25 and risk of acute kidney injury following cardiopulmonary bypass. Clin J Am Soc Nephrol 2011;6:2340–6.

[182] Prowle JR, Ostland V, Calzavacca P, et al. Greater increase in urinary hepcidin predicts protection from acute kidney injury after cardiopulmonary bypass. Nephrol Dial Transplant 2012;27:595–602.

[183] Maddens B, Ghesquiere B, Vanholder R, et al. Chitinase-like proteins are candidate biomarkers for sepsis-induced acute kidney injury. Mol Cell Proteomics 2012;M111(013094):11.

[184] Gygi SP, Rist B, Gerber SA, Turecek F, Gelb MH, Aebersold R. Quantitative analysis of complex protein mixtures using isotope-coded affinity tags. Nat Biotechnol 1999;17:994–9.

[185] Gygi SP, Rist B, Aebersold R. Measuring gene expression by quantitative proteome analysis. Curr Opin Biotechnol 2000;11:396–401.

[186] Unwin RD, Knowles MA, Selby PJ, Banks RE. Urological malignancies and the proteomic-genomic interface. Electrophoresis 1999;20:3629–37.

[187] Beck Jr LH, Bonegio RG, Lambeau G, et al. M-type phospholipase A2 receptor as target antigen in idiopathic membranous nephropathy. N Engl J Med 2009;361:11–21.

[188] Tomas NM, Beck Jr LH, Meyer-Schwesinger C, et al. Thrombospondin type-1 domain-containing 7A in idiopathic membranous nephropathy. N Engl J Med 2014;371:2277–87.

[189] Andeen NK, Yang HY, Dai DF, MacCoss MJ, Smith KD. DnaJ homolog subfamily B member 9 is a putative autoantigen in fibrillary GN. J Am Soc Nephrol 2018;29:231–9.

[190] Dasari S, Alexander MP, Vrana JA, et al. DnaJ heat shock protein family B member 9 is a novel biomarker for fibrillary GN. J Am Soc Nephrol 2018;29:51–6.

[191] Larsen CP, Trivin-Avillach C, Coles P, et al. LDL receptor-related protein 2 (Megalin) as a target antigen in human kidney anti-brush border antibody disease. J Am Soc Nephrol 2018;29:644–53.

[192] Imbert L, Gaudin M, Libong D, et al. Comparison of electrospray ionization, atmospheric pressure chemical ionization and atmospheric pressure photoionization for a lipidomic analysis of Leishmania donovani. J Chromatogr A 2012;1242:75–83.

[193] Dethy JM, Ackermann BL, Delatour C, Henion JD, Schultz GA. Demonstration of direct bioanalysis of drugs in plasma using nanoelectrospray infusion from a silicon chip coupled with tandem mass spectrometry. Anal Chem 2003;75:805–11.

[194] Gibson GT, Mugo SM, Oleschuk RD. Nanoelectrospray emitters: trends and perspective. Mass Spectrom Rev 2009;28:918–36.

[195] Kelly RT, Page JS, Luo Q, et al. Chemically etched open tubular and monolithic emitters for nanoelectrospray ionization mass spectrometry. Anal Chem 2006;78:7796–801.

[196] Lopes F, Cowan DA, Thevis M, Thomas A, Parkin MC. Quantification of intact human insulin-like growth factor-I in serum by nano-ultrahigh-performance liquid chromatography/tandem mass spectrometry. Rapid Commun Mass Spectrom 2014;28:1426–32.

[197] Shui W, Yu Y, Xu X, Huang Z, Xu G, Yang P. Micro-electrospray with stainless steel emitters. Rapid Commun Mass Spectrom 2003;17:1541–7.

[198] Xiong W, Glick J, Lin Y, Vouros P. Separation and sequencing of isomeric oligonucleotide adducts using monolithic columns by ion-pair reversed-phase nano-HPLC coupled to ion trap mass spectrometry. Anal Chem 2007;79:5312–21.

[199] Blum BC, Mousavi F, Emili A. Single-platform 'multi-omic' profiling: unified mass spectrometry and computational workflows for integrative proteomics-metabolomics analysis. Mol Omics 2018;14:307–19.

[200] Steiner S, Aicher L, Raymackers J, et al. Cyclosporine A decreases the protein level of the calcium-binding protein calbindin-D 28 kDa in rat kidney. Biochem Pharmacol 1996;51:253–8.

[201] Mann M. Functional and quantitative proteomics using SILAC. Nat Rev Mol Cell Biol 2006;7:952–8.

[202] Le Bihan T, Goh T, Stewart II, et al. Differential analysis of membrane proteins in mouse fore- and hindbrain using a label-free approach. J Proteome Res 2006;5:2701–10.

[203] Ong SE, Mann M. Mass spectrometry-based proteomics turns quantitative. Nat Chem Biol 2005;1:252–62.

[204] Ideker T, Thorsson V, Ranish JA, et al. Integrated genomic and proteomic analyses of a systematically perturbed metabolic network. Science 2001;292:929–34.

[205] Oda Y, Nagasu T, Chait BT. Enrichment analysis of phosphorylated proteins as a tool for probing the phosphoproteome. Nat Biotechnol 2001;19:379–82.

[206] Zhou H, Watts JD, Aebersold R. A systematic approach to the analysis of protein phosphorylation. Nat Biotechnol 2001;19:375–8.

[207] Fields S, Song O. A novel genetic system to detect protein-protein interactions. Nature 1989;340:245–6.

[208] Staub O, Dho S, Henry P, et al. WW domains of Nedd4 bind to the proline-rich PY motifs in the epithelial Na^+ channel deleted in Liddle's syndrome. EMBO J 1996;15:2371–80.

[209] Debonneville C, Flores SY, Kamynina E, et al. Phosphorylation of Nedd4-2 by Sgk1 regulates epithelial Na(+) channel cell surface expression. EMBO J 2001;20:7052–9.

[210] Snyder PM, Olson DR, Thomas BC. Serum and glucocorticoid-regulated kinase modulates Nedd4-2-mediated inhibition of the epithelial Na^+ channel. J Biol Chem 2002;277:5–8.

[211] Gavin AC, Bosche M, Krause R, et al. Functional organization of the yeast proteome by systematic analysis of protein complexes. Nature 2002;415:141–7.

[212] Ho Y, Gruhler A, Heilbut A, et al. Systematic identification of protein complexes in Saccharomyces cerevisiae by mass spectrometry. Nature 2002;415:180–3.

[213] Jenkins RE, Pennington SR. Arrays for protein expression profiling: towards a viable alternative to two-dimensional gel electrophoresis? Proteomics 2001;1:13–29.

[214] Knezevic V, Leethanakul C, Bichsel VE, et al. Proteomic profiling of the cancer microenvironment by antibody arrays. Proteomics 2001;1:1271–8.

[215] Sreekumar A, Nyati MK, Varambally S, et al. Profiling of cancer cells using protein microarrays: discovery of novel radiation-regulated proteins. Cancer Res 2001;61:7585–93.

[216] Brooks HL, Sorensen AM, Terris J, et al. Profiling of renal tubule Na^+ transporter abundances in NHE3 and NCC null mice using targeted proteomics. J Physiol 2001;530:359–66.

[217] Knepper MA, Masilamani S. Targeted proteomics in the kidney using ensembles of antibodies. Acta Physiol Scand 2001;173:11–21.

[218] Wang XY, Masilamani S, Nielsen J, et al. The renal thiazide-sensitive Na-Cl cotransporter as mediator of the aldosterone-escape phenomenon. J Clin Invest 2001;108:215–22.

[219] Sanchez JC, Wirth P, Jaccoud S, et al. Simultaneous analysis of cyclin and oncogene expression using multiple monoclonal antibody immunoblots. Electrophoresis 1997;18:638–41.

[220] Gingrich JC, Davis DR, Nguyen Q. Multiplex detection and quantitation of proteins on western blots using fluorescent probes. Biotechniques 2000;29:636–42.

[221] Brooks HL, Allred AJ, Beutler KT, Coffman TM, Knepper MA. Targeted proteomic profiling of renal Na(+) transporter and channel abundances in angiotensin II type 1a receptor knockout mice. Hypertension 2002;39:470–3.

[222] Kim GH, Martin SW, Fernandez-Llama P, Masilamani S, Packer RK, Knepper MA. Long-term regulation of renal Na-dependent cotransporters and ENaC: response to altered acid-base intake. Am J Physiol Renal Physiol 2000;279:F459–67.

[223] Kwon TH, Frokiaer J, Han JS, Knepper MA, Nielsen S. Decreased abundance of major Na(+) transporters in kidneys of rats with ischemia-induced acute renal failure. Am J Physiol Renal Physiol 2000;278:F925–39.

[224] Fernandez-Llama P, Jimenez W, Bosch-Marce M, Arroyo V, Nielsen S, Knepper MA. Dysregulation of renal aquaporins and Na-Cl cotransporter in CCl4-induced cirrhosis. Kidney Int 2000;58:216–28.

[225] Kwon TH, Laursen UH, Marples D, et al. Altered expression of renal AQPs and Na(+) transporters in rats with lithium-induced NDI. Am J Physiol Renal Physiol 2000;279:F552–64.

[226] Ecelbarger CA, Knepper MA, Verbalis JG. Increased abundance of distal sodium transporters in rat kidney during vasopressin escape. J Am Soc Nephrol 2001;12:207–17.

[227] Wang W, Kwon TH, Li C, Frokiaer J, Knepper MA, Nielsen S. Reduced expression of Na-K-2Cl cotransporter in medullary TAL in vitamin D-induced hypercalcemia in rats. Am J Physiol Renal Physiol 2002;282:F34–44.

[228] Smith AH, Vrtis JM, Kodadek T. The potential of protein-detecting microarrays for clinical diagnostics. Adv Clin Chem 2004;38:217–38.

[229] Kodadek T, Reddy MM, Olivos HJ, Bachhawat-Sikder K, Alluri PG. Synthetic molecules as antibody replacements. Acc Chem Res 2004;37:711–8.

[230] Kirby R, Cho EJ, Gehrke B, et al. Aptamer-based sensor arrays for the detection and quantitation of proteins. Anal Chem 2004;76:4066–75.

[231] Popper HH, Kothmaier H. Proteomics—tissue and protein microarrays and antibody array: what information is provided? Arch Pathol Lab Med 2008;132:1570–2.

[232] Haab BB, Dunham MJ, Brown PO. Protein microarrays for highly parallel detection and quantitation of specific proteins and antibodies in complex solutions. Genome Biol 2001;2.

[233] Carlsson AC, Ingelsson E, Sundstrom J, et al. Use of proteomics to investigate kidney function decline over 5 years. Clin J Am Soc Nephrol 2017;12:1226–35.

[234] Xu Y, Yang X, Wang E. Review: aptamers in microfluidic chips. Anal Chim Acta 2010;683:12–20.

[235] Elias JE, Haas W, Faherty BK, Gygi SP. Comparative evaluation of mass spectrometry platforms used in large-scale proteomics investigations. Nat Methods 2005;2:667–75.

[236] Adam BL, Qu Y, Davis JW, et al. Serum protein fingerprinting coupled with a pattern-matching algorithm distinguishes prostate cancer from benign prostate hyperplasia and healthy men. Cancer Res 2002;62:3609–14.

[237] Li J, Zhang Z, Rosenzweig J, Wang YY, Chan DW. Proteomics and bioinformatics approaches for identification of serum biomarkers to detect breast cancer. Clin Chem 2002;48:1296–304.

[238] UniProt C. UniProt: a worldwide hub of protein knowledge. Nucleic Acids Res 2019;47: D506–15.

[239] Brosch M, Swamy S, Hubbard T, Choudhary J. Comparison of Mascot and X!Tandem performance for low and high accuracy mass spectrometry and the development of an adjusted Mascot threshold. Mol Cell Proteomics 2008;7:962–70.

[240] Craig R, Beavis RC. TANDEM: matching proteins with tandem mass spectra. Bioinformatics 2004;20:1466–7.

[241] Eng JK, Jahan TA, Hoopmann MR. Comet: an open-source MS/MS sequence database search tool. Proteomics 2013;13:22–4.

[242] Hirosawa M, Hoshida M, Ishikawa M, Toya T. MASCOT: multiple alignment system for protein sequences based on three-way dynamic programming. Comput Appl Biosci 1993;9:161–7.

[243] Lam H, Deutsch EW, Eddes JS, Eng JK, Stein SE, Aebersold R. Building consensus spectral libraries for peptide identification in proteomics. Nat Methods 2008;5:873–5.

[244] Tabb DL, Fernando CG, Chambers MC. MyriMatch: highly accurate tandem mass spectral peptide identification by multivariate hypergeometric analysis. J Proteome Res 2007;6:654–61.

[245] Vaudel M, Barsnes H, Berven FS, Sickmann A, Martens L. SearchGUI: an open-source graphical user interface for simultaneous OMSSA and X!Tandem searches. Proteomics 2011;11:996–9.

[246] Park GW, Hwang H, Kim KH, et al. Integrated proteomic pipeline using multiple search engines for a proteogenomic study with a controlled protein false discovery rate. J Proteome Res 2016;15:4082–90.

[247] Zhang Z, Burke M, Mirokhin YA, et al. Reverse and random decoy methods for false discovery rate estimation in high mass accuracy peptide spectral library searches. J Proteome Res 2018;17:846–57.

[248] Deutsch EW, Mendoza L, Shteynberg D, Slagel J, Sun Z, Moritz RL. Trans-proteomic pipeline, a standardized data processing pipeline for large-scale reproducible proteomics informatics. Proteomics Clin Appl 2015;9:745–54.

[249] Fournier F, Joly Beauparlant C, Paradis R, Droit A. rTANDEM, an R/Bioconductor package for MS/MS protein identification. Bioinformatics 2014;30:2233–4.

[250] Kim S, Pevzner PA. MS-GF+ makes progress towards a universal database search tool for proteomics. Nat Commun 2014;5.

[251] Pfeuffer J, Sachsenberg T, Alka O, et al. OpenMS—a platform for reproducible analysis of mass spectrometry data. J Biotechnol 2017;261:142–8.

[252] Tyanova S, Temu T, Cox J. The MaxQuant computational platform for mass spectrometry-based shotgun proteomics. Nat Protoc 2016;11:2301–19.

[253] Mischak H, Delles C, Klein J, Schanstra JP. Urinary proteomics based on capillary electrophoresis-coupled mass spectrometry in kidney disease: discovery and validation of biomarkers, and clinical application. Adv Chronic Kidney Dis 2010;17:493–506.

[254] Mullen W, Delles C, Mischak H, KUPCa E. Urinary proteomics in the assessment of chronic kidney disease. Curr Opin Nephrol Hypertens 2011;20:654–61.

[255] Spasovski G, Ortiz A, Vanholder R, El Nahas M. Proteomics in chronic kidney disease: the issues clinical nephrologists need an answer for. Proteomics Clin Appl 2011;5:233–40.

[256] Dai H, Zhang H, He Y. Diagnostic accuracy of PLA2R autoantibodies and glomerular staining for the differentiation of idiopathic and secondary membranous nephropathy: an updated meta-analysis. Sci Rep 2015;5.

[257] Cattran DC, Brenchley PE. Membranous nephropathy: integrating basic science into improved clinical management. Kidney Int 2017;91:566–74.

[258] Debiec H, Ronco P. PLA2R autoantibodies and PLA2R glomerular deposits in membranous nephropathy. N Engl J Med 2011;364:689–90.

[259] Haubitz M, Good DM, Woywodt A, et al. Identification and validation of urinary biomarkers for differential diagnosis and evaluation of therapeutic intervention in anti-neutrophil cytoplasmic antibody-associated vasculitis. Mol Cell Proteomics 2009;8:2296–307.

[260] Julian BA, Wittke S, Novak J, et al. Electrophoretic methods for analysis of urinary polypeptides in IgA-associated renal diseases. Electrophoresis 2007;28:4469–83.

[261] Clarke W, Silverman BC, Zhang Z, Chan DW, Klein AS, Molmenti EP. Characterization of renal allograft rejection by urinary proteomic analysis. Ann Surg 2003;237:660–4. discussion 4-5.

[262] Quintana LF, Banon-Maneus E, Sole-Gonzalez A, Campistol JM. Urine proteomics biomarkers in renal transplantation: an overview. Transplantation 2009;88:S45–9.

[263] Quintana LF, Sole-Gonzalez A, Kalko SG, et al. Urine proteomics to detect biomarkers for chronic allograft dysfunction. J Am Soc Nephrol 2009;20:428–35.

[264] Schaub S, Rush D, Wilkins J, et al. Proteomic-based detection of urine proteins associated with acute renal allograft rejection. J Am Soc Nephrol 2004;15:219–27.

[265] Klawitter J, Bendrick-Peart J, Rudolph B, et al. Urine metabolites reflect time-dependent effects of cyclosporine and sirolimus on rat kidney function. Chem Res Toxicol 2009;22:118–28.

[266] Dai Y, Lv T, Wang K, Huang Y, Li D, Liu J. Detection of acute renal allograft rejection by analysis of renal tissue proteomics in rat models of renal transplantation. Saudi J Kidney Dis Transpl 2008;19:952–9.

[267] Reuter S, Reiermann S, Worner R, et al. IF/TA-related metabolic changes—proteome analysis of rat renal allografts. Nephrol Dial Transplant 2010;25:2492–501.

[268] Argiles A, Siwy J, Duranton F, et al. CKD273, a new proteomics classifier assessing CKD and its prognosis. PLoS One 2013;8.

[269] Good DM, Zurbig P, Argiles A, et al. Naturally occurring human urinary peptides for use in diagnosis of chronic kidney disease. Mol Cell Proteomics 2010;9:2424–37.

[270] Merchant ML, Perkins BA, Boratyn GM, et al. Urinary peptidome may predict renal function decline in type 1 diabetes and microalbuminuria. J Am Soc Nephrol 2009;20:2065–74.

[271] Roscioni SS, de Zeeuw D, Hellemons ME, et al. A urinary peptide biomarker set predicts worsening of albuminuria in type 2 diabetes mellitus. Diabetologia 2013;56:259–67.

[272] Schanstra JP, Zurbig P, Alkhalaf A, et al. Diagnosis and prediction of CKD progression by assessment of urinary peptides. J Am Soc Nephrol 2015;26:1999–2010.

[273] Zurbig P, Jerums G, Hovind P, et al. Urinary proteomics for early diagnosis in diabetic nephropathy. Diabetes 2012;61:3304–13.

[274] Siwy J, Schanstra JP, Argiles A, et al. Multicentre prospective validation of a urinary peptidome-based classifier for the diagnosis of type 2 diabetic nephropathy. Nephrol Dial Transplant 2014;29:1563–70.

[275] Andersen S, Mischak H, Zurbig P, Parving HH, Rossing P. Urinary proteome analysis enables assessment of renoprotective treatment in type 2 diabetic patients with microalbuminuria. BMC Nephrol 2010;11.

[276] Rossing K, Mischak H, Parving HH, et al. Impact of diabetic nephropathy and angiotensin II receptor blockade on urinary polypeptide patterns. Kidney Int 2005;68:193–205.

[277] Mischak H, Schanstra JP. CE-MS in biomarker discovery, validation, and clinical application. Proteomics Clin Appl 2011;5:9–23.

[278] Pontillo C, Jacobs L, Staessen JA, et al. A urinary proteome-based classifier for the early detection of decline in glomerular filtration. Nephrol Dial Transplant 2017;32:1510–6.

[279] Pontillo C, Mischak H. Urinary peptide-based classifier CKD273: towards clinical application in chronic kidney disease. Clin Kidney J 2017;10:192–201.

[280] Nandal S, Burt T. integrating pharmacoproteomics into early-phase clinical development: state-of-the-art, challenges, and recommendations. Int J Mol Sci 2017;18.

[281] Dongre AR, Opiteck G, Cosand WL, Hefta SA. Proteomics in the post-genome age. Biopolymers 2001;60:206–11.

[282] Wishart DS. Proteomics and the human metabolome project. Expert Rev Proteomics 2007;4:333–5.

[283] Deidda M, Piras C, Dessalvi CC, et al. Metabolomic approach to profile functional and metabolic changes in heart failure. J Transl Med 2015;13.

[284] Dettmer K, Hammock BD. Metabolomics—a new exciting field within the "omics" sciences. Environ Health Perspect 2004;112:A396–7.

[285] Nicholson JK, Lindon JC, Holmes E. 'Metabonomics': understanding the metabolic responses of living systems to pathophysiological stimuli via multivariate statistical analysis of biological NMR spectroscopic data. Xenobiotica 1999;29:1181–9.

[286] Grams ME, Shafi T, Rhee EP. Metabolomics research in chronic kidney disease. J Am Soc Nephrol 2018;29:1588–90.

[287] Portilla D, Schnackenberg L, Beger RD. Metabolomics as an extension of proteomic analysis: study of acute kidney injury. Semin Nephrol 2007;27:609–20.

[288] Christians U, Klawitter J, Hornberger A, Klawitter J. How unbiased is non-targeted metabolomics and is targeted pathway screening the solution? Curr Pharm Biotechnol 2011;12:1053–66.

[289] Psychogios N, Hau DD, Peng J, et al. The human serum metabolome. PLoS One 2011;6.

[290] Wishart DS, Feunang YD, Marcu A, et al. HMDB 4.0: the human metabolome database for 2018. Nucleic Acids Res 2018;46:D608–17.

[291] Davies R. The metabolomic quest for a biomarker in chronic kidney disease. Clin Kidney J 2018;11:694–703.

[292] Dettmer K, Aronov PA, Hammock BD. Mass spectrometry-based metabolomics. Mass Spectrom Rev 2007;26:51–78.

[293] Bouatra S, Aziat F, Mandal R, et al. The human urine metabolome. PLoS One 2013;8.

[294] Assfalg M, Bertini I, Colangiuli D, et al. Evidence of different metabolic phenotypes in humans. Proc Natl Acad Sci U S A 2008;105:1420–4.

[295] Pinto J, Domingues MR, Galhano E, et al. Human plasma stability during handling and storage: impact on NMR metabolomics. Analyst 2014;139:1168–77.

[296] Sitnikov DG, Monnin CS, Vuckovic D. Systematic assessment of seven solvent and solid-phase extraction methods for metabolomics analysis of human plasma by LC-MS. Sci Rep 2016;6.

[297] Matuszewski BK, Constanzer ML, Chavez-Eng CM. Matrix effect in quantitative LC/MS/MS analyses of biological fluids: a method for determination of finasteride in human plasma at picogram per milliliter concentrations. Anal Chem 1998;70:882–9.

[298] Stahnke H, Kittlaus S, Kempe G, Alder L. Reduction of matrix effects in liquid chromatography-electrospray ionization-mass spectrometry by dilution of the sample extracts: how much dilution is needed? Anal Chem 2012;84:1474–82.

[299] Theodoridis GA, Gika HG, Want EJ, Wilson ID. Liquid chromatography-mass spectrometry based global metabolite profiling: a review. Anal Chim Acta 2012;711:7–16.

[300] Tiziani S, Emwas AH, Lodi A, et al. Optimized metabolite extraction from blood serum for ^{1}H nuclear magnetic resonance spectroscopy. Anal Biochem 2008;377:16–23.

[301] Wang L, Pi Z, Liu S, Liu Z, Song F. Targeted metabolome profiling by dual-probe microdialysis sampling and treatment using Gardenia jasminoides for rats with type 2 diabetes. Sci Rep 2017;7.

[302] Sapcariu SC, Kanashova T, Weindl D, Ghelfi J, Dittmar G, Hiller K. Simultaneous extraction of proteins and metabolites from cells in culture. MethodsX 2014;1:74–80.

[303] David A, Abdul-Sada A, Lange A, Tyler CR, Hill EM. A new approach for plasma (xeno)metabolomics based on solid-phase extraction and nanoflow liquid chromatography-nanoelectrospray ionisation mass spectrometry. J Chromatogr A 2014;1365:72–85.

[304] Zhang Y, Fonslow BR, Shan B, Baek MC, Yates 3rd JR. Protein analysis by shotgun/bottom-up proteomics. Chem Rev 2013;113:2343–94.

[305] Vorreiter F, Richter S, Peter M, Baumann S, von Bergen M, Tomm JM. Comparison and optimization of methods for the simultaneous extraction of DNA, RNA, proteins, and metabolites. Anal Biochem 2016;508:25–33.

[306] Hanna MH, Brophy PD. Metabolomics in pediatric nephrology: emerging concepts. Pediatr Nephrol 2015;30:881–7.

[307] Lindon JC, Nicholson JK. Spectroscopic and statistical techniques for information recovery in metabonomics and metabolomics. Annu Rev Anal Chem (Palo Alto, Calif) 2008;1:45–69.

[308] Syggelou A, Iacovidou N, Atzori L, Xanthos T, Fanos V. Metabolomics in the developing human being. Pediatr Clin N Am 2012;59:1039–58.

[309] Rai RK, Sinha N. Fast and accurate quantitative metabolic profiling of body fluids by nonlinear sampling of ^{1}H-^{13}C two-dimensional nuclear magnetic resonance spectroscopy. Anal Chem 2012;84:10005–11.

[310] Rhee EP. Metabolomics and renal disease. Curr Opin Nephrol Hypertens 2015;24:371–9.

[311] Schnackenberg LK, Beger RD. Monitoring the health to disease continuum with global metabolic profiling and systems biology. Pharmacogenomics 2006;7:1077–86.

[312] Adams RW, Holroyd CM, Aguilar JA, Nilsson M, Morris GA. "Perfecting" WATERGATE: clean proton NMR spectra from aqueous solution. Chem Commun (Camb) 2013;49:358–60.

[313] Hu H, Bradley SA, Krishnamurthy K. Extending the limits of the selective 1D NOESY experiment with an improved selective TOCSY edited preparation function. J Magn Reson 2004;171:201–6.

[314] Piotto M, Saudek V, Sklenar V. Gradient-tailored excitation for single-quantum NMR spectroscopy of aqueous solutions. J Biomol NMR 1992;2:661–5.

[315] Emwas AH, Saccenti E, Gao X, et al. Recommended strategies for spectral processing and post-processing of 1D (1)H-NMR data of biofluids with a particular focus on urine. Metabolomics 2018;14.

[316] Cloarec O, Dumas ME, Craig A, et al. Statistical total correlation spectroscopy: an exploratory approach for latent biomarker identification from metabolic ^{1}H NMR data sets. Anal Chem 2005;77:1282–9.

[317] Fonville JM, Maher AD, Coen M, Holmes E, Lindon JC, Nicholson JK. Evaluation of full-resolution J-resolved ^{1}H NMR projections of biofluids for metabonomics information retrieval and biomarker identification. Anal Chem 2010;82:1811–21.

[318] Feng X, Liu X, Luo Q, Liu BF. Mass spectrometry in systems biology: an overview. Mass Spectrom Rev 2008;27:635–60.

[319] Pan Z, Raftery D. Comparing and combining NMR spectroscopy and mass spectrometry in metabolomics. Anal Bioanal Chem 2007;387:525–7.

[320] Fischer R, Bowness P, Kessler BM. Two birds with one stone: doing metabolomics with your proteomics kit. Proteomics 2013;13:3371–86.

[321] Becker S, Kortz L, Helmschrodt C, Thiery J, Ceglarek U. LC-MS-based metabolomics in the clinical laboratory. J Chromatogr B Analyt Technol Biomed Life Sci 2012;883-884:68–75.

[322] Fischer R, Trudgian DC, Wright C, et al. Discovery of candidate serum proteomic and metabolomic biomarkers in ankylosing spondylitis. Mol Cell Proteomics 2012;11. M111 013904.

[323] Zhou B, Xiao JF, Tuli L, Ressom HW. LC-MS-based metabolomics. Mol BioSyst 2012;8:470–81.

[324] Metz TO, Page JS, Baker ES, et al. High resolution separations and improved ion production and transmission in metabolomics. Trends Anal Chem 2008;27:205–14.

[325] Chervet JP, Ursem M, Salzmann JP. Instrumental requirements for nanoscale liquid chromatography. Anal Chem 1996;68:1507–12.

[326] Wickremsinhe ER, Singh G, Ackermann BL, Gillespie TA, Chaudhary AK. A review of nanoelectrospray ionization applications for drug metabolism and pharmacokinetics. Curr Drug Metab 2006;7:913–28.

[327] Marginean I, Tang K, Smith RD, Kelly RT. Picoelectrospray ionization mass spectrometry using narrow-bore chemically etched emitters. J Am Soc Mass Spectrom 2014;25:30–6.

[328] Wilm M, Mann M. Analytical properties of the nanoelectrospray ion source. Anal Chem 1996;68:1–8.

[329] Chetwynd AJ, David A, Hill EM, Abdul-Sada A. Evaluation of analytical performance and reliability of direct nanoLC-nanoESI-high resolution mass spectrometry for profiling the (xeno) metabolome. J Mass Spectrom 2014;49:1063–9.

[330] Garcia-Villalba R, Carrasco-Pancorbo A, Zurek G, et al. Nano and rapid resolution liquid chromatography-electrospray ionization-time of flight mass spectrometry to identify and quantify phenolic compounds in olive oil. J Sep Sci 2010;33:2069–78.

[331] Lu W, Bennett BD, Rabinowitz JD. Analytical strategies for LC-MS-based targeted metabolomics. J Chromatogr B Analyt Technol Biomed Life Sci 2008;871:236–42.

[332] Myint KT, Aoshima K, Tanaka S, Nakamura T, Oda Y. Quantitative profiling of polar cationic metabolites in human cerebrospinal fluid by reversed-phase nanoliquid chromatography/ mass spectrometry. Anal Chem 2009;81:1121–9.

[333] Smith RD, Shen Y, Tang K. Ultrasensitive and quantitative analyses from combined separations-mass spectrometry for the characterization of proteomes. Acc Chem Res 2004;37:269–78.

[334] Boernsen KO, Gatzek S, Imbert G. Controlled protein precipitation in combination with chip-based nanospray infusion mass spectrometry. An approach for metabolomics profiling of plasma. Anal Chem 2005;77:7255–64.

[335] Page JS, Kelly RT, Tang K, Smith RD. Ionization and transmission efficiency in an electrospray ionization-mass spectrometry interface. J Am Soc Mass Spectrom 2007;18:1582–90.

[336] Shi X, Wahlang B, Wei X, et al. Metabolomic analysis of the effects of polychlorinated biphenyls in nonalcoholic fatty liver disease. J Proteome Res 2012;11:3805–15.

[337] Southam AD, Payne TG, Cooper HJ, Arvanitis TN, Viant MR. Dynamic range and mass accuracy of wide-scan direct infusion nanoelectrospray fourier transform ion cyclotron resonance mass spectrometry-based metabolomics increased by the spectral stitching method. Anal Chem 2007;79:4595–602.

[338] Wu H, Southam AD, Hines A, Viant MR. High-throughput tissue extraction protocol for NMR- and MS-based metabolomics. Anal Biochem 2008;372:204–12.

[339] Hop CE, Chen Y, Yu LJ. Uniformity of ionization response of structurally diverse analytes using a chip-based nanoelectrospray ionization source. Rapid Commun Mass Spectrom 2005;19:3139–42.

[340] Noga M, Sucharski F, Suder P, Silberring J. A practical guide to nano-LC troubleshooting. J Sep Sci 2007;30:2179–89.

[341] Sestak J, Moravcova D, Kahle V. Instrument platforms for nano liquid chromatography. J Chromatogr A 2015;1421:2–17.

[342] Chetwynd AJ, Abdul-Sada A, Hill EM. Solid-phase extraction and nanoflow liquid chromatography-nanoelectrospray ionization mass spectrometry for improved global urine metabolomics. Anal Chem 2015;87:1158–65.

[343] Jones DR, Wu Z, Chauhan D, Anderson KC, Peng J. A nano ultra-performance liquid chromatography-high resolution mass spectrometry approach for global metabolomic profiling and case study on drug-resistant multiple myeloma. Anal Chem 2014;86:3667–75.

[344] Li Z, Tatlay J, Li L. Nanoflow LC-MS for high-performance chemical isotope labeling quantitative metabolomics. Anal Chem 2015;87:11468–74.

[345] Luo X, Li L. Metabolomics of small numbers of cells: metabolomic profiling of 100, 1000, and 10000 human breast cancer cells. Anal Chem 2017;89:11664–71.

[346] Hernandez-Borges J, Aturki Z, Rocco A, Fanali S. Recent applications in nanoliquid chromatography. J Sep Sci 2007;30:1589–610.

[347] Valaskovic GA, Murphy 3rd JP, Lee MS. Automated orthogonal control system for electrospray ionization. J Am Soc Mass Spectrom 2004;15:1201–15.

[348] European Renal Association-European Dialysis and Transplant Association. ERA-EDTA annual report 2015. ERA-EDTA Registry.

[349] Collins DA, Nesterenko EP, Paull B. Porous layer open tubular columns in capillary liquid chromatography. Analyst 2014;139:1292–302.

[350] Nazario CE, Silva MR, Franco MS, Lancas FM. Evolution in miniaturized column liquid chromatography instrumentation and applications: an overview. J Chromatogr A 2015;1421:18–37.

[351] Kiefer P, Delmotte N, Vorholt JA. Nanoscale ion-pair reversed-phase HPLC-MS for sensitive metabolome analysis. Anal Chem 2011;83:850–5.

[352] Peironcely JE, Reijmers T, Coulier L, Bender A, Hankemeier T. Understanding and classifying metabolite space and metabolite-likeness. PLoS One 2011;6.

[353] Zhao Y, Jensen ON. Modification-specific proteomics: strategies for characterization of post-translational modifications using enrichment techniques. Proteomics 2009;9:4632–41.

[354] Coman C, Solari FA, Hentschel A, Sickmann A, Zahedi RP, Ahrends R. Simultaneous metabolite, protein, lipid extraction (SIMPLEX): a combinatorial multimolecular omics approach for systems biology. Mol Cell Proteomics 2016;15:1453–66.

[355] Crutchfield CA, Lu W, Melamud E, Rabinowitz JD. Mass spectrometry-based metabolomics of yeast. Methods Enzymol 2010;470:393–426.

[356] Kelstrup CD, Bekker-Jensen DB, Arrey TN, Hogrebe A, Harder A, Olsen JV. Performance evaluation of the Q exactive HF-X for shotgun proteomics. J Proteome Res 2018;17:727–38.

[357] Xia J, Psychogios N, Young N, Wishart DS. MetaboAnalyst: a web server for metabolomic data analysis and interpretation. Nucleic Acids Res 2009;37:W652–60.

[358] Chadeau-Hyam M, Campanella G, Jombart T, et al. Deciphering the complex: methodological overview of statistical models to derive OMICS-based biomarkers. Environ Mol Mutagen 2013;54:542–57.

[359] Smith CA, O'Maille G, Want EJ, et al. METLIN: a metabolite mass spectral database. Ther Drug Monit 2005;27:747–51.

[360] Blazenovic I, Kind T, Ji J, Fiehn O. Software tools and approaches for compound identification of LC-MS/MS data in metabolomics. Metabolites 2018;8.

[361] Mahieu NG, Patti GJ. Systems-level annotation of a metabolomics data set reduces 25000 features to fewer than 1000 unique metabolites. Anal Chem 2017;89:10397–406.

[362] Fernandez-Albert F, Llorach R, Andres-Lacueva C, Perera A. An R package to analyse LC/MS metabolomic data: MAIT (Metabolite Automatic Identification Toolkit). Bioinformatics 2014;30:1937–9.

[363] Pluskal T, Castillo S, Villar-Briones A, Oresic M. MZmine 2: modular framework for processing, visualizing, and analyzing mass spectrometry-based molecular profile data. BMC Bioinformatics 2010;11.

[364] Tsugawa H, Cajka T, Kind T, et al. MS-DIAL: data-independent MS/MS deconvolution for comprehensive metabolome analysis. Nat Methods 2015;12:523–6.

[365] Smith CA, Want EJ, O'Maille G, Abagyan R, Siuzdak G. XCMS: processing mass spectrometry data for metabolite profiling using nonlinear peak alignment, matching, and identification. Anal Chem 2006;78:779–87.

[366] Kenar E, Franken H, Forcisi S, et al. Automated label-free quantification of metabolites from liquid chromatography-mass spectrometry data. Mol Cell Proteomics 2014;13:348–59.

[367] Fukui Y, Kato M, Inoue Y, Matsubara A, Itoh K. A metabonomic approach identifies human urinary phenylacetylglutamine as a novel marker of interstitial cystitis. J Chromatogr B Analyt Technol Biomed Life Sci 2009;877:3806–12.

[368] Gao X, Chen W, Li R, et al. Systematic variations associated with renal disease uncovered by parallel metabolomics of urine and serum. BMC Syst Biol 2012;6(Suppl 1):S14.

[369] Gronwald W, Klein MS, Zeltner R, et al. Detection of autosomal dominant polycystic kidney disease by NMR spectroscopic fingerprinting of urine. Kidney Int 2011;79:1244–53.

[370] Nevedomskaya E, Pacchiarotta T, Artemov A, et al. (1)H NMR-based metabolic profiling of urinary tract infection: combining multiple statistical models and clinical data. Metabolomics 2012;8:1227–35.

[371] Sato E, Kohno M, Yamamoto M, Fujisawa T, Fujiwara K, Tanaka N. Metabolomic analysis of human plasma from haemodialysis patients. Eur J Clin Investig 2011;41:241–55.

[372] Zivkovic AM, Yang J, Georgi K, et al. Serum oxylipin profiles in IgA nephropathy patients reflect kidney functional alterations. Metabolomics 2012;8:1102–13.

[373] Nkuipou-Kenfack E, Duranton F, Gayrard N, et al. Assessment of metabolomic and proteomic biomarkers in detection and prognosis of progression of renal function in chronic kidney disease. PLoS One 2014;9.

[374] Wishart DS. Metabolomics in monitoring kidney transplants. Curr Opin Nephrol Hypertens 2006;15:637–42.

[375] Bell JD, Lee JA, Lee HA, Sadler PJ, Wilkie DR, Woodham RH. Nuclear magnetic resonance studies of blood plasma and urine from subjects with chronic renal failure: identification of trimethylamine-N-oxide. Biochim Biophys Acta 1991;1096:101–7.

[376] Beger RD, Sun J, Schnackenberg LK. Metabolomics approaches for discovering biomarkers of drug-induced hepatotoxicity and nephrotoxicity. Toxicol Appl Pharmacol 2010;243: 154–166.

[377] Boudonck KJ, Mitchell MW, Nemet L, et al. Discovery of metabolomics biomarkers for early detection of nephrotoxicity. Toxicol Pathol 2009;37:280–92.

[378] Sieber M, Hoffmann D, Adler M, et al. Comparative analysis of novel noninvasive renal biomarkers and metabonomic changes in a rat model of gentamicin nephrotoxicity. Toxicol Sci 2009;109:336–49.

[379] Xu EY, Perlina A, Vu H, et al. Integrated pathway analysis of rat urine metabolic profiles and kidney transcriptomic profiles to elucidate the systems toxicology of model nephrotoxicants. Chem Res Toxicol 2008;21:1548–61.

[380] Klawitter J, Haschke M, Kahle C, et al. Toxicodynamic effects of ciclosporin are reflected by metabolite profiles in the urine of healthy individuals after a single dose. Br J Clin Pharmacol 2010;70:241–51.

[381] Zheng Y, Yu B, Alexander D, Couper DJ, Boerwinkle E. Medium-term variability of the human serum metabolome in the Atherosclerosis Risk in Communities (ARIC) study. OMICS 2014;18:364–73.

[382] Tin A, Nadkarni G, Evans AM, et al. Serum 6-bromotryptophan levels identified as a risk factor for CKD progression. J Am Soc Nephrol 2018;29:1939–47.

[383] Barretina J, Caponigro G, Stransky N, et al. The Cancer Cell Line Encyclopedia enables predictive modelling of anticancer drug sensitivity. Nature 2012;483:603–7.

[384] Rudnick PA, Markey SP, Roth J, et al. A description of the clinical proteomic tumor analysis consortium (CPTAC) common data analysis pipeline. J Proteome Res 2016;15:1023–32.

[385] Efron B, Tibshirani R. Empirical bayes methods and false discovery rates for microarrays. Genet Epidemiol 2002;23:70–86.

[386] Tusher VG, Tibshirani R, Chu G. Significance analysis of microarrays applied to the ionizing radiation response. Proc Natl Acad Sci U S A 2001;98:5116–21.

[387] Wang Y, Ma Y, Carroll RJ. Variance estimation in the analysis of microarray data. J R Stat Soc Series B Stat Methodol 2009;71:425–45.

[388] Zhang HH. Discussion of "Sure independence screening for ultra-high dimensional feature space" J R Stat Soc Series B Stat Methodol 2008;70:849–911.

[389] Wishart DS. Computational approaches to metabolomics. Methods Mol Biol 2010;593:283–313.

[390] Gieser G, Harigaya H, Colangelo PM, Burckart G. Biomarkers in solid organ transplantation. Clin Pharmacol Ther 2011;90:217–20.

[391] Zurbig P, Decramer S, Dakna M, et al. The human urinary proteome reveals high similarity between kidney aging and chronic kidney disease. Proteomics 2009;9:2108–17.

[392] Coresh J, Inker LA, Sang Y, et al. Metabolomic profiling to improve glomerular filtration rate estimation: a proof-of-concept study. Nephrol Dial Transplant 2018.

[393] Sekula P, Goek ON, Quaye L, et al. A metabolome-wide association study of kidney function and disease in the general population. J Am Soc Nephrol 2016;27:1175–88.

[394] Wells DK, Kath WL, Motter AE. Control of stochastic and induced switching in biophysical networks. Phys Rev X 2015;5.

[395] Canadas-Garre M, Anderson K, McGoldrick J, Maxwell AP, McKnight AJ. Genomic approaches in the search for molecular biomarkers in chronic kidney disease. J Transl Med 2018;16.

[396] Consortium EP. An integrated encyclopedia of DNA elements in the human genome. Nature 2012;489:57–74.

[397] Lander ES, Linton LM, Birren B, et al. Initial sequencing and analysis of the human genome. Nature 2001;409:860–921.

[398] Venter JC, Adams MD, Myers EW, et al. The sequence of the human genome. Science 2001;291:1304–51.

[399] Goretti E, Wagner DR, Devaux Y. miRNAs as biomarkers of myocardial infarction: a step forward towards personalized medicine? Trends Mol Med 2014;20:716–25.

[400] Brandenburger T, Salgado Somoza A, Devaux Y, Lorenzen JM. Noncoding RNAs in acute kidney injury. Kidney Int 2018;94:870–81.

[401] Lorenzen JM, Thum T. Circulating and urinary microRNAs in kidney disease. Clin J Am Soc Nephrol 2012;7:1528–33.

[402] Nassirpour R, Mathur S, Gosink MM, et al. Identification of tubular injury microRNA biomarkers in urine: comparison of next-generation sequencing and qPCR-based profiling platforms. BMC Genomics 2014;15:485.

[403] Weber JA, Baxter DH, Zhang S, et al. The microRNA spectrum in 12 body fluids. Clin Chem 2010;56:1733–41.

[404] Cheng L, Sun X, Scicluna BJ, Coleman BM, Hill AF. Characterization and deep sequencing analysis of exosomal and non-exosomal miRNA in human urine. Kidney Int 2014;86:433–44.

[405] Moll AG, Lindenmeyer MT, Kretzler M, Nelson PJ, Zimmer R, Cohen CD. Transcript-specific expression profiles derived from sequence-based analysis of standard microarrays. PLoS One 2009;4.

[406] Keller B, Martini S, Sedor J, Kretzler M. Linking variants from genome-wide association analysis to function via transcriptional network analysis. Semin Nephrol 2010;30:177–84.

[407] Vuylsteke M, Peleman JD, van Eijk MJ. AFLP-based transcript profiling (cDNA-AFLP) for genome-wide expression analysis. Nat Protoc 2007;2:1399–413.

[408] Mortazavi A, Williams BA, McCue K, Schaeffer L, Wold B. Mapping and quantifying mammalian transcriptomes by RNA-Seq. Nat Methods 2008;5:621–8.

[409] Burgos KL, Javaherian A, Bomprezzi R, et al. Identification of extracellular miRNA in human cerebrospinal fluid by next-generation sequencing. RNA 2013;19:712–22.

[410] Git A, Dvinge H, Salmon-Divon M, et al. Systematic comparison of microarray profiling, real-time PCR, and next-generation sequencing technologies for measuring differential microRNA expression. RNA 2010;16:991–1006.

[411] Kolbert CP, Feddersen RM, Rakhshan F, et al. Multi-platform analysis of microRNA expression measurements in RNA from fresh frozen and FFPE tissues. PLoS One 2013;8.

[412] Llorens F, Hummel M, Pantano L, et al. Microarray and deep sequencing cross-platform analysis of the mirRNome and isomiR variation in response to epidermal growth factor. BMC Genomics 2013;14:371.

[413] Meng F, Hackenberg M, Li Z, Yan J, Chen T. Discovery of novel microRNAs in rat kidney using next generation sequencing and microarray validation. PLoS One 2012;7.

[414] Tam S, de Borja R, Tsao MS, McPherson JD. Robust global microRNA expression profiling using next-generation sequencing technologies. Lab Investig 2014;94:350–8.

[415] Wu Q, Wang C, Lu Z, Guo L, Ge Q. Analysis of serum genome-wide microRNAs for breast cancer detection. Clin Chim Acta 2012;413:1058–65.

[416] Spoto B, Leonardis D, Parlongo RM, et al. Plasma cytokines, glomerular filtration rate and adipose tissue cytokines gene expression in chronic kidney disease (CKD) patients. Nutr Metab Cardiovasc Dis 2012;22:981–8.

[417] Szeto CC, Chow KM, Lai KB, et al. mRNA expression of target genes in the urinary sediment as a noninvasive prognostic indicator of CKD. Am J Kidney Dis 2006;47:578–86.

[418] Zehnder D, Quinkler M, Eardley KS, et al. Reduction of the vitamin D hormonal system in kidney disease is associated with increased renal inflammation. Kidney Int 2008;74: 1343–1353.

[419] Zhai YL, Zhu L, Shi SF, Liu LJ, Lv JC, Zhang H. Increased APRIL expression induces IgA1 aberrant glycosylation in IgA nephropathy. Medicine (Baltimore) 2016;95.

[420] Zhou LT, Qiu S, Lv LL, et al. Integrative bioinformatics analysis provides insight into the molecular mechanisms of chronic kidney disease. Kidney Blood Press Res 2018;43:568–81.

[421] Perco P, Muhlberger I, Mayer G, Oberbauer R, Lukas A, Mayer B. Linking transcriptomic and proteomic data on the level of protein interaction networks. Electrophoresis 2010;31:1780–9.

[422] Rudnicki M, Eder S, Perco P, et al. Gene expression profiles of human proximal tubular epithelial cells in proteinuric nephropathies. Kidney Int 2007;71:325–35.

[423] Zheng M, Lv LL, Cao YH, et al. A pilot trial assessing urinary gene expression profiling with an mRNA array for diabetic nephropathy. PLoS One 2012;7.

[424] Zheng M, Lv LL, Cao YH, et al. Urinary mRNA markers of epithelial-mesenchymal transition correlate with progression of diabetic nephropathy. Clin Endocrinol 2012;76:657–64.

[425] Lepenies J, Eardley KS, Kienitz T, et al. Renal TLR4 mRNA expression correlates with inflammatory marker MCP-1 and profibrotic molecule TGF-beta(1) in patients with chronic kidney disease. Nephron Clin Pract 2011;119:c97–c104.

[426] Liu Y. Cellular and molecular mechanisms of renal fibrosis. Nat Rev Nephrol 2011;7:684–96.

[427] Lopez-Hernandez FJ, Lopez-Novoa JM. Role of TGF-beta in chronic kidney disease: an integration of tubular, glomerular and vascular effects. Cell Tissue Res 2012;347:141–54.

[428] Tachaudomdach C, Kantachuvesiri S, Changsirikulchai S, Wimolluck S, Pinpradap K, Kitiyakara C. Connective tissue growth factor gene expression and decline in renal function in lupus nephritis. Exp Ther Med 2012;3:713–8.

[429] Alvarez ML, Distefano JK. The role of non-coding RNAs in diabetic nephropathy: potential applications as biomarkers for disease development and progression. Diabetes Res Clin Pract 2013;99:1–11.

[430] Heggermont WA, Heymans S. MicroRNAs are involved in end-organ damage during hypertension. Hypertension 2012;60:1088–93.

[431] Kasinath BS, Feliers D. The complex world of kidney microRNAs. Kidney Int 2011;80:334–7.

[432] Pandey P, Qin S, Ho J, Zhou J, Kreidberg JA. Systems biology approach to identify transcriptome reprogramming and candidate microRNA targets during the progression of polycystic kidney disease. BMC Syst Biol 2011;5:56.

[433] Redova M, Svoboda M, Slaby O. MicroRNAs and their target gene networks in renal cell carcinoma. Biochem Biophys Res Commun 2011;405:153–6.

[434] White NM, Yousef GM. MicroRNAs: exploring a new dimension in the pathogenesis of kidney cancer. BMC Med 2010;8.

[435] Yi Z, Fu Y, Zhao S, Zhang X, Ma C. Differential expression of miRNA patterns in renal cell carcinoma and nontumorous tissues. J Cancer Res Clin Oncol 2010;136:855–62.

[436] Zampetaki A, Kiechl S, Drozdov I, et al. Plasma microRNA profiling reveals loss of endothelial miR-126 and other microRNAs in type 2 diabetes. Circ Res 2010;107:810–7.

[437] Chen J, Zhang X, Zhang H, et al. Elevated Klotho promoter methylation is associated with severity of chronic kidney disease. PLoS One 2013;8.

[438] Neal CS, Michael MZ, Pimlott LK, Yong TY, Li JY, Gleadle JM. Circulating microRNA expression is reduced in chronic kidney disease. Nephrol Dial Transplant 2011;26:3794–802.

[439] Zawada AM, Rogacev KS, Muller S, et al. Massive analysis of cDNA Ends (MACE) and miRNA expression profiling identifies proatherogenic pathways in chronic kidney disease. Epigenetics 2014;9:161–72.

[440] Zhang W, Shi L, Zhang H, et al. Effect of alprostadil on serum level of miRNA-155 in uremic patients. Zhong Nan Da Xue Xue Bao Yi Xue Ban 2015;40:735–41.

[441] Hu YY, Dong WD, Xu YF, et al. Elevated levels of miR-155 in blood and urine from patients with nephrolithiasis. Biomed Res Int 2014;2014:295651.

[442] Wang G, Kwan BC, Lai FM, Chow KM, Li PK, Szeto CC. Elevated levels of miR-146a and miR-155 in kidney biopsy and urine from patients with IgA nephropathy. Dis Markers 2011;30:171–9.

[443] Ramezani A, Devaney JM, Cohen S, et al. Circulating and urinary microRNA profile in focal segmental glomerulosclerosis: a pilot study. Eur J Clin Investig 2015;45:394–404.

[444] Wang G, Tam LS, Kwan BC, et al. Expression of miR-146a and miR-155 in the urinary sediment of systemic lupus erythematosus. Clin Rheumatol 2012;31:435–40.

[445] Wang G, Tam LS, Li EK, et al. Serum and urinary cell-free MiR-146a and MiR-155 in patients with systemic lupus erythematosus. J Rheumatol 2010;37:2516–22.

[446] Nandakumar P, Tin A, Grove ML, et al. MicroRNAs in the miR-17 and miR-15 families are downregulated in chronic kidney disease with hypertension. PLoS One 2017;12.

[447] Argyropoulos C, Wang K, Bernardo J, et al. Urinary MicroRNA profiling predicts the development of microalbuminuria in patients with type 1 diabetes. J Clin Med 2015;4:1498–517.

[448] Lv LL, Cao YH, Ni HF, et al. MicroRNA-29c in urinary exosome/microvesicle as a biomarker of renal fibrosis. Am J Physiol Renal Physiol 2013;305:F1220–7.

[449] Borges FT, Reis LA, Schor N. Extracellular vesicles: structure, function, and potential clinical uses in renal diseases. Braz J Med Biol Res 2013;46:824–30.

[450] Dear JW, Street JM, Bailey MA. Urinary exosomes: a reservoir for biomarker discovery and potential mediators of intrarenal signalling. Proteomics 2013;13:1572–80.

[451] Salih M, Zietse R, Hoorn EJ. Urinary extracellular vesicles and the kidney: biomarkers and beyond. Am J Physiol Renal Physiol 2014;306:F1251–9.

[452] van Balkom BW, Pisitkun T, Verhaar MC, Knepper MA. Exosomes and the kidney: prospects for diagnosis and therapy of renal diseases. Kidney Int 2011;80:1138–45.

[453] Trams EG, Lauter CJ, Salem Jr N, Heine U. Exfoliation of membrane ecto-enzymes in the form of micro-vesicles. Biochim Biophys Acta 1981;645:63–70.

[454] Johnstone RM, Adam M, Hammond JR, Orr L, Turbide C. Vesicle formation during reticulocyte maturation. Association of plasma membrane activities with released vesicles (exosomes). J Biol Chem 1987;262:9412–20.

[455] Johnstone RM, Bianchini A, Teng K. Reticulocyte maturation and exosome release: transferrin receptor containing exosomes shows multiple plasma membrane functions. Blood 1989;74:1844–51.

[456] Pisitkun T, Shen RF, Knepper MA. Identification and proteomic profiling of exosomes in human urine. Proc Natl Acad Sci U S A 2004;101:13368–73.

[457] Stoorvogel W, Kleijmeer MJ, Geuze HJ, Raposo G. The biogenesis and functions of exosomes. Traffic 2002;3:321–30.

[458] Mathivanan S, Simpson RJ. ExoCarta: a compendium of exosomal proteins and RNA. Proteomics 2009;9:4997–5000.

[459] Admyre C, Grunewald J, Thyberg J, et al. Exosomes with major histocompatibility complex class II and co-stimulatory molecules are present in human BAL fluid. Eur Respir J 2003;22:578–83.

[460] Bard MP, Hegmans JP, Hemmes A, et al. Proteomic analysis of exosomes isolated from human malignant pleural effusions. Am J Respir Cell Mol Biol 2004;31:114–21.

[461] Bobrie A, Colombo M, Raposo G, Thery C. Exosome secretion: molecular mechanisms and roles in immune responses. Traffic 2011;12:1659–68.

[462] Caby MP, Lankar D, Vincendeau-Scherrer C, Raposo G, Bonnerot C. Exosomal-like vesicles are present in human blood plasma. Int Immunol 2005;17:879–87.

[463] Fevrier B, Vilette D, Archer F, et al. Cells release prions in association with exosomes. Proc Natl Acad Sci U S A 2004;101:9683–8.

[464] Heijnen HF, Schiel AE, Fijnheer R, Geuze HJ, Sixma JJ. Activated platelets release two types of membrane vesicles: microvesicles by surface shedding and exosomes derived from exocytosis of multivesicular bodies and alpha-granules. Blood 1999;94:3791–9.

[465] Sullivan R, Saez F, Girouard J, Frenette G. Role of exosomes in sperm maturation during the transit along the male reproductive tract. Blood Cells Mol Dis 2005;35:1–10.

[466] Thery C, Ostrowski M, Segura E. Membrane vesicles as conveyors of immune responses. Nat Rev Immunol 2009;9:581–93.

[467] Urbanelli L, Buratta S, Sagini K, Ferrara G, Lanni M, Emiliani C. Exosome-based strategies for diagnosis and therapy. Recent Pat CNS Drug Discov 2015;10:10–27.

[468] Utleg AG, Yi EC, Xie T, et al. Proteomic analysis of human prostasomes. Prostate 2003;56:150–61.

[469] Mirzakhani M, Shahbazi M, Oliaei F, Mohammadnia-Afrouzi M. Immunological biomarkers of tolerance in human kidney transplantation: an updated literature review. J Cell Physiol 2019;234:5762–74.

[470] Buschow SI, van Balkom BW, Aalberts M, Heck AJ, Wauben M, Stoorvogel W. MHC class II-associated proteins in B-cell exosomes and potential functional implications for exosome biogenesis. Immunol Cell Biol 2010;88:851–6.

[471] Wubbolts R, Leckie RS, Veenhuizen PT, et al. Proteomic and biochemical analyses of human B cell-derived exosomes. Potential implications for their function and multivesicular body formation. J Biol Chem 2003;278:10963–72.

[472] Sabolic I, Valenti G, Verbavatz JM, et al. Localization of the CHIP28 water channel in rat kidney. Am J Phys 1992;263:C1225–33.

[473] Camussi G, Deregibus MC, Bruno S, Cantaluppi V, Biancone L. Exosomes/microvesicles as a mechanism of cell-to-cell communication. Kidney Int 2010;78:838–48.

[474] Clayton A, Turkes A, Dewitt S, Steadman R, Mason MD, Hallett MB. Adhesion and signaling by B cell-derived exosomes: the role of integrins. FASEB J 2004;18:977–9.

[475] Denzer K, Kleijmeer MJ, Heijnen HF, Stoorvogel W, Geuze HJ. Exosome: from internal vesicle of the multivesicular body to intercellular signaling device. J Cell Sci 2000;113(Pt 19):3365–74.

[476] Fevrier B, Raposo G. Exosomes: endosomal-derived vesicles shipping extracellular messages. Curr Opin Cell Biol 2004;16:415–21.

[477] Nolte-'t Hoen EN, Buschow SI, Anderton SM, Stoorvogel W, Wauben MH. Activated T cells recruit exosomes secreted by dendritic cells via LFA-1. Blood 2009;113:1977–81.

[478] Skog J, Wurdinger T, van Rijn S, et al. Glioblastoma microvesicles transport RNA and proteins that promote tumour growth and provide diagnostic biomarkers. Nat Cell Biol 2008;10: 1470–1476.

[479] Smalheiser NR. Exosomal transfer of proteins and RNAs at synapses in the nervous system. Biol Direct 2007;2:35.

[480] Valadi H, Ekstrom K, Bossios A, Sjostrand M, Lee JJ, Lotvall JO. Exosome-mediated transfer of mRNAs and microRNAs is a novel mechanism of genetic exchange between cells. Nat Cell Biol 2007;9:654–9.

[481] Collino F, Deregibus MC, Bruno S, et al. Microvesicles derived from adult human bone marrow and tissue specific mesenchymal stem cells shuttle selected pattern of miRNAs. PLoS One 2010;5.

[482] Deregibus MC, Tetta C, Camussi G. The dynamic stem cell microenvironment is orchestrated by microvesicle-mediated transfer of genetic information. Histol Histopathol 2010;25: 397–404.

[483] Pegtel DM, Cosmopoulos K, Thorley-Lawson DA, et al. Functional delivery of viral miRNAs via exosomes. Proc Natl Acad Sci U S A 2010;107:6328–33.

[484] Quesenberry PJ, Aliotta JM. Cellular phenotype switching and microvesicles. Adv Drug Deliv Rev 2010;62:1141–8.

[485] Thery C, Zitvogel L, Amigorena S. Exosomes: composition, biogenesis and function. Nat Rev Immunol 2002;2:569–79.

[486] Utsugi-Kobukai S, Fujimaki H, Hotta C, Nakazawa M, Minami M. MHC class I-mediated exogenous antigen presentation by exosomes secreted from immature and mature bone marrow derived dendritic cells. Immunol Lett 2003;89:125–31.

[487] Iero M, Valenti R, Huber V, et al. Tumour-released exosomes and their implications in cancer immunity. Cell Death Differ 2008;15:80–8.

[488] Janowska-Wieczorek A, Wysoczynski M, Kijowski J, et al. Microvesicles derived from activated platelets induce metastasis and angiogenesis in lung cancer. Int J Cancer 2005;113:752–60.

[489] Salih M, Demmers JA, Bezstarosti K, et al. Proteomics of urinary vesicles links plakins and complement to polycystic kidney disease. J Am Soc Nephrol 2016;27:3079–92.

[490] Jiang H, Guan G, Zhang R, et al. Identification of urinary soluble E-cadherin as a novel biomarker for diabetic nephropathy. Diabetes Metab Res Rev 2009;25:232–41.

[491] Adachi J, Kumar C, Zhang Y, Olsen JV, Mann M. The human urinary proteome contains more than 1500 proteins, including a large proportion of membrane proteins. Genome Biol 2006;7:R80.

[492] Gonzales PA, Pisitkun T, Hoffert JD, et al. Large-scale proteomics and phosphoproteomics of urinary exosomes. J Am Soc Nephrol 2009;20:363–79.

[493] Floege J. Moderator's view: will 'modern' urine proteomics replace 'old-fashioned' renal biopsy? Nephrol Dial Transplant 2015;30:538–40.

[494] Han WK, Bailly V, Abichandani R, Thadhani R, Bonventre JV. Kidney Injury Molecule-1 (KIM-1): a novel biomarker for human renal proximal tubule injury. Kidney Int 2002;62:237–44.

[495] Barrera-Chimal J, Perez-Villalva R, Cortes-Gonzalez C, et al. Hsp72 is an early and sensitive biomarker to detect acute kidney injury. EMBO Mol Med 2011;3:5–20.

[496] Aiello S, Noris M. Klotho in acute kidney injury: biomarker, therapy, or a bit of both? Kidney Int 2010;78:1208–10.

[497] Dennen P, Altmann C, Kaufman J, et al. Urine interleukin-6 is an early biomarker of acute kidney injury in children undergoing cardiac surgery. Crit Care 2010;14:R181.

[498] Bolignano D, Donato V, Coppolino G, et al. Neutrophil gelatinase-associated lipocalin (NGAL) as a marker of kidney damage. Am J Kidney Dis 2008;52:595–605.

[499] Matsui K, Kamijo-Ikemori A, Hara M, et al. Clinical significance of tubular and podocyte biomarkers in acute kidney injury. Clin Exp Nephrol 2011;15:220–5.

[500] Ramesh G, Krawczeski CD, Woo JG, Wang Y, Devarajan P. Urinary netrin-1 is an early predictive biomarker of acute kidney injury after cardiac surgery. Clin J Am Soc Nephrol 2010;5:395–401.

[501] Alvarez S, Suazo C, Boltansky A, et al. Urinary exosomes as a source of kidney dysfunction biomarker in renal transplantation. Transplant Proc 2013;45:3719–23.

[502] Peake PW, Pianta TJ, Succar L, et al. A comparison of the ability of levels of urinary biomarker proteins and exosomal mRNA to predict outcomes after renal transplantation. PLoS One 2014;9.

[503] Park J, Lin HY, Assaker JP, et al. Integrated kidney exosome analysis for the detection of kidney transplant rejection. ACS Nano 2017;11:11041–6.

[504] Hogan MC, Manganelli L, Woollard JR, et al. Characterization of PKD protein-positive exosome-like vesicles. J Am Soc Nephrol 2009;20:278–88.

[505] Hogan MC, Bakeberg JL, Gainullin VG, et al. Identification of biomarkers for PKD1 using urinary exosomes. J Am Soc Nephrol 2015;26:1661–70.

[506] Conde-Vancells J, Rodriguez-Suarez E, Gonzalez E, et al. Candidate biomarkers in exosome-like vesicles purified from rat and mouse urine samples. Proteomics Clin Appl 2010;4:416–25.

[507] Esteva-Font C, Wang X, Ars E, et al. Are sodium transporters in urinary exosomes reliable markers of tubular sodium reabsorption in hypertensive patients? Nephron Physiol 2010;114:p25–34.

[508] Zhou H, Cheruvanky A, Hu X, et al. Urinary exosomal transcription factors, a new class of biomarkers for renal disease. Kidney Int 2008;74:613–21.

[509] Deregibus MC, Cantaluppi V, Calogero R, et al. Endothelial progenitor cell derived microvesicles activate an angiogenic program in endothelial cells by a horizontal transfer of mRNA. Blood 2007;110:2440–8.

[510] Michael A, Bajracharya SD, Yuen PS, et al. Exosomes from human saliva as a source of microRNA biomarkers. Oral Dis 2010;16:34–8.

[511] Miranda KC, Bond DT, McKee M, et al. Nucleic acids within urinary exosomes/microvesicles are potential biomarkers for renal disease. Kidney Int 2010;78:191–9.

[512] Navarro-Munoz M, Ibernon M, Perez V, et al. Messenger RNA expression of B7-1 and NPHS1 in urinary sediment could be useful to differentiate between minimal-change disease and focal segmental glomerulosclerosis in adult patients. Nephrol Dial Transplant 2011;26: 3914–3923.

[513] van Ham SM, Heutinck KM, Jorritsma T, et al. Urinary granzyme A mRNA is a biomarker to diagnose subclinical and acute cellular rejection in kidney transplant recipients. Kidney Int 2010;78:1033–40.

[514] Chun-Yan L, Zi-Yi Z, Tian-Lin Y, et al. Liquid biopsy biomarkers of renal interstitial fibrosis based on urinary exosome. Exp Mol Pathol 2018;105:223–8.

[515] Gholaminejad A, Abdul Tehrani H, Gholami Fesharaki M. Identification of candidate microRNA biomarkers in renal fibrosis: a meta-analysis of profiling studies. Biomarkers 2018;23:713–24.

[516] Zununi Vahed S, Omidi Y, Ardalan M, Samadi N. Dysregulation of urinary miR-21 and miR-200b associated with interstitial fibrosis and tubular atrophy (IFTA) in renal transplant recipients. Clin Biochem 2017;50:32–9.

[517] Zununi Vahed S, Poursadegh Zonouzi A, Mahmoodpoor F, Samadi N, Ardalan M, Omidi Y. Circulating miR-150, miR-192, miR-200b, and miR-423-3p as non-invasive biomarkers of chronic allograft dysfunction. Arch Med Res 2017;48:96–104.

[518] Wang G, Kwan BC, Lai FM, Chow KM, Kam-Tao Li P, Szeto CC. Expression of microRNAs in the urinary sediment of patients with IgA nephropathy. Dis Markers 2010;28:79–86.

[519] Gelderman MP, Simak J. Flow cytometric analysis of cell membrane microparticles. Methods Mol Biol 2008;484:79–93.

[520] Orozco AF, Lewis DE. Flow cytometric analysis of circulating microparticles in plasma. Cytometry A 2010;77:502–14.

[521] Chaput N, Taieb J, Schartz N, et al. The potential of exosomes in immunotherapy of cancer. Blood Cells Mol Dis 2005;35:111–5.

[522] Escudier B, Dorval T, Chaput N, et al. Vaccination of metastatic melanoma patients with autologous dendritic cell (DC) derived-exosomes: results of the first phase I clinical trial. J Transl Med 2005;3:10.

[523] Morse MA, Garst J, Osada T, et al. A phase I study of dexosome immunotherapy in patients with advanced non-small cell lung cancer. J Transl Med 2005;3:9.

[524] Viaud S, Thery C, Ploix S, et al. Dendritic cell-derived exosomes for cancer immunotherapy: what's next? Cancer Res 2010;70:1281–5.

[525] Zhang Y, Luo CL, He BC, Zhang JM, Cheng G, Wu XH. Exosomes derived from IL-12-anchored renal cancer cells increase induction of specific antitumor response in vitro: a novel vaccine for renal cell carcinoma. Int J Oncol 2010;36:133–40.

[526] Zitvogel L, Regnault A, Lozier A, et al. Eradication of established murine tumors using a novel cell-free vaccine: dendritic cell-derived exosomes. Nat Med 1998;4:594–600.

[527] Beauvillain C, Juste MO, Dion S, Pierre J, Dimier-Poisson I. Exosomes are an effective vaccine against congenital toxoplasmosis in mice. Vaccine 2009;27:1750–7.

[528] Kuate S, Cinatl J, Doerr HW, Uberla K. Exosomal vaccines containing the S protein of the SARS coronavirus induce high levels of neutralizing antibodies. Virology 2007;362:26–37.

[529] Bruno S, Grange C, Deregibus MC, et al. Mesenchymal stem cell-derived microvesicles protect against acute tubular injury. J Am Soc Nephrol 2009;20:1053–67.

[530] Gatti S, Bruno S, Deregibus MC, et al. Microvesicles derived from human adult mesenchymal stem cells protect against ischaemia-reperfusion-induced acute and chronic kidney injury. Nephrol Dial Transplant 2011;26:1474–83.

[531] Wu C, Morris JR. Genes, genetics, and epigenetics: a correspondence. Science 2001;293:1103–5.

[532] Dupont C, Armant DR, Brenner CA. Epigenetics: definition, mechanisms and clinical perspective. Semin Reprod Med 2009;27:351–7.

[533] Robertson KD. DNA methylation and human disease. Nat Rev Genet 2005;6:597–610.

[534] Chung AC, Dong Y, Yang W, Zhong X, Li R, Lan HY. Smad7 suppresses renal fibrosis via altering expression of TGF-beta/Smad3-regulated microRNAs. Mol Ther 2013;21:388–98.

[535] Bechtel W, McGoohan S, Zeisberg EM, et al. Methylation determines fibroblast activation and fibrogenesis in the kidney. Nat Med 2010;16:544–50.

[536] Moore LD, Le T, Fan G. DNA methylation and its basic function. Neuropsychopharmacology 2013;38:23–38.

[537] Okano M, Xie S, Li E. Cloning and characterization of a family of novel mammalian DNA (cytosine-5) methyltransferases. Nat Genet 1998;19:219–20.

[538] Yen RW, Vertino PM, Nelkin BD, et al. Isolation and characterization of the cDNA encoding human DNA methyltransferase. Nucleic Acids Res 1992;20:2287–91.

[539] Bird AP. DNA methylation and the frequency of CpG in animal DNA. Nucleic Acids Res 1980;8:1499–504.

[540] Kulis M, Esteller M. DNA methylation and cancer. Adv Genet 2010;70:27–56.

[541] Chen J, Zhang X, Zhang H, et al. Indoxyl sulfate enhance the hypermethylation of klotho and promote the process of vascular calcification in chronic kidney disease. Int J Biol Sci 2016;12:1236–46.

[542] Ghattas M, El-Shaarawy F, Mesbah N, Abo-Elmatty D. DNA methylation status of the methylenetetrahydrofolate reductase gene promoter in peripheral blood of end-stage renal disease patients. Mol Biol Rep 2014;41:683–8.

[543] Hu Y, Mou L, Yang F, Tu H, Lin W. Curcumin attenuates cyclosporine A induced renal fibrosis by inhibiting hypermethylation of the klotho promoter. Mol Med Rep 2016;14:3229–36.

[544] Ko YA, Mohtat D, Suzuki M, et al. Cytosine methylation changes in enhancer regions of core pro-fibrotic genes characterize kidney fibrosis development. Genome Biol 2013;14:R108.

[545] Sapienza C, Lee J, Powell J, et al. DNA methylation profiling identifies epigenetic differences between diabetes patients with ESRD and diabetes patients without nephropathy. Epigenetics 2011;6:20–8.

[546] Wing MR, Devaney JM, Joffe MM, et al. DNA methylation profile associated with rapid decline in kidney function: findings from the CRIC study. Nephrol Dial Transplant 2014;29:864–72.

[547] National Human Genome Research Institute. NHGRI seeks DNA sequencing technologies fit for routine laboratory and medical use. In: NIH News release. 2008.

[548] Genomes Project C, Auton A, Brooks LD, et al. A global reference for human genetic variation. Nature 2015;526:68–74.

[549] Chial H. DNA sequencing technologies key to the Human Genome Project. Nat Educ 2008;1.

[550] Online Mendelian Inheritance in Man. OMIM gene map statistics. OMIM; 2019.

[551] Schmid H, Henger A, Kretzler M. Molecular approaches to chronic kidney disease. Curr Opin Nephrol Hypertens 2006;15:123–9.

[552] Bockenhauer D, Medlar AJ, Ashton E, Kleta R, Lench N. Genetic testing in renal disease. Pediatr Nephrol 2012;27:873–83.

[553] Devuyst O, Knoers NV, Remuzzi G, et al. Rare inherited kidney diseases: challenges, opportunities, and perspectives. Lancet 2014;383:1844–59.

[554] Jha V, Garcia-Garcia G, Iseki K, et al. Chronic kidney disease: global dimension and perspectives. Lancet 2013;382:260–72.

[555] Wuhl E, van Stralen KJ, Wanner C, et al. Renal replacement therapy for rare diseases affecting the kidney: an analysis of the ERA-EDTA Registry. Nephrol Dial Transplant 2014;29(Suppl. 4):iv1–8.

[556] Ingelfinger JR, Kalantar-Zadeh K, Schaefer F, World Kidney Day Steering C. World Kidney Day 2016: averting the legacy of kidney disease-focus on childhood. Pediatr Nephrol 2016;31:343–8.

[557] Vivante A, Hildebrandt F. Exploring the genetic basis of early-onset chronic kidney disease. Nat Rev Nephrol 2016;12:133–46.

[558] Connaughton DM, Bukhari S, Conlon P, et al. The Irish Kidney Gene Project—prevalence of family history in patients with kidney disease in Ireland. Nephron 2015;130:293–301.

[559] McClellan WM, Satko SG, Gladstone E, Krisher JO, Narva AS, Freedman BI. Individuals with a family history of ESRD are a high-risk population for CKD: implications for targeted surveillance and intervention activities. Am J Kidney Dis 2009;53:S100–6.

[560] Skrunes R, Svarstad E, Reisaeter AV, Vikse BE. Familial clustering of ESRD in the Norwegian population. Clin J Am Soc Nephrol 2014;9:1692–700.

[561] Ayme S, Bockenhauer D, Day S, et al. Common elements in rare kidney diseases: conclusions from a kidney disease: improving global outcomes (KDIGO) controversies conference. Kidney Int 2017;92:796–808.

[562] Groopman EE, Rasouly HM, Gharavi AG. Genomic medicine for kidney disease. Nat Rev Nephrol 2018;14:83–104.

[563] Joly D, Beroud C, Grunfeld JP. Rare inherited disorders with renal involvement-approach to the patient. Kidney Int 2015;87:901–8.

[564] Liapis H, Gaut JP. The renal biopsy in the genomic era. Pediatr Nephrol 2013;28:1207–19.

[565] United States Renal Data System. USRDS annual data report: epidemiology of kidney disease in the United States. Bethesda, MD: National Institutes of Health, National Institute of Diabetes and Digestive and Kidney Diseases; 2018. p. 2018.

[566] Chong JX, Buckingham KJ, Jhangiani SN, et al. The genetic basis of mendelian phenotypes: discoveries, challenges, and opportunities. Am J Hum Genet 2015;97:199–215.

[567] Miller DT, Adam MP, Aradhya S, et al. Consensus statement: chromosomal microarray is a first-tier clinical diagnostic test for individuals with developmental disabilities or congenital anomalies. Am J Hum Genet 2010;86:749–64.

[568] Guttmacher AE, Collins FS. Genomic medicine—a primer. N Engl J Med 2002;347:1512–20.

[569] Dixon-Salazar TJ, Silhavy JL, Udpa N, et al. Exome sequencing can improve diagnosis and alter patient management. Sci Transl Med 2012;4:138ra78.

[570] Lee H, Deignan JL, Dorrani N, et al. Clinical exome sequencing for genetic identification of rare Mendelian disorders. JAMA 2014;312:1880–7.

[571] Valencia CA, Husami A, Holle J, et al. Clinical impact and cost-effectiveness of whole exome sequencing as a diagnostic tool: a pediatric center's experience. Front Pediatr 2015;3:67.

[572] Stokman MF, Renkema KY, Giles RH, Schaefer F, Knoers NV, van Eerde AM. The expanding phenotypic spectra of kidney diseases: insights from genetic studies. Nat Rev Nephrol 2016;12:472–83.

[573] Manolio TA, Fowler DM, Starita LM, et al. Bedside back to bench: building bridges between basic and clinical genomic research. Cell 2017;169:6–12.

[574] Edwards N, Rice SJ, Raman S, et al. A novel LMX1B mutation in a family with end-stage renal disease of 'unknown cause'. Clin Kidney J 2015;8:113–9.

[575] Ellingford JM, Sergouniotis PI, Lennon R, et al. Pinpointing clinical diagnosis through whole exome sequencing to direct patient care: a case of Senior-Loken syndrome. Lancet 2015;385:1916.

[576] Groopman EE, Marasa M, Cameron-Christie S, et al. Diagnostic Utility of Exome Sequencing for Kidney Disease. N Engl J Med 2019;380:142–51.

[577] Munch J, Grohmann M, Lindner TH, Bergmann C, Halbritter J. Diagnosing FSGS without kidney biopsy—a novel INF2-mutation in a family with ESRD of unknown origin. BMC Med Genet 2016;17.

[578] Quaglia M, Musetti C, Ghiggeri GM, et al. Unexpectedly high prevalence of rare genetic disorders in kidney transplant recipients with an unknown causal nephropathy. Clin Transpl 2014;28:995–1003.

[579] Savige J, Colville D, Rheault M, et al. Alport Syndrome in Women and Girls. Clin J Am Soc Nephrol 2016;11:1713–20.

[580] Terryn W, Cochat P, Froissart R, et al. Fabry nephropathy: indications for screening and guidance for diagnosis and treatment by the European Renal Best Practice. Nephrol Dial Transplant 2013;28:505–17.

[581] Wang RY, Lelis A, Mirocha J, Wilcox WR. Heterozygous Fabry women are not just carriers, but have a significant burden of disease and impaired quality of life. Genet Med 2007;9:34–45.

[582] Lentine KL, Kasiske BL, Levey AS, et al. KDIGO Clinical Practice Guideline on the Evaluation and Care of Living Kidney Donors. Transplantation 2017;101:S1–S109.

[583] Genovese G, Friedman DJ, Ross MD, et al. Association of trypanolytic ApoL1 variants with kidney disease in African Americans. Science 2010;329:841–5.

[584] Tzur S, Rosset S, Shemer R, et al. Missense mutations in the APOL1 gene are highly associated with end stage kidney disease risk previously attributed to the MYH9 gene. Hum Genet 2010;128:345–50.

[585] Friedman DJ, Pollak MR. Apolipoprotein L1 and Kidney Disease in African Americans. Trends Endocrinol Metab 2016;27:204–15.

[586] Kruzel-Davila E, Wasser WG, Aviram S, Skorecki K. APOL1 nephropathy: from gene to mechanisms of kidney injury. Nephrol Dial Transplant 2016;31:349–58.

[587] Adalat S, Woolf AS, Johnstone KA, et al. HNF1B mutations associate with hypomagnesemia and renal magnesium wasting. J Am Soc Nephrol 2009;20:1123–31.

[588] Edghill EL, Oram RA, Owens M, et al. Hepatocyte nuclear factor-1beta gene deletions—a common cause of renal disease. Nephrol Dial Transplant 2008;23:627–35.

[589] Mefford HC, Clauin S, Sharp AJ, et al. Recurrent reciprocal genomic rearrangements of 17q12 are associated with renal disease, diabetes, and epilepsy. Am J Hum Genet 2007;81:1057–69.

[590] Moreno-De-Luca D, Consortium S, Mulle JG, et al. Deletion 17q12 is a recurrent copy number variant that confers high risk of autism and schizophrenia. Am J Hum Genet 2010;87:618–30.

[591] Goldstein DB, Allen A, Keebler J, et al. Sequencing studies in human genetics: design and interpretation. Nat Rev Genet 2013;14:460–70.

[592] Matthijs G, Souche E, Alders M, et al. Guidelines for diagnostic next-generation sequencing. Eur J Hum Genet 2016;24:2–5.

[593] Rehm HL. Disease-targeted sequencing: a cornerstone in the clinic. Nat Rev Genet 2013;14:295–300.

[594] Xue Y, Ankala A, Wilcox WR, Hegde MR. Solving the molecular diagnostic testing conundrum for Mendelian disorders in the era of next-generation sequencing: single-gene, gene panel, or exome/genome sequencing. Genet Med 2015;17:444–51.

[595] Richards S, Aziz N, Bale S, et al. Standards and guidelines for the interpretation of sequence variants: a joint consensus recommendation of the American College of Medical Genetics and Genomics and the Association for Molecular Pathology. Genet Med 2015;17:405–24.

[596] Watson CT, Marques-Bonet T, Sharp AJ, Mefford HC. The genetics of microdeletion and microduplication syndromes: an update. Annu Rev Genomics Hum Genet 2014;15:215–44.

[597] Schaaf CP, Wiszniewska J, Beaudet AL. Copy number and SNP arrays in clinical diagnostics. Annu Rev Genomics Hum Genet 2011;12:25–51.

[598] Kearney HM, Thorland EC, Brown KK, Quintero-Rivera F, South ST. Working Group of the American College of Medical Genetics Laboratory Quality Assurance C. American College of Medical Genetics standards and guidelines for interpretation and reporting of postnatal constitutional copy number variants. Genet Med 2011;13:680–5.

[599] Vermeesch JR, Brady PD, Sanlaville D, Kok K, Hastings RJ. Genome-wide arrays: quality criteria and platforms to be used in routine diagnostics. Hum Mutat 2012;33:906–15.

[600] Katsanis SH, Katsanis N. Molecular genetic testing and the future of clinical genomics. Nat Rev Genet 2013;14:415–26.

[601] Shinawi M, Cheung SW. The array CGH and its clinical applications. Drug Discov Today 2008;13:760–70.

[602] Zheng HT, Peng ZH, Li S, He L. Loss of heterozygosity analyzed by single nucleotide polymorphism array in cancer. World J Gastroenterol 2005;11:6740–4.

[603] Hageman GS, Gehrs K, Lejnine S, et al. Clinical validation of a genetic model to estimate the risk of developing choroidal neovascular age-related macular degeneration. Hum Genomics 2011;5:420–40.

[604] Zanke B, Hawken S, Carter R, Chow D. A genetic approach to stratification of risk for age-related macular degeneration. Can J Ophthalmol 2010;45:22–7.

[605] Meschia JF, Singleton A, Nalls MA, et al. Genomic risk profiling of ischemic stroke: results of an international genome-wide association meta-analysis. PLoS One 2011;6.

[606] Tiu RV, Gondek LP, O'Keefe CL, et al. Prognostic impact of SNP array karyotyping in myelodysplastic syndromes and related myeloid malignancies. Blood 2011;117:4552–60.

[607] National Center for Biotechnology Information. dbSNP short genetic variations. NCBI; 2017.

[608] LaFramboise T. Single nucleotide polymorphism arrays: a decade of biological, computational and technological advances. Nucleic Acids Res 2009;37:4181–93.

[609] International HapMap C. The International HapMap Project. Nature 2003;426:789–96.

[610] Walsh AM, Whitaker JW, Huang CC, et al. Integrative genomic deconvolution of rheumatoid arthritis GWAS loci into gene and cell type associations. Genome Biol 2016;17.

[611] Amin Al Olama A, Kote-Jarai Z, Schumacher FR, et al. A meta-analysis of genome-wide association studies to identify prostate cancer susceptibility loci associated with aggressive and non-aggressive disease. Hum Mol Genet 2013;22:408–15.

[612] Billings LK, Florez JC. The genetics of type 2 diabetes: what have we learned from GWAS? Ann N Y Acad Sci 2010;1212:59–77.

[613] Sato-Otsubo A, Sanada M, Ogawa S. Single-nucleotide polymorphism array karyotyping in clinical practice: where, when, and how? Semin Oncol 2012;39:13–25.

[614] Reddy UM, Page GP, Saade GR, et al. Karyotype versus microarray testing for genetic abnormalities after stillbirth. N Engl J Med 2012;367:2185–93.

[615] Wapner RJ, Martin CL, Levy B, et al. Chromosomal microarray versus karyotyping for prenatal diagnosis. N Engl J Med 2012;367:2175–84.

[616] South ST, Lee C, Lamb AN, et al. ACMG Standards and Guidelines for constitutional cytogenomic microarray analysis, including postnatal and prenatal applications: revision 2013. Genet Med 2013;15:901–9.

[617] Caruana G, Wong MN, Walker A, et al. Copy-number variation associated with congenital anomalies of the kidney and urinary tract. Pediatr Nephrol 2015;30:487–95.

[618] Faure A, Bouty A, Caruana G, et al. DNA copy number variants: a potentially useful predictor of early onset renal failure in boys with posterior urethral valves. J Pediatr Urol 2016;12. 227 e1–7.

[619] Fu F, Chen F, Li R, et al. Prenatal diagnosis of fetal multicystic dysplastic kidney via high-resolution whole-genome array. Nephrol Dial Transplant 2016;31:1693–8.

[620] Sanna-Cherchi S, Kiryluk K, Burgess KE, et al. Copy-number disorders are a common cause of congenital kidney malformations. Am J Hum Genet 2012;91:987–97.

[621] Tammimies K, Marshall CR, Walker S, et al. Molecular diagnostic yield of chromosomal microarray analysis and whole-exome sequencing in children with autism spectrum disorder. JAMA 2015;314:895–903.

[622] Verbitsky M, Sanna-Cherchi S, Fasel DA, et al. Genomic imbalances in pediatric patients with chronic kidney disease. J Clin Invest 2015;125:2171–8.

[623] Westland R, Verbitsky M, Vukojevic K, et al. Copy number variation analysis identifies novel CAKUT candidate genes in children with a solitary functioning kidney. Kidney Int 2015;88:1402–10.

[624] Imai K, Kricka LJ, Fortina P. Concordance study of 3 direct-to-consumer genetic-testing services. Clin Chem 2011;57:518–21.

[625] Schouten JP, McElgunn CJ, Waaijer R, Zwijnenburg D, Diepvens F, Pals G. Relative quantification of 40 nucleic acid sequences by multiplex ligation-dependent probe amplification. Nucleic Acids Res 2002;30:e57.

[626] Jeuken J, Cornelissen S, Boots-Sprenger S, Gijsen S, Wesseling P. Multiplex ligation-dependent probe amplification: a diagnostic tool for simultaneous identification of different genetic markers in glial tumors. J Mol Diagn 2006;8:433–43.

[627] Sanger F, Nicklen S, Coulson AR. DNA sequencing with chain-terminating inhibitors. Proc Natl Acad Sci U S A 1977;74:5463–7.

[628] Petersen BS, Fredrich B, Hoeppner MP, Ellinghaus D, Franke A. Opportunities and challenges of whole-genome and -exome sequencing. BMC Genet 2017;18.

[629] Goodwin S, McPherson JD, McCombie WR. Coming of age: ten years of next-generation sequencing technologies. Nat Rev Genet 2016;17:333–51.

[630] Levy SE, Myers RM. Advancements in Next-Generation Sequencing. Annu Rev Genomics Hum Genet 2016;17:95–115.

[631] Schuster SC. Next-generation sequencing transforms today's biology. Nat Methods 2008;5:16–8.

[632] von Bubnoff A. Next-generation sequencing: the race is on. Cell 2008;132:721–3.

[633] Ashley EA. Towards precision medicine. Nat Rev Genet 2016;17:507–22.

[634] Chakravorty S, Hegde M. Gene and variant annotation for mendelian disorders in the era of advanced sequencing technologies. Annu Rev Genomics Hum Genet 2017;18:229–56.

[635] Shashi V, McConkie-Rosell A, Rosell B, et al. The utility of the traditional medical genetics diagnostic evaluation in the context of next-generation sequencing for undiagnosed genetic disorders. Genet Med 2014;16:176–82.

[636] Delio M, Patel K, Maslov A, et al. Development of a targeted multi-disorder high-throughput sequencing assay for the effective identification of disease-causing variants. PLoS One 2015;10.

[637] Saudi Mendeliome G. Comprehensive gene panels provide advantages over clinical exome sequencing for Mendelian diseases. Genome Biol 2015;16:134.

[638] Zemojtel T, Kohler S, Mackenroth L, et al. Effective diagnosis of genetic disease by computational phenotype analysis of the disease-associated genome. Sci Transl Med 2014;6:252ra123.

[639] Braun DA, Lawson JA, Gee HY, et al. Prevalence of monogenic causes in pediatric patients with nephrolithiasis or nephrocalcinosis. Clin J Am Soc Nephrol 2016;11:664–72.

[640] Halbritter J, Baum M, Hynes AM, et al. Fourteen monogenic genes account for 15% of nephrolithiasis/nephrocalcinosis. J Am Soc Nephrol 2015;26:543–51.

[641] Gross O, Kashtan CE, Rheault MN, et al. Advances and unmet needs in genetic, basic and clinical science in Alport syndrome: report from the 2015 International Workshop on Alport Syndrome. Nephrol Dial Transplant 2017;32:916–24.

[642] Savige J, Gregory M, Gross O, Kashtan C, Ding J, Flinter F. Expert guidelines for the management of Alport syndrome and thin basement membrane nephropathy. J Am Soc Nephrol 2013;24:364–75.

[643] McCarthy HJ, Bierzynska A, Wherlock M, et al. Simultaneous sequencing of 24 genes associated with steroid-resistant nephrotic syndrome. Clin J Am Soc Nephrol 2013;8:637–48.

[644] Sadowski CE, Lovric S, Ashraf S, et al. A single-gene cause in 29.5% of cases of steroid-resistant nephrotic syndrome. J Am Soc Nephrol 2015;26:1279–89.

[645] Halbritter J, Diaz K, Chaki M, et al. High-throughput mutation analysis in patients with a nephronophthisis-associated ciliopathy applying multiplexed barcoded array-based PCR amplification and next-generation sequencing. J Med Genet 2012;49:756–67.

[646] Schueler M, Halbritter J, Phelps IG, et al. Large-scale targeted sequencing comparison highlights extreme genetic heterogeneity in nephronophthisis-related ciliopathies. J Med Genet 2016;53:208–14.

[647] Barbour TD, Ruseva MM, Pickering MC. Update on C3 glomerulopathy. Nephrol Dial Transplant 2016;31:717–25.

[648] Bu F, Borsa NG, Jones MB, et al. High-throughput genetic testing for thrombotic microangiopathies and C3 glomerulopathies. J Am Soc Nephrol 2016;27:1245–53.

[649] Riedl M, Thorner P, Licht C. C3 Glomerulopathy. Pediatr Nephrol 2017;32:43–57.

[650] Barbour T, Johnson S, Cohney S, Hughes P. Thrombotic microangiopathy and associated renal disorders. Nephrol Dial Transplant 2012;27:2673–85.

[651] Laurence J. Atypical hemolytic uremic syndrome (aHUS): making the diagnosis. Clin Adv Hematol Oncol 2012;10:1–12.

[652] Lotta LA, Garagiola I, Palla R, Cairo A, Peyvandi F. ADAMTS13 mutations and polymorphisms in congenital thrombotic thrombocytopenic purpura. Hum Mutat 2010;31:11–9.

[653] Nester CM, Barbour T, de Cordoba SR, et al. Atypical aHUS: state of the art. Mol Immunol 2015;67:31–42.

[654] Noris M, Bresin E, Mele C, Remuzzi G. Genetic atypical hemolytic-uremic syndrome. In: Adam MP, Ardinger HH, Pagon RA, et al., editors. GeneReviews((R)). Seattle, WA: University of Washington; 1993.

[655] Botstein D, Risch N. Discovering genotypes underlying human phenotypes: past successes for mendelian disease, future approaches for complex disease. Nat Genet 2003;33 (Suppl):228–37.

[656] Cooper DN, Chen JM, Ball EV, et al. Genes, mutations, and human inherited disease at the dawn of the age of personalized genomics. Hum Mutat 2010;31:631–55.

[657] Bamshad MJ, Ng SB, Bigham AW, et al. Exome sequencing as a tool for Mendelian disease gene discovery. Nat Rev Genet 2011;12:745–55.

[658] Boycott KM, Vanstone MR, Bulman DE, MacKenzie AE. Rare-disease genetics in the era of next-generation sequencing: discovery to translation. Nat Rev Genet 2013;14:681–91.

[659] Ku CS, Naidoo N, Pawitan Y. Revisiting Mendelian disorders through exome sequencing. Hum Genet 2011;129:351–70.

[660] Tarailo-Graovac M, Shyr C, Ross CJ, et al. Exome sequencing and the management of neurometabolic disorders. N Engl J Med 2016;374:2246–55.

[661] Huang KL, Mashl RJ, Wu Y, et al. Pathogenic germline variants in 10,389 adult cancers. Cell 2018;173. 355–70.e14.

[662] Pritchard CC, Mateo J, Walsh MF, et al. Inherited DNA-repair gene mutations in men with metastatic prostate cancer. N Engl J Med 2016;375:443–53.

[663] Zhang J, Walsh MF, Wu G, et al. Germline mutations in predisposition genes in pediatric cancer. N Engl J Med 2015;373:2336–46.

[664] Strande NT, Berg JS. Defining the clinical value of a genomic diagnosis in the era of next-generation sequencing. Annu Rev Genomics Hum Genet 2016;17:303–32.

[665] Yang Y, Muzny DM, Reid JG, et al. Clinical whole-exome sequencing for the diagnosis of mendelian disorders. N Engl J Med 2013;369:1502–11.

[666] Bierzynska A, McCarthy HJ, Soderquest K, et al. Genomic and clinical profiling of a national nephrotic syndrome cohort advocates a precision medicine approach to disease management. Kidney Int 2017;91:937–47.

[667] Braun DA, Schueler M, Halbritter J, et al. Whole exome sequencing identifies causative mutations in the majority of consanguineous or familial cases with childhood-onset increased renal echogenicity. Kidney Int 2016;89:468–75.

[668] Lata S, Marasa M, Li Y, et al. Whole-exome sequencing in adults with chronic kidney disease: a pilot study. Ann Intern Med 2018;168:100–9.

[669] Yao T, Udwan K, John R, et al. Integration of genetic testing and pathology for the diagnosis of adults with FSGS. Clin J Am Soc Nephrol 2019.

[670] Bowling KM, Thompson ML, Amaral MD, et al. Genomic diagnosis for children with intellectual disability and/or developmental delay. Genome Med 2017;9:43.

[671] Eldomery MK, Coban-Akdemir Z, Harel T, et al. Lessons learned from additional research analyses of unsolved clinical exome cases. Genome Med 2017;9:26.

[672] Gilissen C, Hoischen A, Brunner HG, Veltman JA. Disease gene identification strategies for exome sequencing. Eur J Hum Genet 2012;20:490–7.

[673] Need AC, Shashi V, Schoch K, Petrovski S, Goldstein DB. The importance of dynamic re-analysis in diagnostic whole exome sequencing. J Med Genet 2017;54:155–6.

[674] Wenger AM, Guturu H, Bernstein JA, Bejerano G. Systematic reanalysis of clinical exome data yields additional diagnoses: implications for providers. Genet Med 2017;19:209–14.

[675] Mandelker D, Schmidt RJ, Ankala A, et al. Navigating highly homologous genes in a molecular diagnostic setting: a resource for clinical next-generation sequencing. Genet Med 2016;18:1282–9.

[676] Park JY, Clark P, Londin E, Sponziello M, Kricka LJ, Fortina P. Clinical exome performance for reporting secondary genetic findings. Clin Chem 2015;61:213–20.

[677] Carvalho CM, Lupski JR. Mechanisms underlying structural variant formation in genomic disorders. Nat Rev Genet 2016;17:224–38.

[678] Braun DA, Hildebrandt F. Ciliopathies. Cold Spring Harb Perspect Biol 2017;9.

[679] Gee HY, Otto EA, Hurd TW, et al. Whole-exome resequencing distinguishes cystic kidney diseases from phenocopies in renal ciliopathies. Kidney Int 2014;85:880–7.

[680] Renkema KY, Stokman MF, Giles RH, Knoers NV. Next-generation sequencing for research and diagnostics in kidney disease. Nat Rev Nephrol 2014;10:433–44.

[681] Gast C, Pengelly RJ, Lyon M, et al. Collagen (COL4A) mutations are the most frequent mutations underlying adult focal segmental glomerulosclerosis. Nephrol Dial Transplant 2016;31:961–70.

[682] Malone AF, Phelan PJ, Hall G, et al. Rare hereditary COL4A3/COL4A4 variants may be mistaken for familial focal segmental glomerulosclerosis. Kidney Int 2014;86:1253–9.

[683] Porath B, Gainullin VG, Cornec-Le Gall E, et al. Mutations in GANAB, encoding the glucosidase IIα subunit, cause autosomal-dominant polycystic kidney and liver disease. Am J Hum Genet 2016;98:1193–207.

[684] Lu H, Galeano MCR, Ott E, et al. Mutations in DZIP1L, which encodes a ciliary-transition-zone protein, cause autosomal recessive polycystic kidney disease. Nat Genet 2017;49: 1025–34.

[685] Choi M, Scholl UI, Ji W, et al. Genetic diagnosis by whole exome capture and massively parallel DNA sequencing. Proc Natl Acad Sci U S A 2009;106:19096–101.

[686] Isnard P, Rabant M, Labaye J, Antignac C, Knebelmann B, Zaidan M. Karyomegalic interstitial nephritis: a case report and review of the literature. Medicine (Baltimore) 2016;95.

[687] Nakata T, Ishida R, Mihara Y, et al. Steroid-resistant nephrotic syndrome as the initial presentation of nail-patella syndrome: a case of a de novo LMX1B mutation. BMC Nephrol 2017;18.

[688] Wuttke M, Seidl M, Malinoc A, et al. A COL4A5 mutation with glomerular disease and signs of chronic thrombotic microangiopathy. Clin Kidney J 2015;8:690–4.

[689] Smith LD, Willig LK, Kingsmore SF. Whole-exome sequencing and whole-genome sequencing in critically ill neonates suspected to have single-gene disorders. Cold Spring Harb Perspect Med 2015;6:a023168.

[690] Lupski JR, Reid JG, Gonzaga-Jauregui C, et al. Whole-genome sequencing in a patient with Charcot-Marie-Tooth neuropathy. N Engl J Med 2010;362:1181–91.

[691] Saunders CJ, Miller NA, Soden SE, et al. Rapid whole-genome sequencing for genetic disease diagnosis in neonatal intensive care units. Sci Transl Med 2012;4.

[692] Taylor JC, Martin HC, Lise S, et al. Factors influencing success of clinical genome sequencing across a broad spectrum of disorders. Nat Genet 2015;47:717–26.

[693] Yuen RK, Thiruvahindrapuram B, Merico D, et al. Whole-genome sequencing of quartet families with autism spectrum disorder. Nat Med 2015;21:185–91.

[694] Cabezas OR, Flanagan SE, Stanescu H, et al. Polycystic kidney disease with hyperinsulinemic hypoglycemia caused by a promoter mutation in phosphomannomutase 2. J Am Soc Nephrol 2017;28:2529–39.

[695] Carroll C, Hunley TE, Guo Y, Cortez D. A novel splice site mutation in SMARCAL1 results in aberrant exon definition in a child with Schimke immunoosseous dysplasia. Am J Med Genet A 2015;167A:2260–4.

[696] King K, Flinter FA, Nihalani V, Green PM. Unusual deep intronic mutations in the COL4A5 gene cause X linked Alport syndrome. Hum Genet 2002;111:548–54.

[697] Lo YF, Nozu K, Iijima K, et al. Recurrent deep intronic mutations in the SLC12A3 gene responsible for Gitelman's syndrome. Clin J Am Soc Nephrol 2011;6:630–9.

[698] Mele C, Lemaire M, Iatropoulos P, et al. Characterization of a new DGKE intronic mutation in genetically unsolved cases of familial atypical hemolytic uremic syndrome. Clin J Am Soc Nephrol 2015;10:1011–9.

[699] Carss KJ, Arno G, Erwood M, et al. Comprehensive rare variant analysis via whole-genome sequencing to determine the molecular pathology of inherited retinal disease. Am J Hum Genet 2017;100:75–90.

[700] Gilissen C, Hehir-Kwa JY, Thung DT, et al. Genome sequencing identifies major causes of severe intellectual disability. Nature 2014;511:344–7.

[701] Stavropoulos DJ, Merico D, Jobling R, et al. Whole genome sequencing expands diagnostic utility and improves clinical management in pediatric medicine. NPJ Genom Med 2016;1.

[702] Belkadi A, Bolze A, Itan Y, et al. Whole-genome sequencing is more powerful than whole-exome sequencing for detecting exome variants. Proc Natl Acad Sci U S A 2015;112:5473–8.

[703] Lelieveld SH, Spielmann M, Mundlos S, Veltman JA, Gilissen C. Comparison of exome and genome sequencing technologies for the complete capture of protein-coding regions. Hum Mutat 2015;36:815–22.

[704] Mallawaarachchi AC, Hort Y, Cowley MJ, et al. Whole-genome sequencing overcomes pseudogene homology to diagnose autosomal dominant polycystic kidney disease. Eur J Hum Genet 2016;24:1584–90.

[705] Watson CM, Crinnion LA, Berry IR, et al. Enhanced diagnostic yield in Meckel-Gruber and Joubert syndrome through exome sequencing supplemented with split-read mapping. BMC Med Genet 2016;17:1.

[706] Harewood L, Liu M, Keeling J, et al. Bilateral renal agenesis/hypoplasia/dysplasia (BRAHD): postmortem analysis of 45 cases with breakpoint mapping of two de novo translocations. PLoS One 2010;5.

[707] Mansouri MR, Carlsson B, Davey E, et al. Molecular genetic analysis of a de novo balanced translocation t(6;17)(p21.31;q11.2) associated with hypospadias and anorectal malformation. Hum Genet 2006;119:162–8.

[708] Bockenhauer D, Feather S, Stanescu HC, et al. Epilepsy, ataxia, sensorineural deafness, tubulopathy, and KCNJ10 mutations. N Engl J Med 2009;360:1960–70.

[709] Kleta R, Romeo E, Ristic Z, et al. Mutations in SLC6A19, encoding B0AT1, cause Hartnup disorder. Nat Genet 2004;36:999–1002.

[710] Landoure G, Zdebik AA, Martinez TL, et al. Mutations in TRPV4 cause Charcot-Marie-Tooth disease type 2C. Nat Genet 2010;42:170–4.

[711] St Hilaire C, Ziegler SG, Markello TC, et al. NT5E mutations and arterial calcifications. N Engl J Med 2011;364:432–42.

[712] Wellcome Trust Case Control C. Genome-wide association study of 14,000 cases of seven common diseases and 3,000 shared controls. Nature 2007;447:661–78.

[713] Stanescu HC, Arcos-Burgos M, Medlar A, et al. Risk HLA-DQA1 and PLA(2)R1 alleles in idiopathic membranous nephropathy. N Engl J Med 2011;364:616–26.

[714] Cho JH, Gregersen PK. Genomics and the multifactorial nature of human autoimmune disease. N Engl J Med 2011;365:1612–23.

[715] Kleta R, Bernardini I, Ueda M, et al. Long-term follow-up of well-treated nephropathic cystinosis patients. J Pediatr 2004;145:555–60.

[716] Kleta R, Kaskel F, Dohil R, et al. First NIH/Office of rare diseases conference on cystinosis: past, present, and future. Pediatr Nephrol 2005;20:452–4.

[717] Mustafa RA, Yu ASL. Burden of proof for tolvaptan in ADPKD: did REPRISE provide the answer? Clin J Am Soc Nephrol 2018;13:1107–9.

[718] Torres VE, Chapman AB, Devuyst O, et al. Tolvaptan in patients with autosomal dominant polycystic kidney disease. N Engl J Med 2012;367:2407–18.

[719] Torres VE, Chapman AB, Devuyst O, et al. Multicenter, open-label, extension trial to evaluate the long-term efficacy and safety of early versus delayed treatment with tolvaptan in autosomal dominant polycystic kidney disease: the TEMPO 4:4 Trial. Nephrol Dial Transplant 2017;32:1262.

[720] Torres VE, Chapman AB, Devuyst O, et al. Tolvaptan in later-stage autosomal dominant polycystic kidney disease. N Engl J Med 2017;377:1930–42.

[721] Torres VE, Harris PC, Pirson Y. Autosomal dominant polycystic kidney disease. Lancet 2007;369:1287–301.

[722] Horikawa Y, Iwasaki N, Hara M, et al. Mutation in hepatocyte nuclear factor-1 beta gene (TCF2) associated with MODY. Nat Genet 1997;17:384–5.

[723] Bingham C, Bulman MP, Ellard S, et al. Mutations in the hepatocyte nuclear factor-1beta gene are associated with familial hypoplastic glomerulocystic kidney disease. Am J Hum Genet 2001;68:219–24.

[724] Adalat S, Bockenhauer D, Ledermann SE, Hennekam RC, Woolf AS. Renal malformations associated with mutations of developmental genes: messages from the clinic. Pediatr Nephrol 2010;25:2247–55.

[725] Bingham C, Ellard S, van't Hoff WG, et al. Atypical familial juvenile hyperuricemic nephropathy associated with a hepatocyte nuclear factor-1 beta gene mutation. Kidney Int 2003;63:1645–51.

[726] Lindner TH, Njolstad PR, Horikawa Y, Bostad L, Bell GI, Sovik O. A novel syndrome of diabetes mellitus, renal dysfunction and genital malformation associated with a partial deletion of the pseudo-POU domain of hepatocyte nuclear factor-1beta. Hum Mol Genet 1999;8:2001–8.

[727] Ruf RG, Lichtenberger A, Karle SM, et al. Patients with mutations in NPHS2 (podocin) do not respond to standard steroid treatment of nephrotic syndrome. J Am Soc Nephrol 2004;15:722–32.

[728] Buscher AK, Kranz B, Buscher R, et al. Immunosuppression and renal outcome in congenital and pediatric steroid-resistant nephrotic syndrome. Clin J Am Soc Nephrol 2010;5:2075–84.

[729] Rehm HL. Evolving health care through personal genomics. Nat Rev Genet 2017;18:259–67.

[730] Berg JS, Khoury MJ, Evans JP. Deploying whole genome sequencing in clinical practice and public health: meeting the challenge one bin at a time. Genet Med 2011;13:499–504.

[731] Burke W. Genetic tests: clinical validity and clinical utility. Curr Protoc Hum Genet 2014;81. 9.15.1–8.

[732] Green RC, Berg JS, Grody WW, et al. ACMG recommendations for reporting of incidental findings in clinical exome and genome sequencing. Genet Med 2013;15:565–74.

[733] Kalia SS, Adelman K, Bale SJ, et al. Recommendations for reporting of secondary findings in clinical exome and genome sequencing, 2016 update (ACMG SF v2.0): a policy statement of the American College of Medical Genetics and Genomics. Genet Med 2017;19:249–55.

[734] van El CG, Cornel MC, Borry P, et al. Whole-genome sequencing in health care: recommendations of the European Society of Human Genetics. Eur J Hum Genet 2013;21:580–4.

[735] Amendola LM, Dorschner MO, Robertson PD, et al. Actionable exomic incidental findings in 6503 participants: challenges of variant classification. Genome Res 2015;25:305–15.

[736] Olfson E, Cottrell CE, Davidson NO, et al. Identification of medically actionable secondary findings in the 1000 genomes. PLoS One 2015;10.

[737] O'Daniel JM, McLaughlin HM, Amendola LM, et al. A survey of current practices for genomic sequencing test interpretation and reporting processes in US laboratories. Genet Med 2017;19:575–82.

[738] Otten E, Plantinga M, Birnie E, et al. Is there a duty to recontact in light of new genetic technologies? A systematic review of the literature. Genet Med 2015;17:668–78.

[739] Pyeritz RE. The coming explosion in genetic testing—is there a duty to recontact? N Engl J Med 2011;365:1367–9.

[740] Directors ABo. Points to consider in the clinical application of genomic sequencing. Genet Med 2012;14:759–61.

[741] Skirton H, Lewis C, Kent A, Coviello DA, Members of Eurogentest U, Committee EE. Genetic education and the challenge of genomic medicine: development of core competences to support preparation of health professionals in Europe. Eur J Hum Genet 2010;18:972–7.

[742] Kentwell M, Dow E, Antill Y, et al. Mainstreaming cancer genetics: a model integrating germline BRCA testing into routine ovarian cancer clinics. Gynecol Oncol 2017;145:130–6.

[743] Rhodes A, Rosman L, Cahill J, et al. Minding the genes: a multidisciplinary approach towards genetic assessment of cardiovascular disease. J Genet Couns 2017;26:224–31.

[744] National Human Genome Research Institute. The Ethical, Legal and Social Implications (ELSI) Research Program; 2018.

[745] US National Library of Medicine. ClinicalTrials.gov; 2019.

[746] Gross O, Friede T, Hilgers R, et al. Safety and efficacy of the ACE-inhibitor ramipril in Alport syndrome: the double-blind, randomized, placebo-controlled, multicenter phase III eARLY PRO-TECT Alport Trial in pediatric patients. ISRN Pediatr 2012;2012:436046.

[747] US Department of Health and Human Services. How does genetic testing in a research setting differ from clinical genetic testing? Genetics Home Reference; 2019.

[748] Ferreira-Gonzalez A, Teutsch S, Williams MS, et al. US system of oversight for genetic testing: a report from the Secretary's Advisory Committee on Genetics, Health and Society. Per Med 2008;5:521–8.

[749] Jarvik GP, Amendola LM, Berg JS, et al. Return of genomic results to research participants: the floor, the ceiling, and the choices in between. Am J Hum Genet 2014;94:818–26.

[750] Aronson SJ, Rehm HL. Building the foundation for genomics in precision medicine. Nature 2015;526:336–42.

[751] Lyon GJ. Personalized medicine: bring clinical standards to human-genetics research. Nature 2012;482:300–1.

[752] Webb BD, Brandt T, Liu L, et al. A founder mutation in COL4A3 causes autosomal recessive Alport syndrome in the Ashkenazi Jewish population. Clin Genet 2014;86:155–60.

[753] Verlander PC, Kaporis A, Liu Q, Zhang Q, Seligsohn U, Auerbach AD. Carrier frequency of the IVS4 + 4 A --> T mutation of the Fanconi anemia gene FAC in the Ashkenazi Jewish population. Blood 1995;86:4034–8.

[754] Fedick A, Jalas C, Treff NR. A deleterious mutation in the PEX2 gene causes Zellweger syndrome in individuals of Ashkenazi Jewish descent. Clin Genet 2014;85:343–6.

[755] Kramer HJ, Stilp AM, Laurie CC, et al. African ancestry-specific alleles and kidney disease risk in Hispanics/Latinos. J Am Soc Nephrol 2017;28:915–22.

[756] Naik RP, Derebail VK, Grams ME, et al. Association of sickle cell trait with chronic kidney disease and albuminuria in African Americans. JAMA 2014;312:2115–25.

[757] National Institutes of Health. NIH funds precision medicine research with a focus on health disparities. NIH; 2016.

[758] Bredenoord AL, de Vries MC, van Delden H. The right to an open future concerning genetic information. Am J Bioeth 2014;14:21–3.

[759] Wilfond BS, Fernandez CV, Green RC. Disclosing secondary findings from pediatric sequencing to families: considering the "benefit to families" J Law Med Ethics 2015;43:552–8.

[760] Committee on Bioethics, Committee on Genetics, The American College of Medical Genetics and Genomics, Social, Ethical, and Legal Issues Committee. Ethical and policy issues in genetic testing and screening of children. Pediatrics 2013;131:620–2.

[761] Ross LF, Saal HM, David KL, et al. Technical report: ethical and policy issues in genetic testing and screening of children. Genet Med 2013;15:234–45.

[762] Hufnagel SB, Martin LJ, Cassedy A, Hopkin RJ, Antommaria AH. Adolescents' preferences regarding disclosure of incidental findings in genomic sequencing that are not medically actionable in childhood. Am J Med Genet A 2016;170:2083–8.

[763] Joly Y, Ngueng Feze I, Simard J. Genetic discrimination and life insurance: a systematic review of the evidence. BMC Med 2013;11:25.

[764] Otlowski M, Taylor S, Bombard Y. Genetic discrimination: international perspectives. Annu Rev Genomics Hum Genet 2012;13:433–54.

[765] Yoshizawa G, Ho CW, Zhu W, et al. ELSI practices in genomic research in East Asia: implications for research collaboration and public participation. Genome Med 2014;6:39.

[766] Council of Europe. Recommendation of the Committee of Ministers to the member of the States on the processing of personal health-related data for insurance purposes, including data resulting from genetic tests. Quotidiano Sanita; 2016.

[767] Illumina. H3 Africa consortium array available soon. Illumina; 2016.

[768] Association of British Insurers. Concordat and moratorium on genetics and insurance. ABI; 2014.

[769] Canadian Life and Health Insurance Association. Industry code on genetics testing information for insurance underwriting. CLHIA; 2017.

[770] Hudson KL, Pollitz K. Undermining genetic privacy? Employee wellness programs and the law. N Engl J Med 2017;377:1–3.

[771] Pavlova A, Stuart RO, Pohl M, Nigam SK. Evolution of gene expression patterns in a model of branching morphogenesis. Am J Phys 1999;277:F650–63.

[772] Iyengar SK, Schelling JR, Sedor JR. Approaches to understanding susceptibility to nephropathy: from genetics to genomics. Kidney Int 2002;61:S61–7.

[773] Roy S, Coldren C, Karunamurthy A, et al. Standards and guidelines for validating next-generation sequencing bioinformatics pipelines: a joint recommendation of the association for molecular pathology and the College of American Pathologists. J Mol Diagn 2018;20:4–27.

[774] Roy S, LaFramboise WA, Nikiforov YE, et al. Next-generation sequencing informatics: challenges and strategies for implementation in a clinical environment. Arch Pathol Lab Med 2016;140:958–75.

[775] World Health Organization. Community genetics services: report of a WHO consultation on community genetics in low- and middle-income countries. WHO; 2010.

[776] Tekola-Ayele F, Rotimi CN. Translational genomics in low- and middle-income countries: opportunities and challenges. Public Health Genomics 2015;18:242–7.

[777] Maltese PE, Poplavskaia E, Malyutkina I, et al. Genetic tests for low- and middle-income countries: a literature review. Genet Mol Res 2017;16.

[778] Bogershausen N, Altunoglu U, Beleggia F, et al. An unusual presentation of Kabuki syndrome with orbital cysts, microphthalmia, and cholestasis with bile duct paucity. Am J Med Genet A 2016;170:3282–8.

[779] Moosa S, Obregon MG, Altmuller J, et al. Novel IFT122 mutations in three Argentinian patients with cranioectodermal dysplasia: expanding the mutational spectrum. Am J Med Genet A 2016;170A:1295–301.

[780] Osafo C, Raji YR, Burke D, et al. Human heredity and health (H3) in Africa kidney disease research network: a focus on methods in sub-saharan Africa. Clin J Am Soc Nephrol 2015;10:2279–87.

Further reading

Kingsmore SF, Lantos JD, Dinwiddie DL, et al. Next-generation community genetics for low- and middle-income countries. Genome Med 2012;4:25.

CHAPTER 3

Biomarkers in acute kidney disease

Melissa Fang[a], Kavitha Ganta[b,c], Soraya Arzhan[d], and Brent Wagner[b,c,d]

[a]University of New Mexico School of Medicine, Albuquerque, NM, United States, [b]University of New Mexico Health Science Center, Albuquerque, NM, United States, [c]New Mexico Veterans Administration Health Care System, Albuquerque, NM, United States, [d]Kidney Institute of New Mexico, Albuquerque, NM, United States

Introduction

Acute kidney injury (AKI) is a term introduced recently and is a substitute for acute renal failure. This term refers to a continuum of compromised renal function that is characterized by the accumulation of creatinine, products of nitrogen metabolism like urea, dysregulation of the electrolyte balance, acid-base status, and volume status. It can occur within hours to days and carries an increased risk for morbidity and mortality.

Acute kidney injury is an increasing and underrecognized global issue. Unquestionably, this condition imparts significant morality, comorbidity, and financial costs. A vast international database study conducted in 2017 revealed that acute kidney injury is frequent and severe, present in 8%–16% of hospital admissions and increasing the likelihood of hospital mortality [1]. The overall incidence of acute kidney injury in the United States is estimated at 2147 cases per million population per year [2]. Approximately 1% of admitted patients have acute kidney injury at the time of admission, and another 2%–5% are diagnosed during hospitalization. In recent years, perioperative acute kidney injury has been linked to high-risk patients, especially elderly patients with chronic comorbidities such as chronic kidney disease, undergoing high-risk procedures [3]. Acute kidney injury is common in some specific settings; a staggering 67% of intensive care unit patients develop the state [4]. The condition also develops in 15% of patients undergoing cardiopulmonary bypass, a procedure involving multiple renal stressors, and commonly in liver transplant recipients and gastric bypass surgical patients [5–7]. In the United States alone, acute kidney injury hospitalization costs ranged from $5.4 to $24.0 billion in 2017. The burden falls heavily on patients with severe acute kidney injury requiring dialysis, with hospitalization costing over $42,000 on average [8].

Seema S. Ahuja and Brian Castillo: Kidney Biomarkers. https://doi.org/10.1016/B978-0-12-815923-1.00003-1

Definition of AKI

Numerous definitions have been proposed. In 2002, the Acute Dialysis Quality Initiative declared over 30 separate definitions [9]. Understandably, the lack of a universal definition translated into confusion and controversy with respect to epidemiologic data, validation of diagnostic tests, recruitment for prospective trials, and the evaluation of effectiveness of therapies. The need for a comprehensive and standardized definition of acute kidney injury was clear. In 2004, the introduction of the Risk, Injury, Failure, Loss, and End Stage Renal Disease (RIFLE) classification signified an attempt at consensus on a definition. More recently, the AKI Network (AKIN) criteria and Kidney Disease: Improving Global Outcomes (KDIGO) have further attempted a clear definition of this syndrome. All three classifications rely on measuring changes in creatinine and urine output, functions, which are routinely and easily measured. While the limitations of these variable have been made clear in the literature, it should also be acknowledged that the classifications strive to improve the understanding and recognition of acute kidney injury.

RIFLE classification

The Acute Dialysis Quality Initiative (ADQI) was founded in 2002 with the primary goal of developing a definition and evidence-based guidelines for the treatment and prevention of acute kidney injury. Focused on developing a standard, the task force considered the following clinical needs: ease of use and comparability across different centers; utility in different patient populations; and significance of initial baseline measurements and changes to it. In 2004, the RIFLE (risk, injury, failure, loss, and end-stage renal disease) classification was proposed. This relies on changes of the plasma creatinine and/or urine output from baseline to stratify patients into three severity classes and two outcome classes. Risk is defined as a 1.5-fold increase in plasma creatinine, a decrease in glomerular filtration rate >25%, or a urine output <0.5 mL/kg/h for 6 h. Injury is defined as a twofold increase in plasma creatinine, a decrease in glomerular filtration rate >50%, or a urine output <0.5 mL/kg/h for 12 h. Lastly, failure is defined as a threefold increase in plasma creatinine, a decrease in glomerular filtration rate >75%. If the baseline plasma creatinine is ≥4 mg/dL, failure is defined by an increase of >0.5 mg/dL, or a urine output of <0.3 mL/kg/h for 24 h, or anuria for 12 h. These criteria define loss of function as a complete loss for more than 4 weeks that requires dialysis, and end-stage renal disease as complete loss of kidney function for more than 3 months requiring renal replacement therapy.

A notable achievement of RIFLE classification is the construction of a risk stratification model for acute kidney injury. After the implementation of the RIFLE classification, a systematic review of data from 71,000 patients in the various RIFLE categories revealed that the overall mortality rate was 18.9% in the Risk category, 36.1% in the Injury category, and 46.5% in the Failure category. In comparison, healthy patients without acute kidney injury had 6.9% mortality [10]. Not only did the RIFLE classification introduce a standardized definition of acute kidney injury and the various stages of disease progression, but it also reliably demonstrated a stepwise increase in the risk of mortality from each category to the next.

AKIN classification

Even small increases in plasma creatinine are associated with poor outcome. In 2005, the Acute Kidney Injury Network (AKIN) proposed modifications to RIFLE classification that aimed to increase its sensitivity and specificity for acute kidney injury. The new AKIN classification recommended that the diagnosis of acute kidney injury is to be considered only after achieving *adequate hydration and after excluding urinary obstruction*. AKIN also proposed that only plasma creatinine and not glomerular filtration rate changes were to be tracked. Absolute changes in plasma creatinine over the first 48 h of hospitalization are emphasized over baseline creatinine. According to AKIN, acute kidney injury is defined by the sudden decrease in renal function over the period of 48 h as evidenced by an increase in absolute plasma creatinine of at least 0.3 mg/dL, an increase in plasma creatinine $\geq$50% (1.5 $\times$ baseline value), or urine output $<$0.5 mL/kg/h for 6 h. AKIN stages 1, 2, and 3 primarily corresponded to the Risk, Injury, and Failure categories of the RIFLE classification. Stage 3 also includes patients who need renal replacement therapy, regardless of measures of plasma creatinine and urine output. Lastly, the Loss and ESRD categories were removed.

The theoretical benefit of AKIN over RIFLE is in the identification of patients with acute kidney injury earlier in the disease process based on smaller increases in absolute creatinine. However, the literature does not support the ability of the AKIN criteria over RIFLE to identify and stratify hospitalized patients with an acute kidney injury. A head-to-head comparison of the classifications reported that of nearly 50,000 admissions studied, 11.0% were diagnosed with acute kidney injury based on RIFLE criteria, and only 4.8% were diagnosed using AKIN criteria. Reiterating the other studies, the comparison also showed that AKIN had less prognostic value for in-hospital mortality (area under the curve, *AUC*=0.69) compared to RIFLE (*AUC*=0.77) [11–15]. Some research groups, concerned with the possibility that the AKIN classification would exclude a particular population of acute kidney injury patients whose creatinine only rises after the 48-h window, continue to rely on the RIFLE classification which monitors changes within the first 7 days of hospitalization.

KDIGO classification

The Kidney Disease: Improving Global Outcomes (KDIGO) Acute Kidney Injury Work Group proposed a new classification in 2012, which combined RIFLE and AKIN [16]. The new KDIGO model re-examined the criterion of time, ultimately including both changes in creatinine within 48 h as first established by the AKIN model and a decline in glomerular filtration rate over 7 days per the RIFLE model. Like the other classifications, KDIGO defined acute kidney injury as an increase in plasma creatinine by $\geq$0.3 mg/dL within 48 h, increase in creatinine to $\geq$1.5 times baseline within 7 days, or reduction in urine volume to <0.5 mL/kg/h for 6 h. The KIDGO classification also made modifications to the severity of staging. KDIGO stage 1 is defined by plasma creatinine that is 1.5–1.9 times baseline, a 0.3 mg/dL increase of plasma creatinine, or a urine output <0.5 mL/kg/h for 6–12 h. Stage 2 is defined by plasma creatinine that is 2.0–2.9 times baseline or <0.5 mL/kg/h urine output for $\geq$12 h. Stage 3 is defined as plasma creatinine that is three times baseline, plasma creatinine $\geq$4 mg/dL, initiation of RRT, a decrease in *estimated* (not *true*) glomerular filtration rate to <35 mL/min/1.73 m^2 in a patient younger than 18 years old, urine output <0.3 mL/kg/h for $\geq$24 h, or anuria for $\geq$12 h.

As a combination of RIFLE and AKIN, the KDIGO classification benefits from the widened criteria. Like RIFLE, the KDIGO criteria demonstrates a stepwise increase in the risk of mortality from each category to the next. The associations have odds ratios (OR) of 0.58 for stage 1; an OR of 2.2 for stage 2; and 9.55 for stage 3 [17]. In a retrospective study of 1881 cardiac surgery patients, the incidence of acute kidney injury according to the AKIN and KDIGO classification were identical at 25%, and both predicted in-hospital mortality more accurately than RIFLE. KDIGO classification outperformed RIFLE and AKIN classifications in predicting in-hospital mortality in patients on ECMO support [18]. In the context of critical cardiac patients, all three classifications performed equally well in predicting 30-day adverse outcome as defined by readmission for heart failure, the requirement of renal replacement therapy, or death [19]. These studies suggest the KDIGO classification is a reliable evaluation tool with high prognostic value.

These three models provide the field a common language with which to diagnose, prognosticate, study, and treat acute kidney injury. One problem is that all three rely on the assessment of plasma creatinine and oliguria degree. By waiting for particular changes to occur in the urine and blood tests, the current methods of detection and diagnosis can be delayed for several days.

Limitations of acute kidney injury markers

There are numerous concerns and limitations with using plasma creatinine. Plasma creatinine roughly reflects glomerular filtration rate in a steady state, but certainly not when renal function is fluctuating. It takes several days for plasma creatinine to reach a new steady state. There is variability in the endogenous production and equilibration of creatinine into the plasma compartment determined by age, sex, pregnancy status, cirrhosis, diet, and muscle mass. Individuals with lower plasma creatinine, such as children, pregnant women, and those afflicted with cirrhosis, may not be identified in models which use plasma creatinine. About 10%–40% of creatinine elimination occurs via tubular secretion [20]. With diminishing glomerular filtration rate, this tubular secretion increases, potentially overestimating renal function in acute kidney injury patients.

Many common medications, including trimethoprim, salicylates, and cimetidine, can inhibit tubular secretion of creatinine, leading to an elevation of plasma creatinine, which may be falsely interpreted as a marker of severe acute kidney injury. Acute kidney injury may result from mechanical obstruction of the urinary collecting system, including the renal pelvis, ureters, bladder, or urethra. Causes can include stone disease, stricture, tumors, fibrosis, and thrombosis. It is important to recognize that if there is a unilateral obstruction, the preserved function of the contralateral kidney may be sufficient to delay or prevent the rise of plasma creatinine.

In classifications such as RIFLE that require baseline plasma creatinine to define acute kidney injury, stratification and early intervention may be stalled if this baseline value is not known, often the case in clinical practice. One option has been to use the Modification of Diet in Renal Disease (MDRD) equation to estimate the baseline plasma creatinine, but this presents multiple problems. The equation assumes a baseline glomerular filtration rate of either 75–100 mL/min/1.73 m^2, which is potentially problematic in classifications where the magnitude of change of plasma creatinine from baseline is key to stratifying patients [21]. For example, using an assumed glomerular filtration rate of 100 mL/min/1.73 m^2 reduces baseline creatinine estimates, leading to false inflation in the magnitude of change of plasma creatinine throughout hospitalization that increases the rate and severity of acute kidney injury diagnosis [22]. The slower rise in plasma creatinine in patients with chronic kidney disease compared to patients without chronic kidney disease can also impede measurement of percentage increase in plasma creatinine over the defined observation window, thus decreasing the sensitivity of acute kidney injury diagnosis and stratification in such patients [23].

Using urine output is also problematic. It is less specific for the syndrome except when it is severely decreased or absent. Nonetheless, it is a real-time measurement that can be assessed at the hospital bedside, especially in patients who already have a urinary catheter inserted. The downfall of urine output rests in how it is impacted by many factors, including the patient's blood volume status and diuretic use. These factors are not considered in the RIFLE classification.

The number of studies published in the field of acute kidney injury continues to surge annually, indicating a sustained interest in the topic. Mainly due to the standard definitions, researchers have gained the ability to move beyond small single-site studies on well-characterized cohorts to large-scale analyses of heterogeneous populations [24–28]. As part of its ambitious goal to prevent all avoidable death from acute kidney injury internationally by 2025, the International Society of Nephrology carried out a 10-week study in 2014 to assess acute kidney injury prevalence across 289 centers and 72 countries [29]. The ability of such a study to be conducted highlights the widespread acceptance of current acute kidney injury models as well as the potential for global data collection to inform international strategy. Acknowledging the limitations of current acute kidney injury models, the next major step in the field will be to re-examine the available biomarkers, both traditional and novel, and determine which have the potential for clinical implementation and will allow for the most accurate diagnosis, prognostication, and early intervention for acute kidney injury.

The quest for the ideal acute kidney injury biomarkers

For an acute kidney injury biomarker to be clinically applicable, it must exhibit specific desirable characteristics. It should rely on minimally invasive collection techniques at the bedside, such as urine and blood tests. It should be measurable, rapid, and inexpensive to perform in a standard clinical laboratory. The biomarker should be sensitive for acute kidney injury so the phenomenon can be detected early, with a high gradient to assess the degree of damage, guide risk stratification, and determine intervention strategies. Once an effective therapy is initiated, changes in the biomarker level should predict clinical outcomes, not limited only to the need for renal replacement therapy or death, but also including positive results such as renal recovery. Critically, there needs to be biologic plausibility. Centrality of the biomarker with known mechanisms will clarify the settings and scenarios in which it will likely be sensitive and specific for acute kidney injury. If the biomarker can reflect certain etiologies of acute kidney injury, there exists the potential for a panel of select biomarkers to divine a patient's underlying disease processes.

Numerous candidates have been identified, including neutrophil gelatinase-associated lipocalin (NGAL), cystatin C, interleukin 18 (IL-18), kidney injury molecule 1 (KIM-1), liver-type fatty acid-binding protein (L-FABP), tissue inhibitor of metalloproteinase 2 (TIMP-2), and insulin-like growth factor-binding protein 7 (IGFBP7).

Neutrophil gelatinase-associated lipocalin

Also known as siderocalin, lipocalin 2, or oncogene 24p, NGAL is a 25-kDa protein of the family of lipocalins expressed at low concentrations in epithelial cells of the kidney, lungs, gastrointestinal tract, the uterus, and prostate, and additionally stored in specific granules of neutrophils [30, 31]. It filters freely through the glomerulus, and then it is almost entirely resorbed by healthy tubular cells. On average, only 20 ng/mL of NGAL can be detected in the urine of healthy adults [32]. NGAL has many functions. In animal studies, NGAL is induced as part of the innate immunity response in damaged epithelial cells during inflammation [33]. Specifically, it complexes with iron siderophores, which then can be internalized in the proximal tubules via a megalin-cubulin receptor complex and in the distal tubule via the 24p3 receptor [34, 35]. Increasing the intracellular iron concentration regulates iron-responsive genes [36]. Upregulated NGAL expression also correlates with the induction of apoptosis of infiltrating neutrophils, which allows resident cells to escape destruction by the inflammatory response. Other studies suggest NGAL expressed by the damaged tubule may aid renal regeneration and repair after ischemic injury by triggering nephrogenesis and stimulating conversion of mesenchymal cells into kidney epithelia in the mature kidney [37, 38]. During nephrogenesis, NGAL is complementary to transferrin. Delivery of iron into cells is crucial for cell growth and development in postischemic renal regeneration, and it removes iron from the site of tissue injury, limiting iron-mediated cytotoxicity [39].

NGAL is highly upregulated in structurally damaged epithelial cells of the proximal renal tubule early after ischemic and toxic acute kidney injury. Following a 45-min occlusion of the left renal pedicle of mice, a substantial increase in urinary NGAL was detected within 2 h of the procedure that persisted throughout observation of 24 h. In comparison, β2-microglobulin became detectable in the same urine samples 12 h after the procedure. Plasma creatinine did not rise until 24 h after the procedure. A rise, albeit delayed, in urinary NGAL was also observed with mild renal ischemia precipitated by shorter periods of renal pedicle occlusion.

In contrast, plasma creatinine remained within normal limits by the end of the 24-h observation. In other etiologies of tubule cell injury, NGAL increased before the established markers rose as well. In a mouse model of cisplatin nephrotoxicity, urinary NGAL rose after 24 h after administration while urinary β2-microglobulin and urinary NAG were not reliably detected until days 4–5 after administration. By day 3, cisplatin-induced tubule cell necrosis and apoptosis and elevated blood urea nitrogen were already evident [38]. Together, these animal studies indicate that the rise of NGAL significantly precedes changes in other biomarkers.

Studies of human populations have demonstrated the sensitivity, specificity, and prognostic value of NGAL in early acute kidney injury evolving from a variety of etiologies. Several randomized control trials have also been published on the subject, ranging in focus from acute myeloid leukemia (AML), acute lymphoblastic leukemia (ALL), and myelodysplastic syndrome (MDS) patients receiving allogeneic hematopoietic stem cell transplantation [40], stable chronic kidney disease patients undergoing off-pump coronary artery bypass grafting (OP-CABG) [41], women with ovarian cancer undergoing laparotomy [42], and CKD patients incurring contrast-induced acute kidney injury via intra-arterial angiography [43]. Many groups observed positive correlations between urinary NGAL level and diagnosis of acute kidney injury. In a group of AML, ALL, and MDS patients who underwent allogeneic hematopoietic stem cell transplantation, the ratio of urinary NGAL 9 days after hematopoietic stem cell transplantation to baseline urinary NGAL predicted acute kidney injury 2 days before it was diagnosed, with each unit increase of this ratio increased odds of acute kidney injury occurrence by 8%. Likewise, significantly elevated NGAL concentrations were reported at 4 h after Off-pump coronary artery bypass grafting (OP-CABG) in patients who developed acute kidney injury versus those who did not. Similarly, plasma NGAL drawn 6 h after laparotomy was found to be predictive for acute kidney injury occurring within the first 3 days of the surgery. When a threshold on the low side of the gray zone was applied, the plasma NGAL gave high negative predictive values, allowing clinicians to rule out acute kidney injury earlier.

Many studies reported that NGAL can predict contrast-induced nephropathy,[a] a possible leading cause of hospital-acquired acute kidney injury. Serum and urinary NGAL levels increase in patients who developed contrast-induced nephropathy compared to those who did not in the first 6 h following administration of contrast and decreased back to baseline by 12 h [44–48]. Many of

---

[a] Contrast-induced nephropathy, a hobgoblin of medicine, is difficult to assess. Rates have declined to a level where hospitalized patients without any exposure to iodinated contrast have the same rate of acute kidney injury as those exposed to the contrast. Interested readers can find more in an issued dedicated to these controversies, Advances in Chronic Disease May 2017; 24(3).

these studies also determined that no cases were diagnosed within the first 24 h using plasma creatinine changes, demonstrating the role of NGAL as an early marker of contrast-induced nephropathy after invasive cardiac procedures.

Many controlled trials without randomization support the value of NGAL as an early biomarker of acute kidney injury. In children receiving cardiopulmonary bypass, urinary and serum NGAL showed a 92-fold and 19-fold increase, respectively, within 2 h of the procedure. Urinary NGAL levels higher than a threshold of 50 μg/L 2 h after surgery had 100% sensitivity and 98% specificity for the subsequent diagnosis of acute kidney injury. Urinary NGAL levels >25 μg/L had moderate sensitivity, specificity, and positive and negative predictive values [49]. The elevation persisted for 24 h [50]. Similar results were found in adults in intensive care units, with a urinary NGAL threshold of 142.0 ng/mL providing moderate sensitivity and specificity, a high negative predictive value, and yet a low positive predictive value [51]. Adults who underwent cardiac surgery demonstrated a rapid increase in urinary NGAL levels within 3 h postoperatively only providing an improvement in sensitivity and specificity for the development of acute kidney injury over the following 15 h [52]. Defining thresholds obviously modulate the effectiveness of urinary NGAL in risk prediction for the development of acute kidney injury. It is unknown whether a single measured threshold for NGAL can be established for universal prognostication of acute kidney injury.

NGAL expression correlates with severity of kidney injury and predicts clinical consequences ranging from length of hospital stay, length of intensive care unit stay, the risk for dialysis dependency, and death [53]. Acute kidney injury in patients in the highest quintile of plasma NGAL is five times as likely compared to those in the lowest quintile. In this study, 18% of patients with acute kidney injury died, whereas only 1% of patients without acute kidney injury died [53]. The predictive value of plasma NGAL rises with increasing severity of acute kidney injury as categorized by both RIFLE and AKIN. Adults with severer classes often required additional intervention, longer stay, or suffered elevated in-hospital mortality [54–56]. Serum NGAL levels in healthy children also differed significantly compared to critically ill children with systemic inflammatory response syndrome, critically ill children who had septic shock, and critically ill children who subsequently developed acute kidney injury. Those developing acute kidney injury had higher pediatric risk of mortality (PRISM) scores, higher rate of organ failure in the first week of hospitalization, and longer length of intensive care unit stay compared to the children who did not develop acute kidney injury [57].

NGAL may also correlate with other comorbidities. Patients with moderate or severe acute kidney injury are frequently diagnosed with sepsis (18.2% vs

5.7%), diabetes mellitus (48.5% vs 27.8%), and chronic kidney disease (33.3% vs 10.4%) compared to patients without acute kidney injury [51]. Questions that remain are whether universal risk-assessment scores will be able to guide treatment algorithms for acute kidney injury [58, 59].

Many have realized that NGAL may be an early biomarker of kidney infection, organ rejection, or drug toxicity [60–66]. Furthermore, NGAL may be useful in earlier detection of delayed graft function—the requirement for dialysis within 1 week of transplantation, particularly problematic with high rates of deceased, marginal donors—for kidney transplant recipients [67]. In such a population, NGAL has been associated with increased lengths of stay, increased hospital costs associated with dialysis, an increased risk of graft loss at 3 years posttransplant, and increased incidence of acute and chronic rejection [68–71]. Urinary NGAL can be predictive using samples collected within 4 h postsurgery from renal transplant recipients with delayed graft function [72]. In kidney transplant recipients, urinary NGAL levels outperformed plasma creatinine in predicting delayed graft function the day after transplant [73]. Therefore, there are promising studies that suggest that urinary NGAL may have utility as an early marker of delayed graft function and a predictor of graft loss.

At the time of this writing, there are no diagnostic NGAL tests approved by the United States Food & Drug Administration. Invitrogen possesses a human NGAL ELISA kit intended for research purposes only. BioPorto submitted an application for "The NGAL Test" in late 2018 [74]. There is a particle-enhanced turbidimetric immunoassay for the quantitative determination of NGAL in human urine, EDTA- or heparinized- plasma to run on automated clinical chemistry analyzers.

Unfortunately, NGAL demonstrates a low predictive value in certain clinical settings. Plasma and urine NGAL have a low predictive power for acute kidney injury in intensive care unit patients with severe sepsis [75]. In intensive care unit-admitted organ transplant recipients, NGAL is unrelated to acute kidney injury [76]. NGAL is not kidney-specific; therefore, other concurrent disease processes (e.g., infections, inflammation, and neoplasia) confound the predictive ability for acute kidney injury [77]. Plasma NGAL is increased in patients with colorectal neoplasia, diverticulitis, inflammatory bowel diseases, and appendicitis [78–80]—not uncommon conditions. Therefore, NGAL is far from ideal for clinical utility regarding acute kidney injury.

NGAL is an early marker of epithelial structural damage. Given the ubiquity of this cell type, numerous etiologies impact both the plasma and urinary levels. Some studies propose that it can be detected much earlier and with much higher sensitivity than plasma creatinine, yet the lack of specificity for acute kidney injury and current lack of diagnostic assays present obstacles to clinical utility.

Cystatin C

Cystatin C is a member of the cysteine proteinase inhibitor protein superfamily. The normal range of serum cystatin C is 0.51–0.98 mg/L in a healthy population [81]. It is produced at a sustained rate by all nucleated cells. Because it is not bound to plasma proteins, cystatin C is freely filtered by the glomerulus and nearly completely reabsorbed in the proximal tubule, where it is uptaken by megalin [82]. Unlike creatinine, cystatin C is not secreted by the renal tubules and is not affected by changing muscle mass, sex, race, weight, or nutritional status. Additionally, whereas plasma creatinine is distributed in total body water, cystatin C is distributed in the extracellular volume. Therefore, in response to a decrease in (*true*) glomerular filtration rate, cystatin C should rapidly rise to a new steady state more quickly than creatinine, which requires more time for a significant shift to be reflected in total body water [83]. The ability of urinary cystatin C to signal glomerular injury and impairment of tubular reabsorption has made it an attractive potential biomarker for acute kidney injury.

Animal studies have demonstrated the applicability of cystatin C for the detection of acute kidney injury sooner than plasma creatinine. In mice, serum cystatin C demonstrated a detectable increase early with sepsis or bilateral nephrectomy than plasma creatinine or blood urea nitrogen. In septic mice, the inulin-measured glomerular filtration rate fell by >50% by the third hour after induction, which was 3 h before any systemic signs of sepsis (e.g., lethargy or piloerection) were noted. Plasma creatinine gradually elevated over several days. Conversely, cystatin C increased rapidly and reached a steady state within 12 h [84].

In a canine model of gentamicin-induced renal injury, cystatin C showed the most correlative (inverse) relationship for glomerular filtration rate among comparative biomarkers (including plasma creatinine) [85]. The high sensitivity for glomerular function was similarly reflected in a rat model of acute nephrotoxicity, where a rise in urinary cystatin C level occurred on day one after cisplatin-induced acute kidney injury, preceding histopathological changes in the proximal and distal tubules apparent on days 3 and 5, respectively. Renal histology already showed tubular cell necrosis, tubular dilatation, and hyaline cast tubules in the proximal tubules, essentially rendering creatinine ineffective for prediction of damage [86]. In total, these studies suggest that urinary cystatin C could be a practical biomarker, which may reflect and potentially predict kidney injury earlier than plasma creatinine. A randomized, controlled trial to examine the predictive value of cystatin C for acute kidney injury is underway [87].

Prospective studies also suggest that changes in cystatin C precede changes in plasma creatinine. For patients in surgical and medical intensive care units, serum cystatin C was found to have diagnostic value to detect the risk, injury, and failure stages of acute kidney injury [88]. The rise in serum cystatin C preceded that of plasma creatinine by 1.5 days [89]. A substudy from a double-blind, randomized controlled trial of cardiac surgical patients found that an increase in serum cystatin C above a threshold in the first 24 h following surgery was highly predictive of acute kidney injury. As the severity of acute kidney injury increased, so did predictive ability of serum cystatin C. Adding glutathione-*S*-transferase π (π-GST) to the analysis further increased the predictive value [90].

It has been often found that the best combinations involve a single biomarker for tubular injury (NGAL or π-GST) and another for glomerular filtration rate (cystatin C or hepcidin). Such combinations appear to work best if one is measured early and the other measured late in the postoperative period [90]. Using biomarkers derived from different sources of injury and assessing a wider period may have more clinical utility that just a single parameter. Potentially, this could be beneficial in the detection of early and/or subtle kidney injury as it develops.

In the process of cardiopulmonary bypass, elevations of tubular injury biomarkers reflect the damage incurred by the procedure. The marker of glomerular filtration reflects alterations in blood flow secondary to low cardiac output, nonpulsatile flow, or ischemia-reperfusion injury [91, 92]. When renal autoregulation is altered during the hypothermic bypass period, resultant systemic hypotension may cause reduced oxygen delivery to the kidneys, which have some susceptibility to hypoxic injury. When such patients experience hypotension during the procedure, the predictive value of these injury biomarkers increases [93, 94]. Blood samples taken at the end of these procedures demonstrate that serum cystatin C has only moderately predictive value. A persistent elevation of cystatin C at 24 h, conversely, was highly predictive for the need to initiate renal replacement therapy, arguably a surrogate marker of acute kidney injury [95]. While cystatin C does not have significant predictive value immediately after the operation, the evolution in the first 24 h outpaces changes in plasma creatinine, and therefore may have some clinical utility.

Decreases in cystatin C from baseline during the cardiopulmonary bypass may occur from intraoperative hemodilution [95–97], a procedure often used to minimize blood loss [98]. With such large dilutional effects on both serum and urine, estimation of glomerular filtration rates becomes problematic. In such settings, biomarkers should be used with caution due to the significant confounding effects of hemodilution [97].

Cystatin C has shown promise in the evaluation of contrast-induced acute kidney injury. The pathophysiology of contrast-induced nephropathy is not entirely understood [99]. It has been hypothesized to involve renal ischemic injury and tubular epithelial cell damage [100, 101]. Iodinated contrast medium also produces natriuresis and diuresis, which can activate the tubuloglomerular feedback response [102]. Vasoconstriction of the afferent arterioles can decrease the glomerular filtration rate. Reduced filtration entails that cystatin C will increase in the plasma. Several authors have compared levels of cystatin C to creatinine in assessing contrast-induced nephropathy. The analyses suggest that serum cystatin C gains a predictive value for contrast-induced nephropathy by 24 h after administration, 24 h before creatinine becomes indicative. Patients who underwent coronary angiography demonstrated changes within the first postoperative day [103, 104]. By comparison, plasma creatinine elevations were delayed up to 48 h. Serum cystatin C defined thresholds could yield a sensitivity of 94.7% and specificity of 84.8%. For comparison, a plasma creatinine threshold of 1.1 mg/dL provided a sensitivity was 63.2%, and specificity was 78.0% [81].

The United States Food & Drug Administration-approved N-Latex cystatin C assay (Siemens Healthineers, Erlangen, Germany) and its equivalent cystatin C immunoassay (Gentian, Moss, Norway) have been deemed reliable and accurate [105, 106]. LabCorp (Burlington, North Carolina, United States) also performs immunologic cystatin C tests [107].

Plasma cystatin C increases when glomerular filtration rate falls. It may prove to have predictive value for acute kidney injury with improved acuity compared to plasma creatinine. As a corollary, cystatin C may predict contrast-induced nephropathy. The applicability of cystatin C does not encompass every clinical scenario, as evidenced by postcardiopulmonary bypass acute kidney injury. Urinary cystatin C levels in this situation have yet to be thoroughly vetted.

Interleukin 18

IL-18, a member of the proinflammatory IL-1 cytokine family, regulates innate and adaptive immunity [108]. Also known as interferon-gamma inducing factor. Caspase-1 cleaves the IL-18 precursor and allows the mature, active molecule to be secreted from monocytes, macrophages, and dendritic cells [109]. When cells becomes damaged, inactive IL-18 is released and processed extracellularly by neutrophil proteases [110]. Activated IL-18 has many functions. It may form a complex with IL-18 alpha chain and beta chain to signal the release of NFκB [108–111]. If there is sufficient IL-12 or IL-15, IL-18 induces interferon gamma production by NK cells, CD4+ T cells, and CD8+ T cells [112, 113]. Interferon gamma is a cytokine that bridges innate and adaptive immunity, activates macrophages, and stimulates natural killer cells and neutrophils. Ultimately, IL-18 in the presence of IL-12 upregulates cytotoxic actions [114, 115]. Numerous inflammatory processes demonstrate increased IL-18 levels such as autoimmune diseases, diabetic nephropathy, inflammatory bowel disease, macrophage activation syndrome, ischemia-reperfusion injury, sepsis, and acute kidney injury [116–120].

Animal models have primarily focused on building our understanding of the source and functions of caspase-1, an initiator caspase responsible for generating active cytokines, and IL-18. Wild-type mice with bilateral total ischemia and reperfusion demonstrated upregulation of IL-18 in the kidneys at 24 h compared to sham controls [121]. Freshly isolated proximal tubules with inhibited hypoxia-induced caspase-1 activity had less necrotic injury [122]. Similarly, IL-18 inhibitors diminished the induction of interferon gamma, protecting animals from ischemia/reperfusion injury [123, 124].

Few prospective comparative studies have assessed the predictive value of IL-18 for acute kidney injury. Some of these suggest that elevated IL-18 concentration is correlated with higher risk of acute kidney injury and poor clinical outcomes. Elevated IL-18 in infants within 12 h after a congenital heart surgery correlated with prolonged time to extubation, extended intensive care unit length of stay, the need for renal replacement therapy, or death. Inversely, infants with IL-18 <0.8 ng/mL had at least a 90% chance of good clinical outcomes [125]. Another study similarly reported that pediatric patients with the highest quintiles of urine IL-18 were associated with higher odds of acute kidney injury, more extended hospital stay, longer intensive care unit stay, and a longer period of mechanical ventilation compared with the lowest quintiles [126].

Other studies did not identify IL-18 as a significant predictor of acute kidney injury. A large prospective study of patients in the intensive care unit concluded that even the highest recordings of IL-18 did not predict either new or severe acute kidney injury. Moreover, neither baseline IL-18 nor changes in the first 24 h of admission predicted diagnosis or progression of acute kidney injury. Correlations with 90-day mortality and the need to initiate renal replacement therapy were weak as well [127]. It is plausible that the methodology of this study failed to account for the timing of initial renal insult. The study did exclude patients with preexisting acute kidney injury on admission day; however, the only other urinary sample that was collected was at 24 h. The timing of development of acute kidney injury within this 24 h window does affect the levels of urinary biomarkers. Urinary samples collected at additional time points within the 24-h window may have improved timing of the renal injury and clarified risk prediction calculations.

Previous research has emphasized the importance of pinpointing the timing of renal injury as biomarkers rise and fall over time. Urine samples obtained at 2, 4, 6, 8, and 12 h after cardiac surgery in adults showed peaks in urine IL-18 between 2 and 6 h after surgery in the acute kidney injury group compared to the nonacute kidney injury group after correcting for urine creatinine

[128]. Fluxes in IL-18 levels are very time dependent, sometimes increasing around 5 h after a surgery and peaking at 12 h [129]. When compared, IL-18 measurements were significant predictors of acute kidney injury, but NGAL measurements were even more discriminatory. Combining IL-18 with other markers (NGAL, IL-18, KIM-1, and L-FABP) may improve the predictive power [130]. It can be stated that IL-18 doesn't correlate with the timing of renal injury perfectly [126, 131–133].

IL-18 assays are available for research purposes; there are human IL-18 cytokine quantification kits (Cisbio, Bedford, Massachusetts, United States) [134], IL-18 immunoassays (Pacific Biomarkers, Seattle, Washington, United States) [135], and a human IL-18 ELISA kit (R&D Systems, Minneapolis, Minnesota, United States) [136]. None of the available IL-18 assays are United States Food & Drug Administration approved for diagnostic use.

Kidney injury molecule 1 (KIM-1)

KIM-1 is a phosphatidylserine receptor glycoprotein typically absent in the urine of healthy subjects. After an ischemia-reperfusion injury, cells of the proximal tubule will increase KIM-1 mRNA and synthesize the protein. This glycoprotein will be expressed on the apical membrane where it assists in the recognition and phagocytosis of cellular material. KIM-1 then directs the phagocytosed material to the lysosome, where it is degraded [137]. By clearing out the luminal debris, the function of KIM-1 may be to delay tubular obstruction. Once recovered, MAP kinase signaling pathways direct the shedding of the extracellular domain of KIM-1 into the urine [138, 139]. The remnant is degraded via metalloproteinases [140]. The timing of the rise and decline of KIM-1 makes it an attractive target of study in both acute kidney injury and subsequent renal recovery. The expectation is that consecutive measurements of urinary KIM-1 may differentiate between the progression and recovery phases of acute kidney injury and direct early intervention strategies. Successive measurements showing an increase in KIM-1 levels may suggest worsening renal injury, and therapies to attenuate injury may be gauged by this marker. Perhaps consecutively high measurements of KIM-1 indicate recovery-focused treatments.

Animal models confirm a low baseline KIM-1 level that rise in response to renal injury. In large-scale gene expression screens, KIM-1 expression was the most highly increased in rats following ischemia and cisplatin administration [141, 142]. KIM-1 was one of three biomarkers, which exhibited significant correlations with vancomycin-induced renal injury [143]. Increasing vancomycin (intraperitoneally) positively correlated with significant increases in urinary KIM-1, and increases in urinary KIM-1 nearly double that of baseline had over

95% specificity for acute kidney injury [144]. Other studies demonstrated that injecting ≥150 mg/kg/day of vancomycin resulted in linear increases in urinary KIM-1 levels and that this threshold marked the transition from tubular degeneration to tubular necrosis on histology [145]. Importantly, changes in the biomarker levels occurred before any histopathologic damage became evident, suggesting the potential for KIM-1 as a harbinger for early vancomycin-associated acute kidney injury.

Randomized controlled trials on the potential relationship between KIM-1 and acute kidney injury have been sparse, and conclusions have been split. One examined the evolution of acute kidney injury following percutaneous nephrolithotripsy and the other applied the marker to contrast-induced nephropathy. The first involved patients with renal calculi >2 cm, with 29/76 undergoing percutaneous nephrolithotripsy [146]. Detectable KIM-1 differences emerged by 24 h after the operation, outpacing changes in plasma creatinine. KIM-1 did show a significant ability to predict early-stage renal injury in this study population. However, this phenomenon may not translate to contrast-induced nephropathy where a study found no detectable differences in KIM-1 or NGAL levels in a group of chronic kidney disease patients undergoing contrast-enhanced CT scans [147]. Not only is the pathophysiology of contrast-induced nephropathy still incompletely understood [148], the incidence has dropped to a level where it is difficult to conduct clinical studies [99, 149]. Studies on contrast-induced nephropathy have relied mainly on the assumption that contrast materials cause tubular injury. If the injury to the tubules did occur, then tubule injury biomarkers should undoubtedly increase. This study reported that KIM-1, a tubule injury biomarker, did not shift significantly over an extended period, possibly suggesting that contrast-induced nephropathy was not accompanied by a significant degree of tubular injury in this patient population. Our understanding of the pathophysiology of contrast-induced nephropathy must improve before we can meaningfully assess the biological plausibility of various biomarkers and comment on their diagnostic and prognostic value. Together, these randomized controlled trials demonstrate that KIM-1 can predict early renal injury but caution that this biomarker must be used in disease processes involving tubular injury to be effective.

Other controlled and prospective studies investigating KIM-1 as an acute kidney injury biomarker in many patient settings likewise yielded mixed results. In patients who underwent cardiac catheterization, KIM-1 levels increased significantly within the first day of the procedure and remained elevated for 48 h, while estimated glomerular filtration rate based on plasma creatinine did not begin to shift until 48 h after contrast administration [150]. The maximum point occurred around 24 h and increased 45-fold from baseline [151]. However, this increase had only moderately predictive value [152]. A case-control prospective study of 72 children demonstrated high specificity and sensitivity using KIM-1 for the diagnosis of acute kidney injury at 12, 24, and 36h after cardiopulmonary bypass, outweighing other proposed acute kidney injury biomarkers like matrix metalloproteinase-9 and *N*-acetyl-β-D-glucosaminidase [153].

KIM-1 does not appear to have any superiority when compared to other markers, such as π-GST, NGAL, cystatin C, and IL-18 in in predicting deterioration of renal function in patients with AKIN stage 1 acute kidney injury following cardiac surgery [154]. A large multicenter prospective cohort involving adults and children undergoing cardiac surgery demonstrated limitations of KIM-1 correlating with acute kidney injury, dialysis, or death [155].

Up to 70% of severely asphyxiated neonates develop acute kidney injury and suffer high mortality rates [156]. Because significant increases in plasma creatinine in this population are typically not seen until 4 days after delivery, several research groups probed the effectiveness of novel biomarkers in predicting acute kidney injury. Case-control studies revealed no significant biomarker differences between asphyxiated and control neonates and no differences between the neonates who developed acute kidney injury versus those who did not [157]. A larger study posting a similar question did report that the asphyxia group and acute kidney injury group had significantly higher KIM-1 levels 48h after delivery compared to healthy controls [158].

There are mixed results with KIM-1 levels in CKD patients. Spot urinary KIM-1 levels failed to predict proteinuria in a heterogeneous study population of 143 patients with stable chronic kidney disease arising from diabetic and/or hypertensive nephropathy, glomerulonephritis, and polycystic kidney disease [159]. The researchers concluded that tubular atrophy, a hallmark of several disease processes leading to chronic kidney disease, has little or no impact on biomarkers of acute tubular injury such as KIM-1. Conversely, a large prospective observational study involving 1982 patients reported that baseline plasma KIM-1 predicted progression to end-stage renal disease [160]. In line with these results, another prospective study reported that serum and plasma KIM-1 levels positively correlated with chronic kidney disease stage [161].

Baseline serum KIM-1 level in patients with chronic kidney disease, especially in the early stages, correlates with the rate of decline of estimated glomerular filtration rate during a 10years follow-up period [161]. One difference in study designs that may have led to these conflicting findings may be due to the sources from which samples are drawn. Plasma KIM-1 concentrations are determined the volume of KIM-1 distribution and clearance. Urinary excretion can vary over time in patients with acute kidney injury, and a spot collection will certainly rarely be representative. The addition of a 24-h urine collection may help determine a time-averaged production of a kidney injury biomarker (with all of the concomitant drawbacks of a 24-h urine collection). KIM-1 is far from clinical application as of this writing.

KIM-1 assays have limited approved by the United States Food & Drug Administration for certain nonclinical and phase I clinical research situations. On the market are a human urinary TIM-1/KIM-1/HAVCR quantikine ELISA kit (R&D Systems, Minneapolis, Minnesota, United States), a high throughput ultrasensitive urine ELISA kit (Enzo Life Sciences), and an ELISA kit for urine or blood (BosterBio, Pleasanton, California, United States).

KIM-1 is a tubular injury marker produced largely by the proximal tubules. Plasma levels rise with renal injury and fall with recovery. There are some conflicting results, perhaps reflecting the paucity of human studies.

Tissue inhibitor of metalloproteinase-2 and insulin-like growth factor binding protein 7

Renal tubular cells enter a period of G1 cell-cycle arrest after an inciting event of ischemia [162] or sepsis [163]. Insulin-like growth factor binding protein 7 and tissue inhibitor of metalloproteinase-2 are two proteins that have been implicated in G1 cell cycle arrest during the initial phases of cell injury. When a tubular cell enters a G1 cycle arrest, the division will not proceed unless the damaged DNA is repaired [164]. The tissue inhibitor of metalloproteinase-2 has many effects, including inhibition of matrix metalloproteinase activity, suppression of endothelial cell proliferation, and the inhibition of angiogenesis [165]. It has also been shown that the tissue inhibitor of metalloproteinase-2 plays a vital role in the pathophysiology of ischemia-reperfusion injury, both in early tubular and interstitial injury and tubular regeneration postinjury.

The insulin-like growth factor (IGF) system consists of IGFs (IGF-I and IGFII), type I and type II IGF receptors, IGF-binding proteins (IGFBPs), and IGFBP proteases [166]. The IGFBPs modulate the interactions of IGFs with the receptors. IGFBP7 is a member of the IGFBP superfamily and is secreted by epithelial cells, endothelial cells, and vascular smooth muscle cells [167]. IGFBP-7 was identified in urine, plasma, and other tissues, including the kidney [168]. Compared to the other IGFs, the IGFBP-7 binds to the IGFs with a lower affinity and induces apoptosis and suppresses tumor growth. When experimental acute kidney injury was induced in the rats, there was increased expression of IGFBP-7 [169]. This is an interesting finding considering that IGFBP-7 binds to IGF-1 and IGF-1 is responsible in the repair and remodeling of the renal tubular cells after an acute kidney injury due to ischemic or toxic injury. Upregulation of IGFBP-7 is reported in leukemia and thyroid cancer, and downregulation is reported in liver and lung carcinoma [170].

In one prospective study from patients after a cardiopulmonary bypass where 26 out of 50 developed acute kidney injury (as per the KDIGO definition), there was a rise in *the product of TIMP-2 and IGFBP7*, "[TIMP-2] × [IGFBP7]," compared to the presurgery levels. This study also showed that the [TIMP-2] × [IGFBP7] panel had a high sensitivity and specificity in predicting the renal recovery after acute kidney injury [171]. Another prospective study investigated 107 patients who were at increased risk of acute kidney injury after major noncardiac surgery. This study showed that the [TIMP-2] × [IGFBP7] panel was a strong predictor of acute kidney injury in this setting. The highest median values of this biomarker were noticed in patients undergoing transplant surgery, hepatic surgery, or who are septic [172].

Urine TIMP2 and IGFBP7 performed better than comparable biomarkers, including NGAL, KIM-1, and IL-18, for diagnosing acute kidney injury (based on RIFLE criteria) within 12–36 h in the Discovery Study. The Sapphire Validation Study [173] recruited 744 patients from multiple clinical sites in the United States and Europe. The AUCs for acute kidney injury stage 2 or 3 within 12 h for IGFBP7, TIMP, and their combination were essentially 0.80. This study demonstrated that TIMP-2 is better at predicting sepsis-induced acute kidney injury, whereas IGFBP7 was superior in predicting the risk in surgical patients.

IGFBP7 levels may predict mortality, renal recovery, and severity/duration of acute kidney injury in critically ill adults [174]. The Topaz Study was a multicenter prospective study of critically ill patients conducted in the United States. The absolute risk for acute kidney injury with [TIMP-2] × [IGFBP7] greater than a specified threshold was seven times higher than the absolute risk among patients with values below this cutoff [175]. Critically, though, the test had low specificity.

The combination of urine [TIMP-2] × [IGFBP7] was approved for marketing by the United States Food & Drug Administration in September 2014 under the brand name NephroCheck (Astute Medical, San Diego, California, United States). This is the first biomarker for risk assessment of acute kidney injury to become available for clinical use in the United States. This test should only be used in patients 21 years or older, admitted in an intensive care unit who currently have or had an acute respiratory or cardiovascular compromise in the last 24 h. It is to be used in conjunction with clinical evaluation, and is not a point-of-care test. Since the test has low specificity, the predictive value depends heavily on the likelihood of the disease, and therefore not recommended in patients at minimal risk of acute kidney injury. The manufacturers emphasize that this test should not be considered as a substitute for plasma creatinine and that serial measurement does not predict acute kidney injury progression. Bilirubinuria and albuminuria interfere with the test result.

A machine (typically $5000) can perform only one of these tests ($85) at a time, therefore limiting extensive use. Remarkably, in a single-center unblinded randomized clinical trial in critically ill patients after major abdominal surgeries, this testing demonstrated a decrease in the length of ICU/hospital stay, reduction in acute kidney injury severity and cost reduction without differences in the mortality or renal replacement therapy [176].

Conclusions

As recently noted [177], because of broad definitions of acute kidney injury, the multitude of clinical settings where it occurs, and methodological weaknesses, creatinine and oliguria remain the gold standards for the diagnosis of acute kidney injury despite limitations. Perhaps panels of biomarkers will provide practical advantages. Importantly, an ideal biomarker would be strictly limited to renal function without influence from conditions such as sepsis, diabetes mellitus, or malignancy, etc. Even if an ideal undiscovered renal biomarker exists, the utility hinges on a clinician's ability to abrogate kidney injury with an effective therapy.

References

[1] Sawhney S, Fraser SD. Epidemiology of AKI: utilizing large databases to determine the burden of AKI. Adv Chronic Kidney Dis 2017;24(4):194–204.

[2] Ali T, Khan I, Simpson W, Prescott G, Townend J, Smith W, et al. Incidence and outcomes in acute kidney injury: a comprehensive population-based study. J Am Soc Nephrol 2007;18 (4):1292–8.

[3] Meersch M, Schmidt C, Zarbock A. Perioperative acute kidney injury: an under-recognized problem. Anesth Analg 2017;125(4):1223–32.

[4] Goldberg R, Dennen P. Long-term outcomes of acute kidney injury. Adv Chronic Kidney Dis 2008;15(3):297–307.

[5] Kheterpal S, Tremper KK, Heung M, Rosenberg AL, Englesbe M, Shanks AM, et al. Development and validation of an acute kidney injury risk index for patients undergoing general surgery: results from a national data set. Anesthesiology 2009;110(3):505–15.

[6] Thakar CV, Kharat V, Blanck S, Leonard AC. Acute kidney injury after gastric bypass surgery. Clin J Am Soc Nephrol 2007;2(3):426–30.

[7] Cabezuelo JB, Ramirez P, Rios A, Acosta F, Torres D, Sansano T, et al. Risk factors of acute renal failure after liver transplantation. Kidney Int 2006;69(6):1073–80.

[8] Silver SA, Chertow GM. The economic consequences of acute kidney injury. Nephron 2017;137(4):297–301.

[9] Kellum JA, Levin N, Bouman C, Lameire N. Developing a consensus classification system for acute renal failure. Curr Opin Crit Care 2002;8(6):509–14.

[10] Ricci Z, Cruz D, Ronco C. The RIFLE criteria and mortality in acute kidney injury: a systematic review. Kidney Int 2008;73(5):538–46.

[11] Robert AM, Kramer RS, Dacey LJ, Charlesworth DC, Leavitt BJ, Helm RE, et al. Cardiac surgery-associated acute kidney injury: a comparison of two consensus criteria. Ann Thorac Surg 2010;90(6):1939–43.

[12] Fujii T, Uchino S, Takinami M, Bellomo R. Validation of the Kidney Disease Improving Global Outcomes criteria for AKI and comparison of three criteria in hospitalized patients. Clin J Am Soc Nephrol 2014;9(5):848–54.

[13] Lopes JA, Fernandes P, Jorge S, Goncalves S, Alvarez A, Costa e Silva Z, et al. Acute kidney injury in intensive care unit patients: a comparison between the RIFLE and the Acute Kidney Injury Network classifications. Crit Care 2008;12(4):R110.

[14] Bagshaw SM, George C, Bellomo R, Committe ADM. A comparison of the RIFLE and AKIN criteria for acute kidney injury in critically ill patients. Nephrol Dial Transplant 2008;23 (5):1569–74.

[15] Haase M, Bellomo R, Matalanis G, Calzavacca P, Dragun D, Haase-Fielitz A. A comparison of the RIFLE and Acute Kidney Injury Network classifications for cardiac surgery-associated acute kidney injury: a prospective cohort study. J Thorac Cardiovasc Surg 2009;138 (6):1370–6.

[16] Section 1: introduction and methodology. Kidney Int Suppl (2011) 2012;2(1):13–8.

[17] Nisula S, Kaukonen KM, Vaara ST, Korhonen AM, Poukkanen M, Karlsson S, et al. Incidence, risk factors and 90-day mortality of patients with acute kidney injury in Finnish intensive care units: the FINNAKI study. Intensive Care Med 2013;39(3):420–8.

[18] Tsai TY, Chien H, Tsai FC, Pan HC, Yang HY, Lee SY, et al. Comparison of RIFLE, AKIN, and KDIGO classifications for assessing prognosis of patients on extracorporeal membrane oxygenation. J Formos Med Assoc 2017;116(11):844–51.

[19] Roy AK, Mc Gorrian C, Treacy C, Kavanaugh E, Brennan A, Mahon NG, et al. A comparison of traditional and novel definitions (RIFLE, AKIN, and KDIGO) of acute kidney injury for the prediction of outcomes in acute decompensated heart failure. Cardiorenal Med 2013;3(1):26–37.

[20] Shemesh O, Golbetz H, Kriss JP, Myers BD. Limitations of creatinine as a filtration marker in glomerulopathic patients. Kidney Int 1985;28(5):830–8.

[21] Pickering JW, Endre ZH. Back-calculating baseline creatinine with MDRD misclassifies acute kidney injury in the intensive care unit. Clin J Am Soc Nephrol 2010;5(7):1165–73.

[22] Pickering JW, Frampton CM, Endre ZH. Evaluation of trial outcomes in acute kidney injury by creatinine modeling. Clin J Am Soc Nephrol 2009;4(11):1705–15.

[23] Waikar SS, Bonventre JV. Creatinine kinetics and the definition of acute kidney injury. J Am Soc Nephrol 2009;20(3):672–9.

[24] Koza Y. Acute kidney injury: current concepts and new insights. J Inj Violence Res 2016;8 (1):58–62.

[25] Maxwell RA, Bell CM. Acute kidney injury in the critically ill. Surg Clin North Am 2017;97 (6):1399–418.

[26] Nada A, Bonachea EM, Askenazi DJ. Acute kidney injury in the fetus and neonate. Semin Fetal Neonatal Med 2017;22(2):90–7.

[27] Ferenbach DA, Bonventre JV. Acute kidney injury and chronic kidney disease: from the laboratory to the clinic. Nephrol Ther 2016;12(Suppl 1):S41–8.

[28] James MT, Hemmelgarn BR, Wiebe N, Pannu N, Manns BJ, Klarenbach SW, et al. Glomerular filtration rate, proteinuria, and the incidence and consequences of acute kidney injury: a cohort study. Lancet 2010;376(9758):2096–103.

[29] Mehta RL, Burdmann EA, Cerda J, Feehally J, Finkelstein F, Garcia-Garcia G, et al. Recognition and management of acute kidney injury in the International Society of Nephrology 0by25 Global Snapshot: a multinational cross-sectional study. Lancet 2016;387(10032):2017–25.

[30] Zappitelli M, Washburn KK, Arikan AA, Loftis L, Ma Q, Devarajan P, et al. Urine neutrophil gelatinase-associated lipocalin is an early marker of acute kidney injury in critically ill children: a prospective cohort study. Crit Care 2007;11(4):R84.

[31] Borregaard N, Cowland JB. Granules of the human neutrophilic polymorphonuclear leukocyte. Blood 1997;89(10):3503–21.

[32] Kuwabara T, Mori K, Mukoyama M, Kasahara M, Yokoi H, Saito Y, et al. Urinary neutrophil gelatinase-associated lipocalin levels reflect damage to glomeruli, proximal tubules, and distal nephrons. Kidney Int 2009;75(3):285–94.

[33] Borregaard N, Cowland JB. Neutrophil gelatinase-associated lipocalin, a siderophore-binding eukaryotic protein. Biometals 2006;19(2):211–5.

[34] Mori K, Lee HT, Rapoport D, Drexler IR, Foster K, Yang J, et al. Endocytic delivery of lipocalin-siderophore-iron complex rescues the kidney from ischemia-reperfusion injury. J Clin Invest 2005;115(3):610–21.

[35] Langelueddecke C, Roussa E, Fenton RA, Wolff NA, Lee WK, Thevenod F. Lipocalin-2 (24p3/neutrophil gelatinase-associated lipocalin (NGAL)) receptor is expressed in distal nephron and mediates protein endocytosis. J Biol Chem 2012;287(1):159–69.

[36] Yang J, Goetz D, Li JY, Wang W, Mori K, Setlik D, et al. An iron delivery pathway mediated by a lipocalin. Mol Cell 2002;10(5):1045–56.

[37] Tong Z, Wu X, Ovcharenko D, Zhu J, Chen CS, Kehrer JP. Neutrophil gelatinase-associated lipocalin as a survival factor. Biochem J 2005;391(Pt 2):441–8.

[38] Mishra J, Ma Q, Prada A, Mitsnefes M, Zahedi K, Yang J, et al. Identification of neutrophil gelatinase-associated lipocalin as a novel early urinary biomarker for ischemic renal injury. J Am Soc Nephrol 2003;14(10):2534–43.

[39] Makris K, Kafkas N. Neutrophil gelatinase-associated lipocalin in acute kidney injury. Adv Clin Chem 2012;58:141–91.

[40] Taghizadeh-Ghehi M, Sarayani A, Ashouri A, Ataei S, Moslehi A, Hadjibabaie M. Urine neutrophil gelatinase associated lipocalin as an early marker of acute kidney injury in hematopoietic stem cell transplantation patients. Ren Fail 2015;37(6):994–8.

[41] Kanchi M, Manjunath R, Massen J, Vincent L, Belani K. Neutrophil gelatinase-associated lipocalin as a biomarker for predicting acute kidney injury during off-pump coronary artery bypass grafting. Ann Card Anaesth 2017;20(3):297–302.

[42] Hunsicker O, Feldheiser A, Weimann A, Liehre D, Sehouli J, Wernecke KD, et al. Diagnostic value of plasma NGAL and intraoperative diuresis for AKI after major gynecological surgery in patients treated within an intraoperative goal-directed hemodynamic algorithm: a substudy of a randomized controlled trial. Medicine (Baltimore) 2017;96(28).

[43] Ribitsch W, Schilcher G, Quehenberger F, Pilz S, Portugaller RH, Truschnig-Wilders M, et al. Neutrophil gelatinase-associated lipocalin (NGAL) fails as an early predictor of contrast induced nephropathy in chronic kidney disease (ANTI-CI-AKI study). Sci Rep 2017;7:41300.

[44] Benzer M, Alpay H, Baykan O, Erdem A, Demir IH. Serum NGAL, cystatin C and urinary NAG measurements for early diagnosis of contrast-induced nephropathy in children. Ren Fail 2016;38(1):27–34.

[45] Padhy M, Kaushik S, Girish MP, Mohapatra S, Shah S, Koner BC. Serum neutrophil gelatinase associated lipocalin (NGAL) and cystatin C as early predictors of contrast-induced acute kidney injury in patients undergoing percutaneous coronary intervention. Clin Chim Acta 2014;435:48–52.

[46] Filiopoulos V, Biblaki D, Vlassopoulos D. Neutrophil gelatinase-associated lipocalin (NGAL): a promising biomarker of contrast-induced nephropathy after computed tomography. Ren Fail 2014;36(6):979–86.

[47] Kafkas N, Liakos C, Zoubouloglou F, Dagadaki O, Dragasis S, Makris K. Neutrophil gelatinase-associated lipocalin as an early marker of contrast-induced nephropathy after elective invasive cardiac procedures. Clin Cardiol 2016;39(8):464–70.

[48] Liebetrau C, Gaede L, Doerr O, Blumenstein J, Rixe J, Teichert O, et al. Neutrophil gelatinase-associated lipocalin (NGAL) for the early detection of contrast-induced nephropathy after percutaneous coronary intervention. Scand J Clin Lab Invest 2014;74(2):81–8.

[49] Mishra J, Dent C, Tarabishi R, Mitsnefes MM, Ma Q, Kelly C, et al. Neutrophil gelatinase-associated lipocalin (NGAL) as a biomarker for acute renal injury after cardiac surgery. Lancet 2005;365(9466):1231–8.

[50] Dent CL, Ma Q, Dastrala S, Bennett M, Mitsnefes MM, Barasch J, et al. Plasma neutrophil gelatinase-associated lipocalin predicts acute kidney injury, morbidity and mortality after pediatric cardiac surgery: a prospective uncontrolled cohort study. Crit Care 2007;11(6): R127.

[51] Tecson KM, Erhardtsen E, Eriksen PM, Gaber AO, Germain M, Golestaneh L, et al. Optimal cut points of plasma and urine neutrophil gelatinase-associated lipocalin for the prediction of acute kidney injury among critically ill adults: retrospective determination and clinical validation of a prospective multicentre study. BMJ Open 2017;7(7).

[52] Wagener G, Jan M, Kim M, Mori K, Barasch JM, Sladen RN, et al. Association between increases in urinary neutrophil gelatinase-associated lipocalin and acute renal dysfunction after adult cardiac surgery. Anesthesiology 2006;105(3):485–91.

[53] Parikh CR, Coca SG, Thiessen-Philbrook H, Shlipak MG, Koyner JL, Wang Z, et al. Postoperative biomarkers predict acute kidney injury and poor outcomes after adult cardiac surgery. J Am Soc Nephrol 2011;22(9):1748–57.

[54] Haase-Fielitz A, Bellomo R, Devarajan P, Bennett M, Story D, Matalanis G, et al. The predictive performance of plasma neutrophil gelatinase-associated lipocalin (NGAL) increases with grade of acute kidney injury. Nephrol Dial Transplant 2009;24(11):3349–54.

[55] Nickolas TL, Schmidt-Ott KM, Canetta P, Forster C, Singer E, Sise M, et al. Diagnostic and prognostic stratification in the emergency department using urinary biomarkers of nephron damage: a multicenter prospective cohort study. J Am Coll Cardiol 2012;59(3):246–55.

[56] de Geus HR, Bakker J, Lesaffre EM, le Noble JL. Neutrophil gelatinase-associated lipocalin at ICU admission predicts for acute kidney injury in adult patients. Am J Respir Crit Care Med 2011;183(7):907–14.

[57] Wheeler DS, Devarajan P, Ma Q, Harmon K, Monaco M, Cvijanovich N, et al. Serum neutrophil gelatinase-associated lipocalin (NGAL) as a marker of acute kidney injury in critically ill children with septic shock. Crit Care Med 2008;36(4):1297–303.

[58] Forni LG, Darmon M, Ostermann M, Oudemans-van Straaten HM, Pettila V, Prowle JR, et al. Renal recovery after acute kidney injury. Intensive Care Med 2017;43(6):855–66.

[59] Gaiao SM, Paiva J. Biomarkers of renal recovery after acute kidney injury. Rev Bras Ter Intensiva 2017;29(3):373–81.

[60] Petrovic S, Bogavac-Stanojevic N, Peco-Antic A, Ivanisevic I, Kotur-Stevuljevic J, Paripovic D, et al. Clinical application neutrophil gelatinase-associated lipocalin and kidney injury molecule-1 as indicators of inflammation persistence and acute kidney injury in children with urinary tract infection. Biomed Res Int 2013;2013:947157.

[61] Rostami Z, Nikpoor M, Einollahi B. Urinary neutrophil gelatinase associated lipocalin (NGAL) for early diagnosis of acute kidney injury in renal transplant recipients. Nephrourol Mon 2013;5(2):745–52.

[62] Shavit L, Manilov R, Wiener-Well Y, Algur N, Slotki I. Urinary neutrophil gelatinase-associated lipocalin for early detection of acute kidney injury in geriatric patients with urinary tract infection treated by colistin. Clin Nephrol 2013;80(6):405–16.

[63] Tsuchimoto A, Shinke H, Uesugi M, Kikuchi M, Hashimoto E, Sato T, et al. Urinary neutrophil gelatinase-associated lipocalin: a useful biomarker for tacrolimus-induced acute kidney injury in liver transplant patients. PLoS ONE 2014;9(10).

[64] Wasilewska A, Zoch-Zwierz W, Taranta-Janusz K, Michaluk-Skutnik J. Neutrophil gelatinase-associated lipocalin (NGAL): a new marker of cyclosporine nephrotoxicity? Pediatr Nephrol 2010;25(5):889–97.

[65] Kashiwagi E, Tonomura Y, Kondo C, Masuno K, Fujisawa K, Tsuchiya N, et al. Involvement of neutrophil gelatinase-associated lipocalin and osteopontin in renal tubular regeneration and interstitial fibrosis after cisplatin-induced renal failure. Exp Toxicol Pathol 2014;66 (7):301–11.

[66] Shukla A, Rai MK, Prasad N, Agarwal V. Short-term non-steroid anti-inflammatory drug use in spondyloarthritis patients induces subclinical acute kidney injury: biomarkers study. Nephron 2017;135(4):277–86.

[67] Halloran PF, Hunsicker LG. Delayed graft function: state of the art, November 10-11, 2000. Summit meeting, Scottsdale, Arizona, USA. Am J Transplant 2001;1(2):115–20.

[68] Hall IE, Yarlagadda SG, Coca SG, Wang Z, Doshi M, Devarajan P, et al. IL-18 and urinary NGAL predict dialysis and graft recovery after kidney transplantation. J Am Soc Nephrol 2010;21(1):189–97.

[69] Yarlagadda SG, Coca SG, Formica Jr RN, Poggio ED, Parikh CR. Association between delayed graft function and allograft and patient survival: a systematic review and meta-analysis. Nephrol Dial Transplant 2009;24(3):1039–47.

[70] Wu WK, Famure O, Li Y, Kim SJ. Delayed graft function and the risk of acute rejection in the modern era of kidney transplantation. Kidney Int 2015;88(4):851–8.

[71] Scolari MP, Cappuccilli ML, Lanci N, La Manna G, Comai G, Persici E, et al. Predictive factors in chronic allograft nephropathy. Transplant Proc 2005;37(6):2482–4.

[72] Pianta TJ, Peake PW, Pickering JW, Kelleher M, Buckley NA, Endre ZH. Clusterin in kidney transplantation: novel biomarkers versus serum creatinine for early prediction of delayed graft function. Transplantation 2015;99(1):171–9.

[73] Fonseca I, Oliveira JC, Almeida M, Cruz M, Malho A, Martins LS, et al. Neutrophil gelatinase-associated lipocalin in kidney transplantation is an early marker of graft dysfunction and is associated with one-year renal function. J Transplant 2013;2013:650123.

[74] Fantel AG, Juchau MR, Burroughs CJ, Person RE. Studies of embryotoxic mechanisms of niridazole: evidence that oxygen depletion plays a role in dysmorphogenicity. Teratology 1989;39(3):243–51.

[75] Hjortrup PB, Haase N, Treschow F, Moller MH, Perner A. Predictive value of NGAL for use of renal replacement therapy in patients with severe sepsis. Acta Anaesthesiol Scand 2015;59 (1):25–34.

[76] Iguchi N, Uchiyama A, Ueta K, Sawa Y, Fujino Y. Neutrophil gelatinase-associated lipocalin and liver-type fatty acid-binding protein as biomarkers for acute kidney injury after organ transplantation. J Anesth 2015;29(2):249–55.

[77] Chakraborty S, Kaur S, Tong Z, Batra SK, Guha S. Neutrophil gelatinase associated lipocalin: Structure, function and role in human pathogenesis. Acute Phase Proteins: inTech; 2011.

[78] Cruz DN, de Cal M, Garzotto F, Perazella MA, Lentini P, Corradi V, et al. Plasma neutrophil gelatinase-associated lipocalin is an early biomarker for acute kidney injury in an adult ICU population. Intensive Care Med 2010;36(3):444–51.

[79] Poniatowski B, Malyszko J, Bachorzewska-Gajewska H, Malyszko JS, Dobrzycki S. Serum neutrophil gelatinase-associated lipocalin as a marker of renal function in patients with chronic heart failure and coronary artery disease. Kidney Blood Press Res 2009;32(2):77–80.

[80] Nielsen BS, Borregaard N, Bundgaard JR, Timshel S, Sehested M, Kjeldsen L. Induction of NGAL synthesis in epithelial cells of human colorectal neoplasia and inflammatory bowel diseases. Gut 1996;38(3):414–20.

[81] Kato K, Sato N, Yamamoto T, Iwasaki YK, Tanaka K, Mizuno K. Valuable markers for contrast-induced nephropathy in patients undergoing cardiac catheterization. Circ J 2008;72(9):1499–505.

[82] Kaseda R, Iino N, Hosojima M, Takeda T, Hosaka K, Kobayashi A, et al. Megalin-mediated endocytosis of cystatin C in proximal tubule cells. Biochem Biophys Res Commun 2007;357(4):1130–4.

[83] Bjornsson TD. Use of serum creatinine concentrations to determine renal function. Clin Pharmacokinet 1979;4(3):200–22.

[84] Leelahavanichkul A, Souza AC, Street JM, Hsu V, Tsuji T, Doi K, et al. Comparison of serum creatinine and serum cystatin C as biomarkers to detect sepsis-induced acute kidney injury and to predict mortality in CD-1 mice. Am J Physiol Ren Physiol 2014;307(8):F939–48.

[85] Sasaki A, Sasaki Y, Iwama R, Shimamura S, Yabe K, Takasuna K, et al. Comparison of renal biomarkers with glomerular filtration rate in susceptibility to the detection of gentamicin-induced acute kidney injury in dogs. J Comp Pathol 2014;151(2–3):264–70.

[86] Togashi Y, Sakaguchi Y, Miyamoto M, Miyamoto Y. Urinary cystatin C as a biomarker for acute kidney injury and its immunohistochemical localization in kidney in the CDDP-treated rats. Exp Toxicol Pathol 2012;64(7–8):797–805.

[87] Ederoth P, Grins E, Dardashti A, Bronden B, Metzsch C, Erdling A, et al. Ciclosporin to Protect Renal function In Cardiac Surgery (CiPRICS): a study protocol for a double-blind, randomised, placebo-controlled, proof-of-concept study. BMJ Open 2016;6(12).

[88] Herget-Rosenthal S, Marggraf G, Husing J, Goring F, Pietruck F, Janssen O, et al. Early detection of acute renal failure by serum cystatin C. Kidney Int 2004;66(3):1115–22.

[89] Nejat M, Pickering JW, Walker RJ, Endre ZH. Rapid detection of acute kidney injury by plasma cystatin C in the intensive care unit. Nephrol Dial Transplant 2010;25(10):3283–9.

[90] Prowle JR, Calzavacca P, Licari E, Ligabo EV, Echeverri JE, Bagshaw SM, et al. Combination of biomarkers for diagnosis of acute kidney injury after cardiopulmonary bypass. Ren Fail 2015;37(3):408–16.

[91] Basile DP, Anderson MD, Sutton TA. Pathophysiology of acute kidney injury. Compr Physiol 2012;2(2):1303–53.

[92] Moran SM, Myers BD. Pathophysiology of protracted acute renal failure in man. J Clin Invest 1985;76(4):1440–8.

[93] Haase VH. Mechanisms of hypoxia responses in renal tissue. J Am Soc Nephrol 2013;24(4):537–41.

[94] Lannemyr L, Bragadottir G, Krumbholz V, Redfors B, Sellgren J, Ricksten SE. Effects of cardiopulmonary bypass on renal perfusion, filtration, and oxygenation in patients undergoing cardiac surgery. Anesthesiology 2017;126(2):205–13.

[95] Kiessling AH, Dietz J, Reyher C, Stock UA, Beiras-Fernandez A, Moritz A. Early postoperative serum cystatin C predicts severe acute kidney injury following cardiac surgery: a post-hoc analysis of a randomized controlled trial. J Cardiothorac Surg 2014;9:10.

[96] Koyner JL, Bennett MR, Worcester EM, Ma Q, Raman J, Jeevanandam V, et al. Urinary cystatin C as an early biomarker of acute kidney injury following adult cardiothoracic surgery. Kidney Int 2008;74(8):1059–69.

[97] Svenmarker S, Haggmark S, Holmgren A, Naslund U. Serum markers are not reliable measures of renal function in conjunction with cardiopulmonary bypass. Interact Cardiovasc Thorac Surg 2011;12(5):713–7.

[98] Schaller Jr RT, Schaller J, Morgan A, Furman EB. Hemodilution anesthesia: a valuable aid to major cancer surgery in children. Am J Surg 1983;146(1):79–84.

[99] Do C. Intravenous contrast: friend or foe? A review on contrast-induced nephropathy. Adv Chronic Kidney Dis 2017;24(3):147–9.

[100] Tepel M, Zidek W. N-Acetylcysteine in nephrology; contrast nephropathy and beyond. Curr Opin Nephrol Hypertens 2004;13(6):649–54.

[101] Brezis M, Rosen S. Hypoxia of the renal medulla—its implications for disease. N Engl J Med 1995;332(10):647–55.

[102] Morcos SK, Brown PW, Oldroyd S, el Nahas AM, Haylor J. Relationship between the diuretic effect of radiocontrast media and their ability to increase renal vascular resistance. Br J Radiol 1995;68(812):850–3.

[103] Wang M, Zhang L, Yue R, You G, Zeng R. Significance of cystatin C for early diagnosis of contrast-induced nephropathy in patients undergoing coronary angiography. Med Sci Monit 2016;22:2956–61.

[104] Shukla AN, Juneja M, Patel H, Shah KH, Konat A, Thakkar BM, et al. Diagnostic accuracy of serum cystatin C for early recognition of contrast induced nephropathy in Western Indians undergoing cardiac catheterization. Indian Heart J 2017;69(3):311–5.

[105] Floege J, Eng E, Young BA, Alpers CE, Barrett TB, Bowen-Pope DF, et al. Infusion of platelet-derived growth factor or basic fibroblast growth factor induces selective glomerular mesangial cell proliferation and matrix accumulation in rats. J Clin Investig 1993;92(6):2952–62.

[106] Gruff ES. 510(k) summary: Summary of safety and effectiveness information supporting a substantially equivalent determination; 2008.

[107] Gorin Y, Block K, Hernandez J, Bhandari B, Wagner B, Barnes JL, et al. Nox4 NAD(P)H oxidase mediates hypertrophy and fibronectin expression in the diabetic kidney. J Biol Chem 2005;280(47):39616–26.

[108] Novick D, Kim S, Kaplanski G, Dinarello CA. Interleukin-18, more than a Th1 cytokine. Semin Immunol 2013;25(6):439–48.

[109] Chang A, Ko K, Clark MR. The emerging role of the inflammasome in kidney diseases. Curr Opin Nephrol Hypertens 2014;23(3):204–10.

[110] Sugawara S, Uehara A, Nochi T, Yamaguchi T, Ueda H, Sugiyama A, et al. Neutrophil proteinase 3-mediated induction of bioactive IL-18 secretion by human oral epithelial cells. J Immunol 2001;167(11):6568–75.

[111] Matsumoto S, Tsuji-Takayama K, Aizawa Y, Koide K, Takeuchi M, Ohta T, et al. Interleukin-18 activates NF-kappaB in murine T helper type 1 cells. Biochem Biophys Res Commun 1997;234(2):454–7.

[112] Robinson D, Shibuya K, Mui A, Zonin F, Murphy E, Sana T, et al. IGIF does not drive Th1 development but synergizes with IL-12 for interferon-gamma production and activates IRAK and NFkappaB. Immunity 1997;7(4):571–81.

[113] Fantuzzi G, Reed DA, Dinarello CA. IL-12-induced IFN-gamma is dependent on caspase-1 processing of the IL-18 precursor. J Clin Invest 1999;104(6):761–7.

[114] Okamura H, Tsutsi H, Komatsu T, Yutsudo M, Hakura A, Tanimoto T, et al. Cloning of a new cytokine that induces IFN-gamma production by T cells. Nature 1995;378(6552):88–91.

[115] Dao T, Ohashi K, Kayano T, Kurimoto M, Okamura H. Interferon-gamma-inducing factor, a novel cytokine, enhances Fas ligand-mediated cytotoxicity of murine T helper 1 cells. Cell Immunol 1996;173(2):230–5.

[116] Mende R, Vincent FB, Kandane-Rathnayake R, Koelmeyer R, Lin E, Chang J, et al. Analysis of serum interleukin (IL)-1beta and IL-18 in systemic lupus erythematosus. Front Immunol 2018;9:1250.

[117] Novick D, Elbirt D, Miller G, Dinarello CA, Rubinstein M, Sthoeger ZM. High circulating levels of free interleukin-18 in patients with active SLE in the presence of elevated levels of interleukin-18 binding protein. J Autoimmun 2010;34(2):121–6.

[118] Zaharieva E, Kamenov Z, Velikova T, Tsakova A, El-Darawish Y, Okamura H. Interleukin-18 serum level is elevated in type 2 diabetes and latent autoimmune diabetes. Endocr Connect 2018;7(1):179–85.

[119] Nowarski R, Jackson R, Gagliani N, de Zoete MR, Palm NW, Bailis W, et al. Epithelial IL-18 equilibrium controls barrier function in colitis. Cell 2015;163(6):1444–56.

[120] Weiss ES, Girard-Guyonvarc'h C, Holzinger D, de Jesus AA, Tariq Z, Picarsic J, et al. Interleukin-18 diagnostically distinguishes and pathogenically promotes human and murine macrophage activation syndrome. Blood 2018;131(13):1442–55.

[121] Wu H, Craft ML, Wang P, Wyburn KR, Chen G, Ma J, et al. IL-18 contributes to renal damage after ischemia-reperfusion. J Am Soc Nephrol 2008;19(12):2331–41.

[122] Edelstein CL, Shi Y, Schrier RW. Role of caspases in hypoxia-induced necrosis of rat renal proximal tubules. J Am Soc Nephrol 1999;10(9):1940–9.

[123] Melnikov VY, Faubel S, Siegmund B, Lucia MS, Ljubanovic D, Edelstein CL. Neutrophil-independent mechanisms of caspase-1- and IL-18-mediated ischemic acute tubular necrosis in mice. J Clin Invest 2002;110(8):1083–91.

[124] Novick D, Kim SH, Fantuzzi G, Reznikov LL, Dinarello CA, Rubinstein M. Interleukin-18 binding protein: a novel modulator of the Th1 cytokine response. Immunity 1999;10 (1):127–36.

[125] Hazle MA, Gajarski RJ, Aiyagari R, Yu S, Abraham A, Donohue J, et al. Urinary biomarkers and renal near-infrared spectroscopy predict intensive care unit outcomes after cardiac surgery in infants younger than 6 months of age. J Thorac Cardiovasc Surg 2013;146 (4):861–7. e1.

[126] Parikh CR, Devarajan P, Zappitelli M, Sint K, Thiessen-Philbrook H, Li S, et al. Postoperative biomarkers predict acute kidney injury and poor outcomes after pediatric cardiac surgery. J Am Soc Nephrol 2011;22(9):1737–47.

[127] Nisula S, Yang R, Poukkanen M, Vaara ST, Kaukonen KM, Tallgren M, et al. Predictive value of urine interleukin-18 in the evolution and outcome of acute kidney injury in critically ill adult patients. Br J Anaesth 2015;114(3):460–8.

[128] Xin C, Yulong X, Yu C, Changchun C, Feng Z, Xinwei M. Urine neutrophil gelatinase-associated lipocalin and interleukin-18 predict acute kidney injury after cardiac surgery. Ren Fail 2008;30(9):904–13.

[129] Parikh CR, Mishra J, Thiessen-Philbrook H, Dursun B, Ma Q, Kelly C, et al. Urinary IL-18 is an early predictive biomarker of acute kidney injury after cardiac surgery. Kidney Int 2006;70 (1):199–203.

[130] Krawczeski CD, Goldstein SL, Woo JG, Wang Y, Piyaphanee N, Ma Q, et al. Temporal relationship and predictive value of urinary acute kidney injury biomarkers after pediatric cardiopulmonary bypass. J Am Coll Cardiol 2011;58(22):2301–9.

[131] Haase M, Bellomo R, Story D, Davenport P, Haase-Fielitz A. Urinary interleukin-18 does not predict acute kidney injury after adult cardiac surgery: a prospective observational cohort study. Crit Care 2008;12(4):R96.

[132] Liangos O, Tighiouart H, Perianayagam MC, Kolyada A, Han WK, Wald R, et al. Comparative analysis of urinary biomarkers for early detection of acute kidney injury following cardiopulmonary bypass. Biomarkers 2009;14(6):423–31.

[133] Liang XL, Liu SX, Chen YH, Yan LJ, Li H, Xuan HJ, et al. Combination of urinary kidney injury molecule-1 and interleukin-18 as early biomarker for the diagnosis and progressive assessment of acute kidney injury following cardiopulmonary bypass surgery: a prospective nested case-control study. Biomarkers 2010;15(4):332–9.

[134] Ghosh Choudhury G, Jin DC, Kim Y, Celeste A, Ghosh-Choudhury N, Abboud HE. Bone morphogenetic protein-2 inhibits MAPK-dependent Elk-1 transactivation and DNA synthesis induced by EGF in mesangial cells. Biochem Biophys Res Commun 1999;258(2):490–6.

[135] Ghosh Choudhury G, Kim YS, Simon M, Wozney J, Harris S, Ghosh-Choudhury N, et al. Bone morphogenetic protein 2 inhibits platelet-derived growth factor-induced c-fos gene transcription and DNA synthesis in mesangial cells. Involvement of mitogen-activated protein kinase. J Biol Chem 1999;274(16):10897–902.

[136] Human total IL-18/IL-1F4 quantikine ELISA kit. Available from: https://www.rndsystems.com/products/human-total-il-18-il-1f4-quantikine-elisa-kit_dl180.

[137] Ichimura T, Asseldonk EJ, Humphreys BD, Gunaratnam L, Duffield JS, Bonventre JV. Kidney injury molecule-1 is a phosphatidylserine receptor that confers a phagocytic phenotype on epithelial cells. J Clin Invest 2008;118(5):1657–68.

[138] Bailly V, Zhang Z, Meier W, Cate R, Sanicola M, Bonventre JV. Shedding of kidney injury molecule-1, a putative adhesion protein involved in renal regeneration. J Biol Chem 2002;277(42):39739–48.

[139] van Timmeren MM, van den Heuvel MC, Bailly V, Bakker SJ, van Goor H, Stegeman CA. Tubular kidney injury molecule-1 (KIM-1) in human renal disease. J Pathol 2007;212(2):209–17.

[140] Zhang Z, Humphreys BD, Bonventre JV. Shedding of the urinary biomarker kidney injury molecule-1 (KIM-1) is regulated by MAP kinases and juxtamembrane region. J Am Soc Nephrol 2007;18(10):2704–14.

[141] Bonventre JV. Kidney injury molecule-1 (KIM-1): a urinary biomarker and much more. Nephrol Dial Transplant 2009;24(11):3265–8.

[142] Amin RP, Vickers AE, Sistare F, Thompson KL, Roman RJ, Lawton M, et al. Identification of putative gene based markers of renal toxicity. Environ Health Perspect 2004;112(4):465–79.

[143] Rhodes NJ, Prozialeck WC, Lodise TP, Venkatesan N, O'Donnell JN, Pais G, et al. Evaluation of vancomycin exposures associated with elevations in novel urinary biomarkers of acute kidney injury in vancomycin-treated rats. Antimicrob Agents Chemother 2016;60(10):5742–51.

[144] Vaidya VS, Ozer JS, Dieterle F, Collings FB, Ramirez V, Troth S, et al. Kidney injury molecule-1 outperforms traditional biomarkers of kidney injury in preclinical biomarker qualification studies. Nat Biotechnol 2010;28(5):478–85.

[145] Fuchs TC, Frick K, Emde B, Czasch S, von Landenberg F, Hewitt P. Evaluation of novel acute urinary rat kidney toxicity biomarker for subacute toxicity studies in preclinical trials. Toxicol Pathol 2012;40(7):1031–48.

[146] Daggulli M, Utangac MM, Dede O, Bodakci MN, Hatipoglu NK, Penbegul N, et al. Potential biomarkers for the early detection of acute kidney injury after percutaneous nephrolithotripsy. Ren Fail 2016;38(1):151–6.

[147] Kooiman J, van de Peppel WR, Sijpkens YW, Brulez HF, de Vries PM, Nicolaie MA, et al. No increase in kidney injury molecule-1 and neutrophil gelatinase-associated lipocalin excretion following intravenous contrast enhanced-CT. Eur Radiol 2015;25(7):1926–34.

[148] Jorgensen AL. Contrast-induced nephropathy: pathophysiology and preventive strategies. Crit Care Nurse 2013;33(1):37–46.

[149] Luk L, Steinman J, Newhouse JH. Intravenous contrast-induced nephropathy-the rise and fall of a threatening idea. Adv Chronic Kidney Dis 2017;24(3):169–75.

[150] Han WK, Wagener G, Zhu Y, Wang S, Lee HT. Urinary biomarkers in the early detection of acute kidney injury after cardiac surgery. Clin J Am Soc Nephrol 2009;4(5):873–82.

[151] Tu Y, Wang H, Sun R, Ni Y, Ma L, Xv F, et al. Urinary netrin-1 and KIM-1 as early biomarkers for septic acute kidney injury. Ren Fail 2014;36(10):1559–63.

[152] Torregrosa I, Montoliu C, Urios A, Andres-Costa MJ, Gimenez-Garzo C, Juan I, et al. Urinary KIM-1, NGAL and L-FABP for the diagnosis of AKI in patients with acute coronary syndrome or heart failure undergoing coronary angiography. Heart Vessel 2015;30(6):703–11.

[153] Han WK, Waikar SS, Johnson A, Betensky RA, Dent CL, Devarajan P, et al. Urinary biomarkers in the early diagnosis of acute kidney injury. Kidney Int 2008;73(7):863–9.

[154] Arthur JM, Hill EG, Alge JL, Lewis EC, Neely BA, Janech MG, et al. Evaluation of 32 urine biomarkers to predict the progression of acute kidney injury after cardiac surgery. Kidney Int 2014;85(2):431–8.

[155] Parikh CR, Thiessen-Philbrook H, Garg AX, Kadiyala D, Shlipak MG, Koyner JL, et al. Performance of kidney injury molecule-1 and liver fatty acid-binding protein and combined biomarkers of AKI after cardiac surgery. Clin J Am Soc Nephrol 2013;8(7):1079–88.

[156] Shah P, Riphagen S, Beyene J, Perlman M. Multiorgan dysfunction in infants with post-asphyxial hypoxic-ischaemic encephalopathy. Arch Dis Child Fetal Neonatal Ed 2004;89 (2):F152–5.

[157] Sarafidis K, Tsepkentzi E, Agakidou E, Diamanti E, Taparkou A, Soubasi V, et al. Serum and urine acute kidney injury biomarkers in asphyxiated neonates. Pediatr Nephrol 2012;27 (9):1575–82.

[158] Cao XY, Zhang HR, Zhang W, Chen B. Diagnostic values of urinary netrin-1 and kidney injury molecule-1 for acute kidney injury induced by neonatal asphyxia. Zhongguo Dang Dai Er Ke Za Zhi 2016;18(1):24–8.

[159] Seibert FS, Sitz M, Passfall J, Haesner M, Laschinski P, Buhl M, et al. Prognostic value of urinary calprotectin, NGAL and KIM-1 in chronic kidney disease. Kidney Blood Press Res 2018;43(4):1255–62.

[160] Alderson HV, Ritchie JP, Pagano S, Middleton RJ, Pruijm M, Vuilleumier N, et al. The associations of blood kidney injury molecule-1 and neutrophil gelatinase-associated lipocalin with progression from CKD to ESRD. Clin J Am Soc Nephrol 2016;11(12):2141–9.

[161] Sabbisetti VS, Waikar SS, Antoine DJ, Smiles A, Wang C, Ravisankar A, et al. Blood kidney injury molecule-1 is a biomarker of acute and chronic kidney injury and predicts progression to ESRD in type I diabetes. J Am Soc Nephrol 2014;25(10):2177–86.

[162] Witzgall R, Brown D, Schwarz C, Bonventre JV. Localization of proliferating cell nuclear antigen, vimentin, c-Fos, and clusterin in the postischemic kidney. Evidence for a heterogenous genetic response among nephron segments, and a large pool of mitotically active and dedifferentiated cells. J Clin Invest 1994;93(5):2175–88.

[163] Yang QH, Liu DW, Long Y, Liu HZ, Chai WZ, Wang XT. Acute renal failure during sepsis: potential role of cell cycle regulation. J Infect 2009;58(6):459–64.

[164] Boonstra J, Post JA. Molecular events associated with reactive oxygen species and cell cycle progression in mammalian cells. Gene 2004;337:1–13.

[165] Vijayan A, Faubel S, Askenazi DJ, Cerda J, Fissell WH, Heung M, et al. Clinical use of the urine biomarker [TIMP-2] × [IGFBP7] for acute kidney injury risk assessment. Am J Kidney Dis 2016;68(1):19–28.

[166] Hwa V, Oh Y, Rosenfeld RG. The insulin-like growth factor-binding protein (IGFBP) superfamily. Endocr Rev 1999;20(6):761–87.

[167] Kashani K, Al-Khafaji A, Ardiles T, Artigas A, Bagshaw SM, Bell M, et al. Discovery and validation of cell cycle arrest biomarkers in human acute kidney injury. Crit Care 2013;17(1):R25.

[168] Degeorges A, Wang F, Frierson Jr HF, Seth A, Sikes RA. Distribution of IGFBP-rP1 in normal human tissues. J Histochem Cytochem 2000;48(6):747–54.

[169] Li HL, Yan Z, Ke ZP, Tian XF, Zhong LL, Lin YT, et al. IGFBP2 is a potential biomarker in acute kidney injury (AKI) and resveratrol-loaded nanoparticles prevent AKI. Oncotarget 2018;9 (93):36551–60.

[170] Ruan W, Zhu S, Wang H, Xu F, Deng H, Ma Y, et al. IGFBP-rP1, a potential molecule associated with colon cancer differentiation. Mol Cancer 2010;9:281.

[171] Meersch M, Schmidt C, Van Aken H, Martens S, Rossaint J, Singbartl K, et al. Urinary TIMP-2 and IGFBP7 as early biomarkers of acute kidney injury and renal recovery following cardiac surgery. PLoS ONE 2014;9(3).

[172] Gocze I, Koch M, Renner P, Zeman F, Graf BM, Dahlke MH, et al. Urinary biomarkers TIMP-2 and IGFBP7 early predict acute kidney injury after major surgery. PLoS ONE 2015;10(3).

[173] Koyner JL, Shaw AD, Chawla LS, Hoste EA, Bihorac A, Kashani K, et al. Tissue inhibitor metalloproteinase-2 (TIMP-2)IGF-binding protein-7 (IGFBP7) levels are associated with adverse long-term outcomes in patients with AKI. J Am Soc Nephrol 2015;26(7):1747–54.

[174] Aregger F, Uehlinger DE, Witowski J, Brunisholz RA, Hunziker P, Frey FJ, et al. Identification of IGFBP-7 by urinary proteomics as a novel prognostic marker in early acute kidney injury. Kidney Int 2014;85(4):909–19.

[175] Bihorac A, Chawla LS, Shaw AD, Al-Khafaji A, Davison DL, Demuth GE, et al. Validation of cell-cycle arrest biomarkers for acute kidney injury using clinical adjudication. Am J Respir Crit Care Med 2014;189(8):932–9.

[176] Gocze I, Jauch D, Gotz M, Kennedy P, Jung B, Zeman F, et al. Biomarker-guided intervention to prevent acute kidney injury after major surgery: the prospective randomized BigpAK study. Ann Surg 2018;267(6):1013–20.

[177] Perico L, Perico N, Benigni A. The incessant search for renal biomarkers: is it really justified? Curr Opin Nephrol Hypertens 2019;28(2):195–202.

CHAPTER 4

Biomarkers in diabetic kidney disease

Parisa Mortaji[a] and Brent Wagner[b,c,d]
[a]*Department of Medicine, University of Colorado, Aurora, CO, United States,* [b]*University of New Mexico Health Science Center, Albuquerque, NM, United States,* [c]*New Mexico Veterans Administration Health Care System, Albuquerque, NM, United States,* [d]*Kidney Institute of New Mexico, Albuquerque, NM, United States*

Type 1 and 2 diabetes mellitus are *multifactorial* chronic diseases involving both genetic and environmental factors (for a review, see Refs. [1, 2]). Type 1 diabetes mellitus is typically considered an autoimmune disease with an unclear pathogenesis affecting 10%–15% of all patients with diabetes worldwide. It is thought that continued exposure of β-cells to pancreatic β-cell autoantigens leads to the production of islet-targeting autoantibodies such as the 65-kDa glutamic acid decarboxylase antibody, insulinoma-associated protein 2 antibody, and zinc transporter 8 antibody [3–5]. These autoantibodies appear months or years before symptom onset. This activates β-cells and dendritic cells, which subsequently present the autoantigens to T cells, leading to T-cell-mediated pancreatic β-cell destruction. Resultant insulin deficiency and concomitant hyperglycemia then promote associated symptoms. Onset of type 1 diabetes typically occurs in childhood or adolescence, with a peak incidence between 12 and 14 years of age.

The main identified genetic risk factor for type 1 diabetes is the human leukocyte antigen (HLA) class II haplotypes: HLA-DR and HLA-DQ [6–8]. Genetic risk factors alone are not sufficient to cause disease. Proposed environmental triggers include viral infections, gestational events, and age at introduction of food [9–17].

Type 1 diabetes mellitus can be classified into stages [1]. One such system has two stages: presymptomatic, in which there are decreased numbers of pancreatic β-cells without symptoms; and symptomatic, in which evidence of hyperglycemia manifests clinically. Another system classifies type 1 diabetes into three stages. Stage 1 is defined by the presence of autoantibodies and the absence of hyperglycemia. Stage 2 is defined by concomitancy of both autoantibodies

Seema S. Ahuja and Brian Castillo: Kidney Biomarkers. https://doi.org/10.1016/B978-0-12-815923-1.00004-3

and hyperglycemia. Stage 3 is the development of overt symptomatic type 1 diabetes mellitus.

Type 2 diabetes mellitus is the most common type of diabetes, accounting for 85%–90% of global cases [2]. Similar to type 1 diabetes, type 2 diabetes is also influenced by both genetic and environmental factors. It tends to cluster in families and thus is considered heritable, although precise genetic markers pertaining to the pathogenesis have yet to be identified. The single most important factor contributing to development of type 2 diabetes is obesity, and the risk is markedly increased with a body mass index of greater than or equal to 30 kg/m^2. Other environmental risk factors include poor diet, physical inactivity, sedentary lifestyle, and cigarette smoking. In fact, up to 90% of cases of type 2 diabetes are considered preventable through a healthy diet, a body mass index less than or equal to 25 kg/m^2, exercising a minimum of 30 min/day, avoidance of cigarette smoking, and moderate alcohol consumption.

The pathogenesis of type 2 diabetes involves at least eight identified pathways; however, the most important are the interplay between insulin resistance and secretion [18, 19]. Increased peripheral insulin resistance precedes the onset of disease by many years. The pancreatic β-cells initially compensate by increasing insulin secretion; however, over time, β-cell function becomes progressively impaired. Additionally, the mass of the β-cells also decreases, likely through apoptosis and dysregulated autophagy. Overt type 2 diabetes mellitus occurs when β-cell insulin secretion is insufficient to offset peripheral insulin resistance.

The microvascular complications of both type 1 and type 2 diabetes mellitus have been attributed mainly to hyperglycemia and directly related to the severity and duration of elevated glucose levels. These include retinopathy, neuropathy, and nephropathy [20–27]. The cells in these tissues are unable to effectively regulate glucose uptake during hyperglycemia leading to increased intracellular glucose levels and subsequent cellular damage resulting from multiple factors, including oxidative stress. Macrovascular complications include increased risk of coronary heart disease, cerebrovascular disease, and peripheral artery disease.

Diabetic kidney disease

Diabetes is one of the most common chronic diseases and is responsible for approximately 50% of end-stage renal disease cases [28]. End-stage renal disease is costly and devastating, demarcating a need for renal replacement therapy either through dialysis or renal transplantation. Diabetic kidney disease is a microvascular complication that develops in approximately one-third of patients with type 1 diabetes and half of patients with type 2 diabetes [29, 30]. Given the heavy disease burden, screening for renal involvement is recommended annually within 5 years after diagnosis of type 1 diabetes and at the time of diagnosis of type 2 diabetes [31]. The definition of a biomarker—i.e., biological marker—is an objective, accurately measurable, and reproducible sign that predict onset of disease or an outcome of interest [32]. The standard test to screen for diabetic kidney disease is urine albumin, with disease progression correlated with an increase in urinary albumin excretion rate through normo- to ("macro-") albuminuria [33].

Diabetic kidney disease is termed diabetic nephropathy after tissue biopsy confirms the diagnosis [34]. The pathological staging of diabetic nephropathy is divided into four classes. Class I is defined by thickening of the glomerular basement membrane. Class IIa represents mild mesangial expansion involving >25% of the mesangium. Class IIb represents severe mesangial expansion involving >25% of the mesangium. Class III is defined by Kimmelstiel Wilson[a] nodules and diffuse glomerulosclerosis in <50% of the glomeruli. Class IV, the final pathological stage, is characterized by advanced diabetic glomerulosclerosis, with >50% glomerular involvement (see Table 1).

Diabetes causes damage to the kidney via multiple mechanisms, including hyperglycemia, glycosylation end-products, inflammation, and microalbuminuria, all of which lead to fibrosis and ultimately renal failure [35, 36]. The regions of the nephron that are particularly susceptible to damage include the glomeruli and the tubules. Thus, biomarkers of kidney damage reflecting either early glomerular or tubular dysfunction would be valuable for monitoring the progression of diabetic kidney disease [37, 38].

The clinical stages of disease progression were classically based on the diagnostic criteria of chronic kidney disease: normoalbuminuria (<30 mg/day), "microalbuminuria" (30–300 mg/day), "macroalbuminuria" (>300 mg/day—also defined as overt proteinuria [39]), a compromise in the glomerular filtration rate, with progression to the need for renal replacement, respectively [40]. Two out of three consecutive urinary albumin samples, ideally at least 3 months apart, must be abnormal for diagnosis [40, 41]. The urinary albumin to creatinine ratio has served as a convenient test for measuring albumin with many favoring a random morning spot sample [41, 42]. Stage 2 represents an early clinically identifiable landmark characterized by "microalbuminuria" and an increase in the (true) glomerular filtration rate, indicating hyperfiltration and glomerular damage [43, 44].

Chronic kidney disease, in practice, is often attributed to diabetes based on a patient's history: the duration of diabetes, the presence of other microvascular complications such as retinopathy, and the degree of control of hyperglycemia

---

[a]Dr. Paul Kimmelstein (1900–1970) was not married to Dr. Clifford Wilson (1906–1997); therefore, one of the authors—BW—prefers to avoid hyphenating their names.

Table 1 Pathological staging of diabetic nephropathy

Class	Description
I	Glomerular basement membrane thickening only
IIa	Mild mesangial expansion (in >25% of the mesangium)
IIb	Severe mesangial expansion (in >25% of the mesangium)
III	Nodular sclerosis (Kimmelstiel Wilson nodules and <50% diffuse glomerulosclerosis)
IV	Advanced diabetic glomerulosclerosis (>50% diffuse glomerulosclerosis)

Adapted from Thomas MC, Brownlee M, Susztak K, Sharma K, Jandeleit-Dahm KA, Zoungas S, et al. Diabetic kidney disease. Nat Rev Dis Primers 2015;1:15018; Tervaert TW, Mooyaart AL, Amann K, Cohen AH, Cook HT, Drachenberg CB, et al. Pathologic classification of diabetic nephropathy. J Am Soc Nephrol 2010;21(4):556–63.

[34, 45]. Hence, findings of *both* albuminuria and retinopathy do favor a diagnosis of diabetic kidney disease [41, 42]. Therefore, retinopathy can be labeled as an important (albeit late) clinical biomarker for a patient's likelihood of having diabetic nephropathy.

Urine albumin

"Microalbuminuria" is generally defined as urinary excretion of 30–300 mg of albumin/day (National Kidney Foundation, 2012).[b] The pathophysiology of albuminuria in relation to diabetes is a correlate with the degree of glomerular damage. Diabetes, in the early stages, causes both glomerular hyperfiltration and damage to all three layers of the glomerular basement membrane: the endothelium, the basement membrane, and the podocyte layer [33, 43]. The culmination of these structural and biochemical processes—increased pore size and loss of barrier charge—contributes to the development of "microalbuminuria." Albuminuria can even act as a stressor for renal proximal tubule epithelial cells, culminating in tubulointerstitial inflammation and progressive renal damage [46, 47].

Albuminuria is commonly used as a predictor of diabetic kidney disease progression in either type 1 or type 2 diabetes mellitus [48–50]. Given that this can vacillate in the former, and is regarded as a relatively late manifestation of kidney damage in the latter, there is a frontier for better biomarkers. Typically, significant renal damage has already occurred by the time albuminuria is detected [51]. An ideal biomarker would predict progression of any kidney disease at the earliest stage where there is an opportunity to impact patient outcomes. Approximately 10%–30% of patients with either type 1 or type 2 diabetes develop renal impairment *prior to the onset of albuminuria,* a phenomenon

[b]Note that this is a generalization as the rates of urinary albumin and creatinine are dependent on sex.

termed normoalbuminuric diabetic kidney disease [52, 53]. These patients are predominantly elderly, female, and have a lower prevalence of comorbid retinopathy [54, 55]. In another subset of patients, "microalbuminuria" resolves with time, making it an overall poor predictor of progression [56, 57]. Furthermore, albuminuria is not specific to diabetic glomerulopathy but is present in a variety of renal and nonglomerular diseases, including urinary tract infections and periods of hemodynamic stress (such as sepsis, congestive heart failure, and exercise) [58–60].

It has been suggested that tubular damage may play an equally important role in the pathophysiology of diabetic kidney disease perhaps preceding the development of glomerular damage [61, 62]. Again, urine albumin is not positioned to be *the* optimal biomarker for the early detection of diabetic kidney disease. Improving the sensitivity and specificity of the tests we currently have has the potential to greatly impact the outcomes for millions. The discovery of such biomarkers may permit the categorization of either tubular or glomerular disease with applications to either research or clinical purposes. Other potential benefits of these undiscovered biomarkers may include the prediction of the natural history of disease, detection at earlier stages, or avenues for targeted therapeutic interventions (while possibly decreasing risks of toxicity or the mis-application of impotent therapies).

Urine protein

n addition to albumin, the increase of other urinary proteins has been shown to predominate in early diabetic kidney disease, often interpreted as the evolution of the glomerular pathology. These proteins may be associated with renal damage, such as alpha-1 microglobulin, cystatin C, transferrin, nephrin, metalloproteinase-9, immunoglobulin G (IgG), and beta-2 macroglobulin [63–67]. At lower levels of proteinuria, the overall contribution of albumin can be extremely variable [68].

In nondiabetic patients, nonalbumin proteins have been considered suitable markers for tubular rather than glomerular damage. Kim et al. demonstrated a similar association in a study of 118 patients with type 2 diabetes with *estimated* (not true) glomerular filtration rates of at least 60 mL/min/1.73 m^2 [69]. Nonalbuminuric proteinuria was associated with several urinary tubular damage markers, again demonstrating that tubular damage is a consequence of diabetic kidney disease. A urinary protein panel may be more sensitive overall compared with each individual protein alone.

Low-molecular-weight urinary proteomics with capillary electrophoresis-coupled mass spectrometry is compelling for clinical use. Urine samples were taken from type 1 ($n = 16$) and type 2 diabetic patients ($n = 19$) and the proteome of each subject was profiled according to a previously generated chronic kidney disease classifier of 273 urinary proteins [70]. Patients with type 1 diabetes had 66/273 significant biomarkers at an early stage, as compared to 85/273 biomarkers for type 2 diabetic patients. Urine albumin was also measured, and the study analyzed the ability of the classifier to predict progression of normoalbuminuric patients to "macroalbuminuria." It was found that urinary proteomics, specifically elevated levels of collagen fragments, predicted overt proteinuria 3–5 years before its onset, and predicted microalbuminuria 1.5 years before its onset. Thus, collagen fragments may be useful as an early biomarker of diabetic kidney disease to assess for risk of progression to "macroalbuminuria."

Plasma creatinine[c]

Plasma creatinine has long been used as a standard for measuring both acute and chronic renal impairment as it roughly and inversely correlates with the true glomerular filtration rate [71, 72]. Creatinine has been (and continues to be) ideally suited as an index of renal function because it is not protein bound, freely filtered, not metabolized by the kidney, and physiologically inactive. However, creatinine has limitations for gauging renal function or (*true*) glomerular filtration rate for many reasons. The magnitude of rise of plasma creatinine is a rather late marker of renal disease, and it is clearly nonspecific in terms of revealing etiology of kidney demise [73]. Plasma creatinine can reflect many variables, including age, sex, ancestry, nutritional status, diet, muscle mass, amputation status, volume depletion, and use of drugs that impact either glomerular hemodynamics or the tubular trafficking [71, 72]. Creatinine is freely filtered by the kidney and also secreted and reabsorbed by renal tubules to a variable and unpredictable degree. In studies of healthy subjects, repeated measurements of serum creatinine do show some variability. Therefore, plasma creatinine has inherent imprecision as a correlate of the glomerular filtration rate and many factors impact its concentration [74]. While (*true*) glomerular filtration rate itself would obviously be a great estimator of the degree of damage due to diabetic kidney disease, by the time severe renal insufficiency is detected, the likelihood of treatment success is limited given the chronicity of the structural and functional damage that has already occurred. There's room for a more specific biomarker that could signal early disease (if it's out there).

[c]At the hospitals where we work, the laboratories often run the test off green-top, heparinized tubes. The laboratory workers who answered the phone on the day of this writing, then, get full blame for error.

Cystatin C

Cystatin C is a cysteine protease inhibitor [75–77], produced by nearly all nucleated cells in the body, and is completely filtered through the glomerulus. In the health state, serum cystatin C concentrations should be negligible [78, 79].

Serum and urine cystatin C were measured in 119 patients with type 2 diabetes and 50 disease-free controls [80]. *Estimated* (not *true*) glomerular filtration rates were used for statistical comparison. Serum cystatin C was higher in the selected patients with type 2 diabetes (1.05 ± 0.34 mg/L) as compared with controls (0.63 ± 0.14 mg/L). In another study, elevated cystatin C levels were used to diagnose diabetic kidney disease in normoalbuminuric patients with diabetes as compared to controls [81, 82]. In this study, serum cystatin C cutoff was determined to be 1.26 mg/L, with a sensitivity of 88.2% and a specificity of 84.8%. Overall, it was concluded that serum cystatin C had appeal as an earlier biomarker of diabetic kidney disease with good sensitivity and specificity. A consecutive cohort study was conducted to evaluate the ability of urine cystatin C to predict microalbuminuria [83]. A total of 146 participants, 42 with diabetic kidney disease and 142 with nondiabetic kidney disease, were enrolled. The ability of urine cystatin C to predict microalbuminuria was determined to have a sensitivity of 71.2% and a specificity of 80.0% at the cutoff value of 0.415 ng/mL.

Neutrophil gelatinase-associated lipocalin (NGAL)

Neutrophil gelatinase-associated lipocalin is a glycoprotein in the lipocalin superfamily that is a commonly studied novel biomarker for renal disease [75, 84, 85]. Neutrophil gelatinase-associated lipocalin is highly expressed in a variety of cell types during periods of neutrophilic or renal tubular cellular damage [86, 87]. Elevated serum levels can correlate with structural tissue damage. Neutrophil gelatinase-associated lipocalin is freely filtered through the glomerulus and nearly completely reabsorbed in the proximal tubule [75, 84, 85]. During periods of renal tubular injury induced by inflammation, ischemia, or hyperglycemia, there is decreased reabsorption of neutrophil gelatinase-associated lipocalin in the proximal tubule leading to increased urinary excretion, thus validating its potential use as a biomarker of renal damage [88].

Urinary neutrophil gelatinase-associated lipocalin is similarly as predictive of microalbuminuria as cystatin C in patients with a history of type 2 diabetes mellitus (over a minimum study duration of 6 years) [83]. The ability of urinary neutrophil gelatinase-associated lipocalin to predict microalbuminuria was determined to have a sensitivity of 65.2% and a specificity of 93.7% at the cutoff value of 75.5 ng/mL. Increased neutrophil gelatinase-associated lipocalin has been correlated with the severity of albuminuria in diabetic kidney disease patients [89]. Therefore, both levels of urinary neutrophil gelatinase-associated lipocalin and urinary cystatin C may be elevated in type 2 diabetic patients with diabetic kidney disease and normal urinary albumin/creatinine ratio of less than 30 mg/g.

Clusterin

Clusterin is a glycoprotein with many biological functions. It has been shown to play a role in sperm maturation, lipid transport, and tissue remodeling [90]. In the kidneys, clusterin functions as a defense factor to prevent renal fibrosis, with levels reflecting proximal tubular damage. Urine clusterin is predictive of "microalbuminuria" with a sensitivity of 69.7% and specificity of 86.2% at the cut-off value of 568.5 ng/mL [83]. Clusterin is upregulated by the transforming growth factor-β signaling pathway in renal tubular epithelial cells after an injurious insult, therefore may lack specificity for diabetic kidney disease.

Transforming growth factor-β1 and bone morphogenetic protein-7

Transforming growth factor-β1 and bone morphogenetic protein-7 are secretory cytokines in the transforming growth factor-β superfamily [91]. In diabetic kidney disease as well as other renal pathologies, transforming growth factor-β1 is a key cytokine promoting fibrosis and has been shown to be elevated in patients with diabetes mellitus [92, 93]. Bone morphogenetic protein-7 is an antagonist to transforming growth factor-β 1 with antifibrotic and antiinflammatory properties [94–98], and its levels were found to be decreased in diabetic kidney disease [99]. When exogenously administered in animal models of diabetes mellitus, it delayed progression of renal disease [100, 101].

Serum transforming growth factor-β 1 was found to be substantially increased in 102 patients with type 2 diabetes compared to 179 controls, with bone morphogenetic protein-7 levels decreased in patients with type 2 diabetes mellitus who subsequently progressed to major renal disease compared to matched controls [102]. These novel biomarkers were shown to predict renal disease at a much higher rate than albuminuria and estimated (*not* true) glomerular filtration rate. In addition, they had additive properties such that testing for both transforming growth factor-β1 and bone morphogenetic protein-7 led to further improvement in detection of risk for diabetic kidney disease. The fact that these two biomarkers represented decent predictors for diabetic kidney disease supports what is now well known regarding the pathogenesis of diabetic kidney disease—i.e., renal fibrosis and vascular sclerosis are important histological predictors of the progression of renal disease.

Urine immunoglobulin G, ceruloplasmin, and transferrin

The role of urinary immunoglobulin G (IgG), ceruloplasmin, and transferrin has been examined as potential early biomarkers for diabetic kidney disease [62]. Patients with type 2 diabetes and normoalbuminuria were recruited ($n = 140$). Overnight urine collections were obtained on three different days and urinary IgG, ceruloplasmin, and transferrin was measured. The protein-to-creatinine ratio for each of these was calculated. At 5 years of follow up, 117 patients remained in the study, of which 17 had progressed to "microalbuminuria." Urine IgG-to-creatinine ratios, ceruloplasmin-to-creatinine ratios, and transferrin-to-creatinine ratios were all significantly elevated in 17 patients who had progressed to "microalbuminuria" compared to those who did not. It was concluded that these proteins can predict progression to diabetic kidney disease at an earlier stage and with higher sensitivity when compared to albumin.

Immunoglobulin G4 and Smad1

Smad1 is a transcription factor involved in the production of type IV collagen that functions downstream of the transforming growth factor-β superfamily proteins [103]. Collagen is a major constituent of mesangial matrix expansion during progression of diabetic nephropathy; therefore, it was hypothesized that Smad1 levels may be increased and represent a potential biomarker for diabetic kidney disease progression. In animal studies, early increases in urinary Smad1 correlated with the development of mesangial matrix expansion, *but not with albuminuria* [104].

Immunoglobulin G4 is a negatively charged protein, a property that typically hinders its excretion across the glomerular capillary. However, under conditions of glomerular basement membrane compromise, the charge barrier is disrupted, leading to increased urinary excretion [105, 106]. Urinary Smad1 and urinary immunoglobulin G4 were examined in 554 patients with type 2 diabetes mellitus, assessed via enzyme-linked immunosorbent assays [107]. There was a *mild* (yet statistically significant) association between increased baseline albuminuria (albumin normalized to creatinine) and higher baseline urinary Smad1 (*also* normalized to urine creatinine). Urinary immunoglobulin G4 concentrations were found to positively correlate with pathological changes involving the surface density of the peripheral glomerular basement membrane, whereas Smad1 was positively correlated with the degree of mesangial expansion. Increased urinary immunoglobulin G4 was also associated with a lower estimated (*not* true) glomerular filtration rate. Increases in both biomarkers significantly predicted later decline of *estimated* glomerular filtration rate in the subjects who did not have underlying "macroalbuminuria." Based on the histopathological correlation and this finding, it was concluded that immunoglobulin G4 and Smad1 reflect early diabetic nephropathy.

Adiponectin

Adiponectin is a protein that regulates energy homeostasis, insulin sensitivity, and acts as an antiinflammatory and antiatherogenic agent [108]. It is highly expressed in adipose cells and abundant in plasma, although plasma levels are decreased in patients with obesity and related diagnoses, such as type 2 diabetes mellitus [109–111]. Conversely, patients with type 1 diabetes mellitus have increased plasma adiponectin levels [112–116]. Positive associations between elevated serum adiponectin, albuminuria, and reduced *estimated* glomerular filtration rates in patients with type 1 diabetes mellitus have been reported [117].

Fibroblast growth factor 23

Fibroblast growth factor 23, a very important phosphaturic protein, has been shown to play a role in mineral metabolism in chronic kidney disease. Fibroblast growth factor 23 was assessed in 56 patients with type 2 diabetes mellitus and "macroalbuminuric" diabetic kidney disease in Brazil [118]. Serum fibroblast growth factor 23 was associated with serum creatinine and proteinuria and an independent predictor of death, doubling of baseline serum creatinine, and/or dialysis need, even after adjusting for creatinine clearance. Fibroblast growth factor 23 may relate to the risk of diabetic kidney disease progression in patients with "macroalbuminuria."

Retinol-binding protein

Retinol-binding protein is a low-molecular-weight protein in the lipocalin superfamily; it is freely filtered at the glomerulus and nearly completely reabsorbed in the proximal tubule [119, 120]. Urinary retinol-binding protein has been previously identified as a biomarker of proximal tubular damage and may have some utility for the diagnosis of several renal diseases. Increased levels of both serum and urine retinol binding protein have been found in patients afflicted with diabetes [121, 122].

Ratios of urine retinol binding protein to urine creatinine, urine neutrophil gelatinase-associated lipocalin to urine creatinine, urine tumor necrosis factor-α to urine creatinine, and urine interleukin-18 to urine creatinine were measured in 293 patients with type 2 diabetes mellitus [89]. Groups were characterized as normoalbuminuria, "microalbuminuria," and "macroalbuminuria." The levels of urine retinol-binding protein: urine creatinine, urine tumor necrosis factor-α: urine creatinine, and urine interleukin-18: urine creatinine were markedly increased in the albuminuric groups. The urine neutrophil gelatinase-associated lipocalin: urine creatinine level was significantly higher in the type 2 diabetes patients with "microalbuminuria" than in control subjects. It was concluded that urinary retinol-binding protein, neutrophil gelatinase-associated lipocalin, tumor necrosis factor-α and interleukin-18 were significantly elevated in the "microalbuminuria" and "macroalbuminuria" groups relative to those with normoalbuminuria. This suggests the potential utility of these as novel biomarkers for progression of diabetic kidney disease.

Vascular endothelial growth factor

Vascular endothelial growth factor has been implicated in the pathogenesis of diabetic kidney disease, as it is likely produced by both glomerular podocytes and tubular epithelial cells during periods of renal injury [123–125]. Vascular endothelial growth factor levels have been shown to be significantly high in many studies of animals with diabetic nephropathy [123]. Patients with type 2 diabetes ($n=107$) were compared to healthy controls ($n=47$) and plasma and urinary vascular endothelial growth factor and soluble vascular endothelial growth factor receptor (sFLT-1) levels assessed [126]. Subjects were divided into four groups based on urinary albumin-to-creatinine ratio. Urinary and plasma vascular endothelial growth factor and sFLT-1 levels were measured by quantitative enzyme-linked immunosorbent assay. Urinary vascular endothelial growth factor was significantly higher in the diabetic groups than the control group, with progressively higher levels with increasing urinary albumin excretion (overt proteinuria > "microalbuminuria" > normoalbuminuria). Urinary sFLT-1 levels were also elevated in the "microalbuminuria" and overt proteinuria groups, although *this increase did not correlate with the degree of albuminuria*. This study demonstrated that urinary vascular endothelial growth factor increased with progression of diabetic kidney disease and strongly correlated with urinary albumin-to-creatinine ratio.

Nephrin

Nephrin is a transmembrane protein component of the slit diaphragm of glomerular podocytes. In one study, 15 biopsies from patients with type 2 diabetic kidney disease were obtained [127]. Fourteen were found to have moderate to severe nodular diabetic glomerulosclerosis, and one subject had diffuse diabetic glomerulosclerosis. Researchers also collected a one-time random urine albumin to creatinine ratio and categorized diabetic patients into subgroups of normoalbuminuria, "microalbuminuria," or "macroalbuminuria." Expression of podocyte specific proteins nephrin, synaptopodin, and podocin in renal biopsies of patients with type 2 diabetes was decreased.

Urinary nephrin was subsequently measured via enzyme-linked immunosorbent assay in 66 diabetic patients with various stages of chronic kidney disease and albuminuria compared to 10 healthy controls. Urine nephrin-to-creatinine ratio was significantly different between the diabetic and control group. Nephrinuria correlated significantly with "microalbuminuria" and "macroalbuminuria," but not normoalbuminuria (although nephrinuria was also detected in 54% of diabetic patients with normoalbuminuria). Nephrinuria was detected in early disease (i.e., normoalbuminuria) and increased with overt proteinuria, affirming its potential use as an early biomarker of diabetic kidney disease.

Glycosaminoglycans

Glycosaminoglycans are components of the extracellular matrix and cellular plasma membranes. Given that compromised glomerular basement membrane permeability in diabetic kidney disease leads to proteinuria, one study involved 86 diabetic patients, 33 with type 1 diabetes and 53 with type 2 diabetes, as well as 30 healthy controls [128]. Total urinary glycosaminoglycan concentrations, which included chondroitin, dermatan, and heparan sulfate, were measured through 24-h urine collections. Albumin was also measured, and diabetic patients were subsequently classified into the normoalbuminuria, "microalbuminuria," or "macroalbuminuria" groups.

All patients with diabetes excreted significantly more glycosaminoglycan than the control group, independent of duration of disease. However, patients with underlying diabetic kidney disease (defined as the presence of either "micro-" or "macro-" albuminuria) had significantly higher levels of glycosaminoglycan excretion, and this positively correlated with duration of diabetic kidney disease, as compared to diabetic patients with intact renal function. In summary, patients with both type 1 and type 2 diabetes, as well as "microalbuminuria" or "macroalbuminuria," have increased urinary glycosaminoglycan excretion, suggesting its role as a potential biomarker.

Tumor necrosis factor-α

Tumor necrosis factor-α is a proinflammatory cytokine. In type 2 diabetes, there is the presence of both a chronic low-grade inflammation and innate immune system activation, which likely underlie much of the pathogenesis [129–133]. Plasma tumor necrosis factor-α can be free or bound to circulating tumor necrosis factor receptors 1 or 2. The tumor necrosis factor pathway markers thus constitute free tumor necrosis factor-α, total tumor necrosis factor-α, tumor necrosis factor receptor 1, and tumor necrosis factor receptor 2.

Correlation exists between elevated markers of the tumor necrosis factor pathway and development of diabetic nephropathy and end-stage renal disease [134]. Patients with type 2 diabetes mellitus ($n=410$) were followed over 12 years and assessed for diabetic nephropathy and end-stage renal disease. Elevated plasma tumor necrosis factor receptor 1 and tumor necrosis factor receptor 2 were significantly higher in the group that developed end-stage renal disease compared to those who did not. This association was present regardless of the presence or absence of "microalbuminuria." Of the tumor necrosis factor pathway markers tested, tumor necrosis factor receptor 1 was the best overall marker for prediction of time to end-stage renal disease. A single measurement of tumor necrosis factor receptor 1 may be able to predict the 12-year risk of end-stage renal disease in type 2 diabetic patients, regardless of the presence or absence of proteinuria. Similarly, in a cross-sectional study of 314 patients with type 2 diabetes and normoalbuminuria elevated levels of circulating tumor necrosis factor-α and tumor necrosis factor receptors were associated with reduced renal function [135].

Podocalyxin

Podocytes are one of the three layers of the glomerular basement membrane. The function and structure of this epithelial layer are clearly compromised in diabetic kidney disease [136]. Urinary podocalyxin is an extracellular podocyte-specific protein. Urinary podocalyxin was measured via quantitative enzyme-linked immunosorbent assay in a study involving 142 participants with glomerular diseases, 71 patients with type 2 diabetes, and 69 healthy controls [137]. Levels of urinary podocalyxin were significantly higher in subjects with type 2 diabetes, regardless of proteinuria, as compared with healthy controls, with 53.8% of normoalbuminuric diabetic patients having urinary podocalyxin values above the cutoff (normal controls had urinary podocalyxin levels of 7.1 ± 0.5 ng/μmol creatinine). Urinary podocalyxin positively correlated with glycated hemoglobin $A1c_c$ and urinary albumin, but no such correlation was observed with respect to creatinine levels and *estimated* (i.e., not *true*) glomerular filtration rate. However, creatinine and *estimated* glomerular filtration rate are typically abnormal in advanced stages of diabetic kidney disease when functional nephrons are no longer present. If podocalyxin represents an earlier biomarker of diabetic nephropathy—as evidenced by its presence in normoalbuminuric patients—it is logical that there such correlation is lost as disease progresses (i.e., as creatinine increases reflecting the decline of the *true* glomerular filtration rate). In summary, urinary podocalyxin may be a useful early biomarker to detect diabetic kidney disease, as podocyte injury certainly precedes development of microalbuminuria.

Wilm's tumor-1

Wilm's tumor-1 is a transcription factor, requisite for nephrogenesis (particularly podocyte development), and can be shed by renal epithelial cells [138, 139]. Urinary Wilm's tumor-1 has been associated with podocyte damage and tubulointerstitial fibrosis in past studies [140, 141]. Podocytopenia is a known structural consequence of diabetic kidney disease, mechanistically resulting in proteinuria [136, 142]. Thus, urinary Wilm's tumor 1 levels were evaluated as a potential early biomarker of podocyte damage, an early pathological finding of diabetic kidney disease. Wilm's tumor-1 was found in the urines of 33 out of 48 patients with type 1 diabetes, but in only 1 healthy control in another study [143]. Wilm's tumor-1 levels were positive in all patients with proteinuria, and half of patients without proteinuria. In addition, levels were significantly higher in patients with proteinuria compared to those without. Wilm's tumor-1 also correlated positively with urine protein-to-creatinine ratio, elevated serum creatinine, and reduced *estimated* glomerular filtration rates. In conclusion, Wilm's tumor-1 may be useful as an early biomarker of diabetic kidney disease, as it also provides insight into the stage of functional damage.

Microribonucleic acids

Microribonucleic acids (miRNAs) are naturally occurring noncoding RNAs that regulate gene expression by altering mRNA expression [144–147]. They have been extensively studied as potential noninvasive biomarkers for diabetic kidney disease. Out of 40 type 1 diabetic patients, 10 subjects who never developed renal disease were matched with 10 subjects who developed overt diabetic nephropathy in an analysis of urinary miRNA profiles [148]. In addition, 10 individuals with intermittent albuminuria (i.e., a subsequent reversion to normoalbuminuria) were matched with 10 subjects who had persistent "microalbuminuria" (the authors noting the persistence of albuminuria at follow-up visits). The patients were followed biennially for 10 years and again at 18 years. Depending on the stage of diabetic kidney disease, 27 miRNAs were identified at significantly different levels in subjects, concluding that urinary miRNA profiling differs based on the stage of diabetic kidney disease.

Differential expression of a urinary miRNA panel was measured in 40 patients with type 2 diabetes mellitus and 12 controls using PCR array [149]. Exosomal miR15-b, miR-34a, and miR-636 were the most significantly expressed miRNAs. These three miRNAs were then measured in urine samples of 136 patients with type 2 diabetes mellitus (90 with albuminuria, 46 without) and 44 healthy controls using quantitative real-time PCR to validate their expression. Cutoff values used for miR-15b, miR-34a, and miR-636 were 1.84, 0.697, and 1.045, respectively. Sensitivities for miR-15b, miR-34a, and miR-636 were 97.8%, 93.3, and 97.8%, respectively, and specificities were 82.2%, 86.7%, and 93.3%, respectively, for detection of type 2 diabetic kidney disease. In addition, normoalbuminuric patients tested positive for miR-15b, miR-34a, and miR-636 at 37%, 13%, and 13%, respectively.

The diagnostic value of the 3-miRNA panel rather than each individual miRNA reached a sensitivity of 100% in diagnosing diabetic kidney disease.[d] These three miRNAs that exhibited significant expression are important factors in the fatty acid metabolism, glucose homeostasis, *and* renal fibrosis. Critically, this urinary miRNA panel was positive in an average of 28.3% of normoalbuminuric patients, suggesting a potential use as an earlier and sensitive biomarker of diabetic kidney disease.

Conclusions

Both diabetes mellitus types 1 and 2 are leading causes of chronic kidney disease, including end-stage renal disease. The lesions are characterized by an increase in extracellular matrix in the glomeruli, mesangial cell proliferation, and interstitial fibrosis. These conditions correspond with a turnover of extracellular matrix, such as collagen type IV. Rational biomarkers for the early detection of diabetic kidney disease include the mechanistic mediators (e.g., transforming growth factor β1, bone morphogenic protein-7, Smad1, adiponectin, fibroblast growth factor 23, vascular endothelial growth factor, tumor necrosis factor-α) and the pathophysiologic sequelae (plasma cystatin C, neutrophil gelatinase-associated lipocalin or urinary clusterin, immunoglobin G, retinol-binding protein, nephrin, glycosaminoglycans, podocalyxin, or Wilm's tumor 1). Perhaps the early detection of the onset of renal disease will permit the clinician to implement kidney-sparing therapies and retard the progression to end-stage renal disease.

---

[d]The reader should be reminded that the pulse of a patient afflicted with diabetes also has a 100% sensitivity for the presence of diabetic kidney disease.

References

[1] Katsarou A, Gudbjornsdottir S, Rawshani A, Dabelea D, Bonifacio E, Anderson BJ, et al. Type 1 diabetes mellitus. Nat Rev Dis Primers 2017;3.

[2] DeFronzo RA, Ferrannini E, Groop L, Henry RR, Herman WH, Holst JJ, et al. Type 2 diabetes mellitus. Nat Rev Dis Primers 2015;1.

[3] Ziegler AG, Hummel M, Schenker M, Bonifacio E. Autoantibody appearance and risk for development of childhood diabetes in offspring of parents with type 1 diabetes: the 2-year analysis of the German BABYDIAB Study. Diabetes 1999;48(3):460–8.

[4] Ilonen J, Hammais A, Laine AP, Lempainen J, Vaarala O, Veijola R, et al. Patterns of beta-cell autoantibody appearance and genetic associations during the first years of life. Diabetes 2013;62(10):3636–40.

[5] Krischer JP, Lynch KF, Schatz DA, Ilonen J, Lernmark A, Hagopian WA, et al. The 6 year incidence of diabetes-associated autoantibodies in genetically at-risk children: the TEDDY study. Diabetologia 2015;58(5):980–7.

[6] Nerup J, Platz P, Andersen OO, Christy M, Lyngsoe J, Poulsen JE, et al. HL-A antigens and diabetes mellitus. Lancet 1974;2(7885):864–6.

[7] Singal DP, Blajchman MA. Histocompatibility (HL-A) antigens, lymphocytotoxic antibodies and tissue antibodies in patients with diabetes mellitus. Diabetes 1973;22(6):429–32.

[8] Cudworth AG, Woodrow JC. Evidence for HL-A-linked genes in "juvenile" diabetes mellitus. Br Med J 1975;3(5976):133–5.

[9] Beyerlein A, Donnachie E, Jergens S, Ziegler AG. Infections in early life and development of type 1 diabetes. JAMA 2016;315(17):1899–901.

[10] Ashton MP, Eugster A, Walther D, Daehling N, Riethausen S, Kuehn D, et al. Incomplete immune response to coxsackie B viruses associates with early autoimmunity against insulin. Sci Rep 2016;6.

[11] Hyoty H. Viruses in type 1 diabetes. Pediatr Diabetes 2016;17(Suppl. 22):56–64.

[12] Knip M, Virtanen SM, Akerblom HK. Infant feeding and the risk of type 1 diabetes. Am J Clin Nutr 2010;91(5):1506S–13S.

[13] La Torre D, Seppanen-Laakso T, Larsson HE, Hyotylainen T, Ivarsson SA, Lernmark A, et al. Decreased cord-blood phospholipids in young age-at-onset type 1 diabetes. Diabetes 2013;62(11):3951–6.

[14] Oresic M, Gopalacharyulu P, Mykkanen J, Lietzen N, Makinen M, Nygren H, et al. Cord serum lipidome in prediction of islet autoimmunity and type 1 diabetes. Diabetes 2013;62(9):3268–74.

[15] Lynch KF, Lernmark B, Merlo J, Cilio CM, Ivarsson SA, Lernmark A, et al. Cord blood islet autoantibodies and seasonal association with the type 1 diabetes high-risk genotype. J Perinatol 2008;28(3):211–7.

[16] Resic Lindehammer S, Honkanen H, Nix WA, Oikarinen M, Lynch KF, Jonsson I, et al. Seroconversion to islet autoantibodies after enterovirus infection in early pregnancy. Viral Immunol 2012;25(4):254–61.

[17] Viskari HR, Roivainen M, Reunanen A, Pitkaniemi J, Sadeharju K, Koskela P, et al. Maternal first-trimester enterovirus infection and future risk of type 1 diabetes in the exposed fetus. Diabetes 2002;51(8):2568–71.

[18] Lecture DRAB. From the triumvirate to the ominous octet: a new paradigm for the treatment of type 2 diabetes mellitus. Diabetes 2009;58(4):773–95.

[19] DeFronzo RA. Insulin resistance, lipotoxicity, type 2 diabetes and atherosclerosis: the missing links. The Claude Bernard Lecture 2009. Diabetologia 2010;53(7):1270–87.

[20] de Ferranti SD, de Boer IH, Fonseca V, Fox CS, Golden SH, Lavie CJ, et al. Type 1 diabetes mellitus and cardiovascular disease: a scientific statement from the American Heart Association and American Diabetes Association. Circulation 2014;130(13):1110–30.

[21] Anderzen J, Samuelsson U, Gudbjornsdottir S, Hanberger L, Akesson K. Teenagers with poor metabolic control already have a higher risk of microvascular complications as young adults. J Diabetes Complications 2016;30(3):533–6.

[22] Raile K, Galler A, Hofer S, Herbst A, Dunstheimer D, Busch P, et al. Diabetic nephropathy in 27,805 children, adolescents, and adults with type 1 diabetes: effect of diabetes duration, A1C, hypertension, dyslipidemia, diabetes onset, and sex. Diabetes Care 2007;30(10):2523–8.

[23] Gross JL, de Azevedo MJ, Silveiro SP, Canani LH, Caramori ML, Zelmanovitz T. Diabetic nephropathy: diagnosis, prevention, and treatment. Diabetes Care 2005;28(1):164–76.

[24] Voulgari C, Psallas M, Kokkinos A, Argiana V, Katsilambros N, Tentolouris N. The association between cardiac autonomic neuropathy with metabolic and other factors in subjects with type 1 and type 2 diabetes. J Diabetes Complications 2011;25(3):159–67.

[25] Inzucchi SE, Bergenstal RM, Buse JB, Diamant M, Ferrannini E, Nauck M, et al. Management of hyperglycemia in type 2 diabetes: a patient-centered approach: position statement of the American Diabetes Association (ADA) and the European Association for the Study of Diabetes (EASD). Diabetes Care 2012;35(6):1364–79.

[26] Handelsman Y, Bloomgarden ZT, Grunberger G, Umpierrez G, Zimmerman RS, Bailey TS, et al. American association of clinical endocrinologists and american college of endocrinology—clinical practice guidelines for developing a diabetes mellitus comprehensive care plan—2015. Endocr Pract 2015;21(Suppl. 1):1–87.

[27] Pozzilli P, Leslie RD, Chan J, De Fronzo R, Monnier L, Raz I, et al. The A1C and ABCD of glycaemia management in type 2 diabetes: a physician's personalized approach. Diabetes Metab Res Rev 2010;26(4):239–44.

[28] Saran R, Li Y, Robinson B, Abbott KC, Agodoa LY, Ayanian J, et al. US Renal Data System 2015 Annual Data Report: epidemiology of kidney disease in the United States. Am J Kidney Dis 2016;67(3 Suppl. 1):Svii. S1–305.

[29] Thomas MC, Weekes AJ, Broadley OJ, Cooper ME, Mathew TH. The burden of chronic kidney disease in Australian patients with type 2 diabetes (the NEFRON study). Med J Aust 2006; 185(3):140–4.

[30] Dwyer JP, Parving HH, Hunsicker LG, Ravid M, Remuzzi G, Lewis JB. Renal dysfunction in the presence of normoalbuminuria in type 2 diabetes: results from the DEMAND Study. Cardiorenal Med 2012;2(1):1–10.

[31] Whaley-Connell AT, Vassalotti JA, Collins AJ, Chen SC, McCullough PA. National Kidney Foundation's Kidney Early Evaluation Program (KEEP) annual data report 2011: executive summary. Am J Kidney Dis 2012;59(3 Suppl. 2):S1–4.

[32] Strimbu K, Tavel JA. What are biomarkers? Curr Opin HIV AIDS 2010;5(6):463–6.

[33] Parving HH, Persson F, Rossing P. Microalbuminuria: a parameter that has changed diabetes care. Diabetes Res Clin Pract 2015;107(1):1–8.

[34] Thomas MC, Brownlee M, Susztak K, Sharma K, Jandeleit-Dahm KA, Zoungas S, et al. Diabetic kidney disease. Nat Rev Dis Primers 2015;1.

[35] Sun YM, Su Y, Li J, Wang LF. Recent advances in understanding the biochemical and molecular mechanism of diabetic nephropathy. Biochem Biophys Res Commun 2013;433(4):359–61.

[36] Arora MK, Singh UK. Molecular mechanisms in the pathogenesis of diabetic nephropathy: an update. Vascul Pharmacol 2013;58(4):259–71.

[37] Hong CY, Chia KS, Ling SL. Urinary protein excretion in type 2 diabetes with complications. J Diabetes Complications 2000;14(5):259–65.

[38] Weitgasser R, Schnoell F, Gappmayer B, Kartnig I. Prospective evaluation of urinary *N*-acetyl-beta-D-glucosaminidase with respect to macrovascular disease in elderly type 2 diabetic patients. Diabetes Care 1999;22(11):1882–6.

[39] Warram JH, Gearin G, Laffel L, Krolewski AS. Effect of duration of type I diabetes on the prevalence of stages of diabetic nephropathy defined by urinary albumin/creatinine ratio. J Am Soc Nephrol 1996;7(6):930–7.

[40] National Kidney Foundation. KDOQI Clinical Practice Guideline for Diabetes and CKD: 2012 update. Am J Kidney Dis 2012;60(5):850–86.

[41] Tuttle KR, Bakris GL, Bilous RW, Chiang JL, de Boer IH, Goldstein-Fuchs J, et al. Diabetic kidney disease: a report from an ADA Consensus Conference. Diabetes Care 2014; 37(10):2864–83.

[42] Standards of Medical Care in Diabetes-2016: summary of revisions. Diabetes Care 2016;-39 Suppl. 1:S4–5.

[43] Zeni L, Norden AGW, Cancarini G, Unwin RJ. A more tubulocentric view of diabetic kidney disease. J Nephrol 2017;30(6):701–17.

[44] Haneda M, Utsunomiya K, Koya D, Babazono T, Moriya T, Makino H, et al. Classification of diabetic nephropathy 2014. Nihon Jinzo Gakkai Shi 2014;56(5):547–52.

[45] Alicic RZ, Rooney MT, Tuttle KR. Diabetic kidney disease: challenges, progress, and possibilities. Clin J Am Soc Nephrol 2017;12(12):2032–45.

[46] Gilbert RE, Cooper ME. The tubulointerstitium in progressive diabetic kidney disease: more than an aftermath of glomerular injury? Kidney Int 1999;56(5):1627–37.

[47] Tang SC, Lai KN. The pathogenic role of the renal proximal tubular cell in diabetic nephropathy. Nephrol Dial Transplant 2012;27(8):3049–56.

[48] Perkins BA, Ficociello LH, Roshan B, Warram JH, Krolewski AS. In patients with type 1 diabetes and new-onset microalbuminuria the development of advanced chronic kidney disease may not require progression to proteinuria. Kidney Int 2010;77(1):57–64.

[49] Adler AI, Stevens RJ, Manley SE, Bilous RW, Cull CA, Holman RR, et al. Development and progression of nephropathy in type 2 diabetes: the United Kingdom Prospective Diabetes Study (UKPDS 64). Kidney Int 2003;63(1):225–32.

[50] Lee SY, Choi ME. Urinary biomarkers for early diabetic nephropathy: beyond albuminuria. Pediatr Nephrol 2015;30(7):1063–75.

[51] Barratt J, Topham P. Urine proteomics: the present and future of measuring urinary protein components in disease. CMAJ 2007;177(4):361–8.

[52] Perkins BA, Ficociello LH, Ostrander BE, Silva KH, Weinberg J, Warram JH, et al. Microalbuminuria and the risk for early progressive renal function decline in type 1 diabetes. J Am Soc Nephrol 2007;18(4):1353–61.

[53] Krolewski AS, Niewczas MA, Skupien J, Gohda T, Smiles A, Eckfeldt JH, et al. Early progressive renal decline precedes the onset of microalbuminuria and its progression to macroalbuminuria. Diabetes Care 2014;37(1):226–34.

[54] Chen C, Wang C, Hu C, Han Y, Zhao L, Zhu X, et al. Normoalbuminuric diabetic kidney disease. Front Med 2017;11(3):310–8.

[55] Yokoyama H, Sone H, Oishi M, Kawai K, Fukumoto Y, Kobayashi M, et al. Prevalence of albuminuria and renal insufficiency and associated clinical factors in type 2 diabetes: the Japan Diabetes Clinical Data Management study (JDDM15). Nephrol Dial Transplant 2009; 24(4):1212–9.

[56] Zachwieja J, Soltysiak J, Fichna P, Lipkowska K, Stankiewicz W, Skowronska B, et al. Normal-range albuminuria does not exclude nephropathy in diabetic children. Pediatr Nephrol 2010;25(8):1445–51.

[57] Fioretto P, Steffes MW, Mauer M. Glomerular structure in nonproteinuric IDDM patients with various levels of albuminuria. Diabetes 1994;43(11):1358–64.

[58] Stamm WE, Hooton TM. Management of urinary tract infections in adults. N Engl J Med 1993;329(18):1328–34.

[59] Bellinghieri G, Savica V, Santoro D. Renal alterations during exercise. J Ren Nutr 2008;18(1): 158–64.

[60] Haffner SM, Stern MP, Gruber MK, Hazuda HP, Mitchell BD, Patterson JK. Microalbuminuria. Potential marker for increased cardiovascular risk factors in nondiabetic subjects? Arteriosclerosis 1990;10(5):727–31.

[61] Kotajima N, Kimura T, Kanda T, Obata K, Kuwabara A, Fukumura Y, et al. Type IV collagen as an early marker for diabetic nephropathy in non-insulin-dependent diabetes mellitus. J Diabetes Complications 2000;14(1):13–7.

[62] Narita T, Hosoba M, Kakei M, Ito S. Increased urinary excretions of immunoglobulin g, ceruloplasmin, and transferrin predict development of microalbuminuria in patients with type 2 diabetes. Diabetes Care 2006;29(1):142–4.

[63] Halimi JM, Matthias B, Al-Najjar A, Laouad I, Chatelet V, Marliere JF, et al. Respective predictive role of urinary albumin excretion and nonalbumin proteinuria on graft loss and death in renal transplant recipients. Am J Transplant 2007;7(12):2775–81.

[64] Cieciura T, Urbanowicz A, Perkowska-Ptasinska A, Nowacka-Cieciura E, Tronina O, Majchrzak J, et al. Tubular and glomerular proteinuria in diagnosing chronic allograft nephropathy with relevance to the degree of urinary albumin excretion. Transplant Proc 2005;37(2):987–90.

[65] Patari A, Forsblom C, Havana M, Taipale H, Groop PH, Holthofer H. Nephrinuria in diabetic nephropathy of type 1 diabetes. Diabetes 2003;52(12):2969–74.

[66] Bernard AM, Amor AA, Goemaere-Vanneste J, Antoine JL, Lauwerys RR, Lambert A, et al. Microtransferrinuria is a more sensitive indicator of early glomerular damage in diabetes than microalbuminuria. Clin Chem 1988;34(9):1920–1.

[67] Langham RG, Kelly DJ, Cox AJ, Thomson NM, Holthofer H, Zaoui P, et al. Proteinuria and the expression of the podocyte slit diaphragm protein, nephrin, in diabetic nephropathy: effects of angiotensin converting enzyme inhibition. Diabetologia 2002;45(11):1572–6.

[68] Ballantyne FC, Gibbons J, O'Reilly DS. Urine albumin should replace total protein for the assessment of glomerular proteinuria. Ann Clin Biochem 1993;30(Pt 1):101–3.

[69] Kim SS, Song SH, Kim IJ, Kim WJ, Jeon YK, Kim BH, et al. Nonalbuminuric proteinuria as a biomarker for tubular damage in early development of nephropathy with type 2 diabetic patients. Diabetes Metab Res Rev 2014;30(8):736–41.

[70] Zurbig P, Jerums G, Hovind P, Macisaac RJ, Mischak H, Nielsen SE, et al. Urinary proteomics for early diagnosis in diabetic nephropathy. Diabetes 2012;61(12):3304–13.

[71] Slocum JL, Heung M, Pennathur S. Marking renal injury: can we move beyond serum creatinine? Transl Res 2012;159(4):277–89.

[72] Perrone RD, Madias NE, Levey AS. Serum creatinine as an index of renal function: new insights into old concepts. Clin Chem 1992;38(10):1933–53.

[73] Doi K, Yuen PS, Eisner C, Hu X, Leelahavanichkul A, Schnermann J, et al. Reduced production of creatinine limits its use as marker of kidney injury in sepsis. J Am Soc Nephrol 2009;20(6): 1217–21.

[74] Inker LA, Schmid CH, Tighiouart H, Eckfeldt JH, Feldman HI, Greene T, et al. Estimating glomerular filtration rate from serum creatinine and cystatin C. N Engl J Med 2012;367 (1):20–9.

[75] Alter ML, Kretschmer A, Von Websky K, Tsuprykov O, Reichetzeder C, Simon A, et al. Early urinary and plasma biomarkers for experimental diabetic nephropathy. Clin Lab 2012; 58(7-8):659–71.

[76] Conti M, Moutereau S, Zater M, Lallali K, Durrbach A, Manivet P, et al. Urinary cystatin C as a specific marker of tubular dysfunction. Clin Chem Lab Med 2006;44(3):288–91.

[77] Dieterle F, Perentes E, Cordier A, Roth DR, Verdes P, Grenet O, et al. Urinary clusterin, cystatin C, beta2-microglobulin and total protein as markers to detect drug-induced kidney injury. Nat Biotechnol 2010;28(5):463–9.

[78] Avinash S, Singh VP, Agarwal AK, Chatterjee S, Identification AV. Stratification of diabetic kidney disease using serum cystatin C and serum creatinine based estimating equations in type 2 diabetes: a comparative analysis. J Assoc Physicians India 2015;63(11):28–35.

[79] Javanmardi M, Azadi NA, Amini S, Abdi M. Diagnostic value of cystatin C for diagnosis of early renal damages in type 2 diabetic mellitus patients: the first experience in Iran. J Res Med Sci 2015;20(6):571–6.

[80] Qamar A, Hayat A, Ahmad TM, Khan A, Hasnat MNU, Tahir S. Serum cystatin C as an early diagnostic biomarker of diabetic kidney disease in type 2 diabetic patients. J Coll Physicians Surg Pak 2018;28(4):288–91.

[81] El-Kafrawy NA, Shohaib AA, El-Deen SMK, El Barbary H, Seleem AS. Evaluation of serum cystatin C as an indicator of early renal function decline in type 2 diabetes. Menoufia Med J 2014;27(1):60–5.

[82] Singla K, Sodhi K, Pandey R, Singh J, Sharma P. The utility of serum cystatin C in the diagnosis of early diabetic nephropathy. J Pharm Biomed Sci 2014;4:84–7.

[83] Zeng XF, Lu DX, Li JM, Tan Y, Li Z, Zhou L, et al. Performance of urinary neutrophil gelatinase-associated lipocalin, clusterin, and cystatin C in predicting diabetic kidney disease and diabetic microalbuminuria: a consecutive cohort study. BMC Nephrol 2017;18(1).

[84] Zeng XF, Li JM, Tan Y, Wang ZF, He Y, Chang J, et al. Performance of urinary NGAL and L-FABP in predicting acute kidney injury and subsequent renal recovery: a cohort study based on major surgeries. Clin Chem Lab Med 2014;52(5):671–8.

[85] Haase M, Bellomo R, Devarajan P, Schlattmann P, Haase-Fielitz A, Group NM-aI. Accuracy of neutrophil gelatinase-associated lipocalin (NGAL) in diagnosis and prognosis in acute kidney injury: a systematic review and meta-analysis. Am J Kidney Dis 2009;54(6):1012–24.

[86] Mori K, Nakao K. Neutrophil gelatinase-associated lipocalin as the real-time indicator of active kidney damage. Kidney Int 2007;71(10):967–70.

[87] Bennett M, Dent CL, Ma Q, Dastrala S, Grenier F, Workman R, et al. Urine NGAL predicts severity of acute kidney injury after cardiac surgery: a prospective study. Clin J Am Soc Nephrol 2008;3(3):665–73.

[88] Wang JY, Yang JH, Xu J, Jia JY, Zhang XR, Yue XD, et al. Renal tubular damage may contribute more to acute hyperglycemia induced kidney injury in non-diabetic conscious rats. J Diabetes Complications 2015;29(5):621–8.

[89] Wu J, Shao X, Lu K, Zhou J, Ren M, Xie X, et al. Urinary RBP and NGAL levels are associated with nephropathy in patients with type 2 diabetes. Cell Physiol Biochem 2017;42(2):594–602.

[90] Khan Z, Pandey M. Role of kidney biomarkers of chronic kidney disease: an update. Saudi J Biol Sci 2014;21(4):294–9.

[91] Allendorph GP, Vale WW, Choe S. Structure of the ternary signaling complex of a TGF-beta superfamily member. Proc Natl Acad Sci U S A 2006;103(20):7643–8.

[92] Sharma K, Ziyadeh FN, Alzahabi B, McGowan TA, Kapoor S, Kurnik BR, et al. Increased renal production of transforming growth factor-beta1 in patients with type II diabetes. Diabetes 1997;46(5):854–9.

[93] Hellmich B, Schellner M, Schatz H, Pfeiffer A. Activation of transforming growth factor-beta1 in diabetic kidney disease. Metabolism 2000;49(3):353–9.

[94] Zeisberg M. Bone morphogenic protein-7 and the kidney: current concepts and open questions. Nephrol Dial Transplant 2006;21(3):568–73.

[95] Gould SE, Day M, Jones SS, Dorai H. BMP-7 regulates chemokine, cytokine, and hemodynamic gene expression in proximal tubule cells. Kidney Int 2002;61(1):51–60.

[96] Wang S, Hirschberg R. BMP7 antagonizes TGF-beta-dependent fibrogenesis in mesangial cells. Am J Physiol Renal Physiol 2003;284(5):F1006–13.

[97] Zeisberg M, Bottiglio C, Kumar N, Maeshima Y, Strutz F, Muller GA, et al. Bone morphogenic protein-7 inhibits progression of chronic renal fibrosis associated with two genetic mouse models. Am J Physiol Renal Physiol 2003;285(6):F1060–7.

[98] Zeisberg M, Hanai J, Sugimoto H, Mammoto T, Charytan D, Strutz F, et al. BMP-7 counteracts TGF-beta1-induced epithelial-to-mesenchymal transition and reverses chronic renal injury. Nat Med 2003;9(7):964–8.

[99] Wang SN, Lapage J, Hirschberg R. Loss of tubular bone morphogenetic protein-7 in diabetic nephropathy. J Am Soc Nephrol 2001;12(11):2392–9.

[100] Wang S, Chen Q, Simon TC, Strebeck F, Chaudhary L, Morrissey J, et al. Bone morphogenic protein-7 (BMP-7), a novel therapy for diabetic nephropathy. Kidney Int 2003;63(6): 2037–49.

[101] Wang S, de Caestecker M, Kopp J, Mitu G, Lapage J, Hirschberg R. Renal bone morphogenetic protein-7 protects against diabetic nephropathy. J Am Soc Nephrol 2006;17(9):2504–12.

[102] Wong MG, Perkovic V, Woodward M, Chalmers J, Li Q, Hillis GS, et al. Circulating bone morphogenetic protein-7 and transforming growth factor-beta1 are better predictors of renal end points in patients with type 2 diabetes mellitus. Kidney Int 2013;83(2):278–84.

[103] Abe H, Matsubara T, Iehara N, Nagai K, Takahashi T, Arai H, et al. Type IV collagen is transcriptionally regulated by Smad1 under advanced glycation end product (AGE) stimulation. J Biol Chem 2004;279(14):14201–6.

[104] Matsubara T, Abe H, Arai H, Nagai K, Mima A, Kanamori H, et al. Expression of Smad1 is directly associated with mesangial matrix expansion in rat diabetic nephropathy. Lab Invest 2006;86(4):357–68.

[105] Deckert T, Feldt-Rasmussen B, Djurup R, Deckert M. Glomerular size and charge selectivity in insulin-dependent diabetes mellitus. Kidney Int 1988;33(1):100–6.

[106] Melvin T, Kim Y, Michael AF. Selective binding of IgG4 and other negatively charged plasma proteins in normal and diabetic human kidneys. Am J Pathol 1984;115(3):443–6.

[107] Doi T, Moriya T, Fujita Y, Minagawa N, Usami M, Sasaki T, et al. Urinary IgG4 and Smad1 are specific biomarkers for renal structural and functional changes in early stages of diabetic nephropathy. Diabetes 2018;67(5):986–93.

[108] Goldstein BJ, Scalia R. Adiponectin: a novel adipokine linking adipocytes and vascular function. J Clin Endocrinol Metab 2004;89(6):2563–8.

[109] Arita Y, Kihara S, Ouchi N, Takahashi M, Maeda K, Miyagawa J, et al. Paradoxical decrease of an adipose-specific protein, adiponectin, in obesity. Biochem Biophys Res Commun 1999;257(1):79–83.

[110] Weyer C, Funahashi T, Tanaka S, Hotta K, Matsuzawa Y, Pratley RE, et al. Hypoadiponectinemia in obesity and type 2 diabetes: close association with insulin resistance and hyperinsulinemia. J Clin Endocrinol Metab 2001;86(5):1930–5.

[111] Hotta K, Funahashi T, Arita Y, Takahashi M, Matsuda M, Okamoto Y, et al. Plasma concentrations of a novel, adipose-specific protein, adiponectin, in type 2 diabetic patients. Arterioscler Thromb Vasc Biol 2000;20(6):1595–9.

[112] Imagawa A, Funahashi T, Nakamura T, Moriwaki M, Tanaka S, Nishizawa H, et al. Elevated serum concentration of adipose-derived factor, adiponectin, in patients with type 1 diabetes. Diabetes Care 2002;25(9):1665–6.

[113] Mannucci E, Ognibene A, Cremasco F, Dicembrini I, Bardini G, Brogi M, et al. Plasma adiponectin and hyperglycaemia in diabetic patients. Clin Chem Lab Med 2003;41(9):1131–5.

[114] Perseghin G, Lattuada G, Danna M, Sereni LP, Maffi P, De Cobelli F, et al. Insulin resistance, intramyocellular lipid content, and plasma adiponectin in patients with type 1 diabetes. Am J Physiol Endocrinol Metab 2003;285(6):E1174–81.

[115] Hadjadj S, Aubert R, Fumeron F, Pean F, Tichet J, Roussel R, et al. Increased plasma adiponectin concentrations are associated with microangiopathy in type 1 diabetic subjects. Diabetologia 2005;48(6):1088–92.

[116] Frystyk J, Tarnow L, Hansen TK, Parving HH, Flyvbjerg A. Increased serum adiponectin levels in type 1 diabetic patients with microvascular complications. Diabetologia 2005;48(9): 1911–8.

[117] Schalkwijk CG, Chaturvedi N, Schram MT, Fuller JH, Stehouwer CD, Group EPCS. Adiponectin is inversely associated with renal function in type 1 diabetic patients. J Clin Endocrinol Metab 2006;91(1):129–35.

[118] Titan SM, Zatz R, Graciolli FG, dos Reis LM, Barros RT, Jorgetti V, et al. FGF-23 as a predictor of renal outcome in diabetic nephropathy. Clin J Am Soc Nephrol 2011;6(2):241–7.

[119] Flower DR. The lipocalin protein family: structure and function. Biochem J 1996;318(Pt 1):1–14.

[120] Fiseha T, Tamir Z. Urinary markers of tubular injury in early diabetic nephropathy. Int J Nephrol 2016;2016.

[121] Hong CY, Chia KS, Ling SL. Urine protein excretion among Chinese patients with type 2 diabetes mellitus. Med J Malaysia 2000;55(2):220–9.

[122] Takebayashi K, Suetsugu M, Wakabayashi S, Aso Y, Inukai T. Retinol binding protein-4 levels and clinical features of type 2 diabetes patients. J Clin Endocrinol Metab 2007;92 (7):2712–9.

[123] Cooper ME, Vranes D, Youssef S, Stacker SA, Cox AJ, Rizkalla B, et al. Increased renal expression of vascular endothelial growth factor (VEGF) and its receptor VEGFR-2 in experimental diabetes. Diabetes 1999;48(11):2229–39.

[124] Tsuchida K, Makita Z, Yamagishi S, Atsumi T, Miyoshi H, Obara S, et al. Suppression of transforming growth factor beta and vascular endothelial growth factor in diabetic nephropathy in rats by a novel advanced glycation end product inhibitor, OPB-9195. Diabetologia 1999;42(5):579–88.

[125] Flyvbjerg A, Dagnaes-Hansen F, De Vriese AS, Schrijvers BF, Tilton RG, Rasch R. Amelioration of long-term renal changes in obese type 2 diabetic mice by a neutralizing vascular endothelial growth factor antibody. Diabetes 2002;51(10):3090–4.

[126] Kim NH, Oh JH, Seo JA, Lee KW, Kim SG, Choi KM, et al. Vascular endothelial growth factor (VEGF) and soluble VEGF receptor FLT-1 in diabetic nephropathy. Kidney Int 2005;67(1): 167–77.

[127] Jim B, Ghanta M, Qipo A, Fan Y, Chuang PY, Cohen HW, et al. Dysregulated nephrin in diabetic nephropathy of type 2 diabetes: a cross sectional study. PLoS One 2012;7(5).

[128] Poplawska-Kita A, Mierzejewska-Iwanowska B, Szelachowska M, Siewko K, Nikolajuk A, Kinalska I, et al. Glycosaminoglycans urinary excretion as a marker of the early stages of diabetic nephropathy and the disease progression. Diabetes Metab Res Rev 2008;24(4): 310–7.

[129] Hotamisligil GS, Spiegelman BM. Tumor necrosis factor alpha: a key component of the obesity-diabetes link. Diabetes 1994;43(11):1271–8.

[130] Pickup JC, Crook MA. Is type II diabetes mellitus a disease of the innate immune system? Diabetologia 1998;41(10):1241–8.

[131] Pradhan AD, Manson JE, Rifai N, Buring JE, Ridker PM. C-reactive protein, interleukin 6, and risk of developing type 2 diabetes mellitus. JAMA 2001;286(3):327–34.

[132] Spranger J, Kroke A, Mohlig M, Hoffmann K, Bergmann MM, Ristow M, et al. Inflammatory cytokines and the risk to develop type 2 diabetes: results of the prospective population-based European Prospective Investigation into Cancer and Nutrition (EPIC)-Potsdam Study. Diabetes 2003;52(3):812–7.

[133] Liu S, Tinker L, Song Y, Rifai N, Bonds DE, Cook NR, et al. A prospective study of inflammatory cytokines and diabetes mellitus in a multiethnic cohort of postmenopausal women. Arch Intern Med 2007;167(15):1676–85.

[134] Niewczas MA, Gohda T, Skupien J, Smiles AM, Walker WH, Rosetti F, et al. Circulating TNF receptors 1 and 2 predict ESRD in type 2 diabetes. J Am Soc Nephrol 2012;23(3): 507–15.

[135] Gohda T, Nishizaki Y, Murakoshi M, Nojiri S, Yanagisawa N, Shibata T, et al. Clinical predictive biomarkers for normoalbuminuric diabetic kidney disease. Diabetes Res Clin Pract 2018;141:62–8.

[136] Wolf G, Chen S, Ziyadeh FN. From the periphery of the glomerular capillary wall toward the center of disease: podocyte injury comes of age in diabetic nephropathy. Diabetes 2005; 54(6):1626–34.

[137] Hara M, Yamagata K, Tomino Y, Saito A, Hirayama Y, Ogasawara S, et al. Urinary podocalyxin is an early marker for podocyte injury in patients with diabetes: establishment of a highly sensitive ELISA to detect urinary podocalyxin. Diabetologia 2012;55(11):2913–9.

[138] Zhou H, Cheruvanky A, Hu X, Matsumoto T, Hiramatsu N, Cho ME, et al. Urinary exosomal transcription factors, a new class of biomarkers for renal disease. Kidney Int 2008;74(5): 613–21.

[139] Lee H, Han KH, Lee SE, Kim SH, Kang HG, Cheong HI. Urinary exosomal WT1 in childhood nephrotic syndrome. Pediatr Nephrol 2012;27(2):317–20.

[140] Michaud JL, Kennedy CR. The podocyte in health and disease: insights from the mouse. Clin Sci (Lond) 2007;112(6):325–35.

[141] Su J, Li SJ, Chen ZH, Zeng CH, Zhou H, Li LS, et al. Evaluation of podocyte lesion in patients with diabetic nephropathy: Wilms' tumor-1 protein used as a podocyte marker. Diabetes Res Clin Pract 2010;87(2):167–75.

[142] Susztak K, Raff AC, Schiffer M, Bottinger EP. Glucose-induced reactive oxygen species cause apoptosis of podocytes and podocyte depletion at the onset of diabetic nephropathy. Diabetes 2006;55(1):225–33.

[143] Kalani A, Mohan A, Godbole MM, Bhatia E, Gupta A, Sharma RK, et al. Wilm's tumor-1 protein levels in urinary exosomes from diabetic patients with or without proteinuria. PLoS One 2013;8(3).

[144] Carthew RW, Sontheimer EJ. Origins and mechanisms of miRNAs and siRNAs. Cell 2009;136(4):642–55.

[145] Lee RC, Ambros V. An extensive class of small RNAs in Caenorhabditis elegans. Science 2001;294(5543):862–4.

[146] Guo H, Ingolia NT, Weissman JS, Bartel DP. Mammalian microRNAs predominantly act to decrease target mRNA levels. Nature 2010;466(7308):835–40.

[147] Bartel DP. MicroRNAs: genomics, biogenesis, mechanism and function. Cell 2004;116(2): 281–97.

[148] Argyropoulos C, Wang K, McClarty S, Huang D, Bernardo J, Ellis D, et al. Urinary microRNA profiling in the nephropathy of type 1 diabetes. PLoS One 2013;8(1).

[149] Eissa S, Matboli M, Aboushahba R, Bekhet MM, Soliman Y. Urinary exosomal microRNA panel unravels novel biomarkers for diagnosis of type 2 diabetic kidney disease. J Diabetes Complications 2016;30(8):1585–92.

Further reading

Tervaert TW, Mooyaart AL, Amann K, Cohen AH, Cook HT, Drachenberg CB, et al. Pathologic classification of diabetic nephropathy. J Am Soc Nephrol 2010;21(4):556–63.

CHAPTER 5

Biomarkers in renal vasculitis

Haiyan Zhang
UC San Diego Health, Anatomic Pathology, La Jolla, CA, United States

Introduction

Primary systemic vasculitides are classified according to their clinical, serological, and pathological features. The Chapel Hill Consensus conference updated the most recent nomenclature criteria of the classification in 2012 [1]. Vasculitis is best classified based on the size of the involved vessels into large, medium, and small vessel disease. Small vessel vasculitis (SVV), which includes glomerulonephritis, is by far the most frequent vasculitic lesion in the kidney, and the defining renal lesion is a necrotizing crescentic glomerulonephritis. Medium vessel vasculitis occasionally involves the kidney as necrotizing arteritis. Large vessel vasculitis only rarely affects the kidney, and does so most often secondarily by ischemia from proximal arterial narrowing [2]. Approximately 75% of patients with renal vasculitis have overt evidence of systemic vasculitis. Renal-limited vasculitis (RLV) is morphologically indistinguishable with those with systemic vasculitis [3].

Renal vasculitis usually clinically presents as rapidly progressive glomerulonephritis, and pathologically presents as necrotizing crescentic glomerulonephritis. Necrotizing crescentic glomerulonephritis is categorized by the deposits in the glomerulus: immune complex, antiglomerular basement membrane (GBM) antibodies, and pauciimmune. In patients with active untreated pauciimmune crescentic glomerulonephritis, approximately 85%–95% are found to be antineutrophilic cytoplasmic antibodies (ANCA) positive [3, 4]. Crescentic glomerulonephritis that is not associated with ANCA, also without evidence of immune-complex or anti-GBM, is rare and accounts for no more than 5% of all crescentic glomerulonephritis [4].

In this chapter, we review the biomarkers in small vessel vasculitis that affect the kidney, with emphasis on ANCA-associated vasculitis (AAV).

Seema S. Ahuja and Brian Castillo: Kidney Biomarkers. https://doi.org/10.1016/B978-0-12-815923-1.00005-5

ANCA-associated vasculitis

The antineutrophil cytoplasmic antibody (ANCA)-associated vasculitides (AAV) are a group of disorders characterized by pauci-immune necrotizing inflammation of the small to medium vessels in association with autoantibodies against the cytoplasmic region of the neutrophil and are differentiated by subtle differences in clinical phenotype [5]. Three entities are included in this definition: granulomatosis with polyangiitis (GPA, formerly known as Wegener's granulomatosis), microscopic polyangiitis (MPA), and eosinophilic granulomatosis with polyangiitis (EGPA, formerly known as Churg-Strauss syndrome). AAV is the major cause of rapidly progressive crescentic glomerulonephritis (>80%) and is responsible for 5% of patients on renal replacement therapy [4].

The disease course of AAV can be variable. In conjunction with the short- and long-term toxicity of current therapies, this makes optimal management of AAV patients highly complex, and therefore is often guided by physician-biased approaches rather than more objective approaches. Customization of treatment is therefore highly desirable but is limited by the lack of markers signaling the likely outcome of disease [5]. Tissue sampling to identify ongoing disease and differentiate it from chronic damage may be helpful but is limited in its use as a longitudinal marker. Kidney biopsies, for example, may be biased by sampling error—especially when considering focal disease—and are invasive, making repeated assessments less appealing. For this reason, there has been a long-standing search for noninvasive biomarkers that can predict treatment responses and disease relapses [5].

GPA is most commonly associated with proteinase-3 (PR3)-ANCA. MPA is typically associated with myeloperoxidase (MPO)-ANCA. However, only half of patients with EGPA have a detectable ANCA. At one extreme are RLV patients with glomerulonephritis as the only expression of vasculitis, with approximately 80% of patients MPO-ANCA positive. At the other extreme are patients with destructive lesions in the nasal septum with 90% positivity for PR3-ANCA. Patients with pulmonary capillaritis and no nodules or cavities have a similar frequency of MPO-ANCA and PR3-ANCA, whereas patients with lung nodules or cavities more often have PR3-ANCA [6]. There are two conditions usually ANCA-negative. One is most children with EGPA, another one is some patients with limited GPA (often confined to respiratory tract). It is uncertain whether these represent variants of AAV or have different etiologies with similar final pathologies. In a recent study [7], among the 114 patients with pauci-immune glomerulonephritis, 29 (25.4%) were ANCA-negative. Compared with the 85 ANCA-positive patients, ANCA-negative patients were younger at the onset, had higher urinary protein excretion, and were more likely to have mesangial proliferation morphologically; however, no differences were found in renal function, need for dialysis, extrarenal involvement, renal and global survival, as well as treatment response and relapse rates between ANCA-negative and positive groups.

Therefore, serologic classification of AAV (e.g., MPO-ANCA, PR3-ANCA, ANCA-negative) as well as clinicopathologic classification (e.g., MPA, GPA, EGPA, RLV), and combinations of both, are useful for characterizing the nature and outcome of the disease in a given patient, and for predicting the prognosis and response to treatment [8].

What are ANCAs?

ANCAs are group of autoantibodies, mainly of IgG type, directed against antigens found in the cytoplasmic granules of neutrophils and monocytes. ANCAs are classified according to the indirect immunofluorescence (IIF) patterns they produce on normal neutrophils and according to their target antigens. ANCA IIF patterns include cytoplasmic granular fluorescence with central interlobular accentuation (the classical "C-ANCA"), flat homogeneous cytoplasmic fluorescence ("C-ANCA (atypical)"), perinuclear fluorescence with nuclear extension (the classical "P-ANCA"), perinuclear fluorescence without nuclear extension (usually described as "P-ANCA" but sometimes called "P-ANCA (atypical)" or even atypical ANCA), and other less usual patterns, including the combination of cytoplasmic and perinuclear staining ("atypical"). The major target antigen of perinuclear staining (P-ANCA) is MPO. Classical C-ANCA is almost always directed against PR3, and very rarely against MPO or against both PR3 and MPO simultaneously [9].

The three most commonly used assays are indirect immunofluorescence (IIF), the direct and "capture" enzyme-linked immunosorbent assays (ELISAs) for ANCA against PR3 and MPO. However, the International Consensus Statement for Testing and Reporting ANCA has recommended that all sera are screened for ANCA by IIF and that IIF-positivity is confirmed by direct ELISAs. Some laboratories test by direct ELISA alone, others screen with direct ELISA and confirm positive sera by IIF, and a few use capture ELISAs. The newest consensus, in 2017, recommended that using antigen-specific immunoassays alone may replace the two-stage method [10].

How do ANCAs cause vasculitis?

A considerable body of evidence has shown that ANCAs directly cause vascular injury. The most accepted model of pathogenesis proposes that ANCAs activate cytokine-primed neutrophils, leading to bystander damage of endothelial cells and an escalation of inflammation with recruitment of mononuclear cells [11, 12]. Agent stimulation, such as lipopolysaccharide (LPS) or the cytokine tumor necrosis factor alpha (TNF-α), results in the translocation of ANCA-antigen (MPO and/or PR3) from intracellular granules to the primed neutrophil surface. Subsequently, ANCA activation of these primed neutrophils leads to a respiratory burst and degranulation, therefore potentially causing endothelial injury [11, 12]. On the other hand, numerous cytotoxic mediators are released following neutrophil stimulation by ANCAs, including reactive oxygen species, chemokines, cytokines, proteolytic enzymes, and nitric oxide (NO). The firm adhesion of activated neutrophils to endothelial cells results in endothelial cell damage and may also recruit other inflammatory cells including monocytes and T cells. Following neutrophil activation by ANCA, neutrophils are driven down an accelerated but aberrant apoptotic pathway [11, 12]. However, there is a failure of cell surface changes, for example, bi-lipid cellular membrane phosphatidylserine (PS) externalization, which is normally an early feature of apoptosis. As a result, apoptotic neutrophils are processed by the phagocytes abnormally [11, 12]. This leads to a common finding of leukocytosis in vasculitis.

Recently, increasing evidence supports the proposal of genetic factors predisposing to AAV development. The first genome-wide association study (GWAS), from the United Kingdom, involved 1233 cases of AAV and 5884 controls. A replication cohort from Northern Europe involved 1454 AAV cases and 1666 controls [13]. The strongest genetic associations were with the antigenic specificity of ANCAs, not with the clinical subtypes of AAV. PR3-ANCA was associated with HLA-DP and the genes encoding alpha(1)-antitrypsin (SERPINA1, the main inhibitor of PR3), and the gene encoding PR3 itself (PRTN3). In contrast, MPO-ANCA was significantly associated with HLA-DQ. This study thus confirmed that the pathogenesis of ANCA-associated vasculitis has a genetic component, and shows genetic distinctions between GPA and MPA in terms of the ANCA serotype, but not the phenotype of AAV per se [11, 13]. Other genetic polymorphisms, such as CTLA4 and PTPN22 (and probably many others), are also likely to be contributory. The haplotype carrying DRB1*13:02 was suggested to be protective against MPA and MPO-ANCA-positive vasculitis in a Japanese study [14]. Even though these studies provide insight into the pathogenesis of the breakdown of immune tolerance in patients with AAV, the strength of these associations does not allow these genetic polymorphisms to be used to identify high-risk populations that could be targeted for screening; nor can they be used for the diagnosis of vasculitis, or to determine prognosis or choice of therapy. Similar to ANCA directed against PR3 or MPO, these markers are helpful for diagnosis in AAV, in conjunction with appropriate clinical and pathological findings; however, they are less useful when assessing disease activity in patients with an established diagnosis, or in predicting relapses [11].

ANCA as a biomarker for diagnosis

Although ANCA-negative AAV cases occur, they have become less common with improvements in ANCA testing methodology (including using both indirect immunofluorescence and enzyme-linked immunosorbent assay for screening, or antigen-specific immunoassays recently). Moreover, it is increasingly recognized that such ANCA-negative AAV cases have a broad differential diagnosis that may include entities such as chronic granulomatous disease, IgG4 disease, monogenic granulomatous diseases including Blau syndrome, or other immune-mediated diseases [11]. As a result, most modern clinicians would be reluctant to make a firm diagnosis of AAV without detectable ANCA. In fact, ANCA positivity is usually a requirement for enrolment into clinical trials involving patients with AAV, and is included in pediatric classification criteria and adult definitions of AAV [11].

The International Consensus Statement for Testing and Reporting ANCA recommended that all sera are screened for ANCA by IIF and that IIF-positivity is confirmed by direct ELISAs. When the results of the IIF test were combined with those of the ELISAs (C-ANCA/anti-PR3 positive, P-ANCA/anti-MPO positive), the diagnostic specificity increased to 99%. The sensitivity of the combination of C-ANCA + anti-PR3 or P-ANCA + anti-MPO for WG, MPA, or iRPGN was 73%, 67%, and 82%, respectively [15].

The availability of reliable antigen-specific immunoassays has raised doubts as to whether the two-stage diagnostic strategy currently recommended for ANCA detection is the best approach. The use of antigen-specific assays as the initial and/or only step has been suggested as an alternative approach to screening by IIF. In a 2016 large multicenter study [16] by the European Vasculitis Study Group (EUVAS), the diagnostic performance of antigen-specific immunoassays was confirmed to equal or even to exceed the diagnostic performance of IIF. New guidelines for ANCA testing have been developed based on this study, and according to the revised 2017 International Consensus recommendations, testing for ANCA in small vessel vasculitis can be done by PR3- and MPO-ANCA immunoassays, without the categorical need for IIF. This Consensus Statement highlights the value of ANCA testing as a tool for diagnosis (but not follow-up) of GPA and MPA and gives a historical perspective of ANCAs in small-vessel vasculitis. The clinical utility of ANCA depends on the type of assay performed and the appropriate ordering of testing the right clinical setting. Accurate identification of all patients with AAV and the avoidance of misdiagnosis can be achieved using a "gating policy" based on clinical information given to the laboratory at the time of request [10].

It is reported that some pediatric patients with vasculitis (particularly single organ disease) were initially ANCA-negative, and subsequently became ANCA-positive. This suggests that the negative predictive value of ANCA testing for patients with single organ disease at first presentation might be suboptimal [16]. Therefore, using ANCA as a diagnostic biomarker is much more meaningful under the right clinical context and, when positive, is especially useful for those with atypical presentations and/or single organ disease [11].

ANCA as a biomarker for disease activity

Although ANCAs are important in the pathogenesis and diagnosis of AAV, it remains debatable whether ANCAs are useful to monitor disease activity. In a metaanalysis [17], nine studies were evaluated and it was found that both rising ANCA and persistently positive ANCA were poorly associated with flares.

On the other hand, Kemna et al. [18] noted that longitudinal ANCA measurements could be useful in a subset of patients with renal involvement. They found that rising ANCA titers were significantly correlated with relapses in patients with renal disease (hazard ratio: 11.09), whereas this association was less clear-cut in those subjects without overt renal disease (hazard ratio: 2.79). However, fewer than half of the patients with a rising ANCA titer experienced a relapse within a year, which makes it difficult to use these data to treat patients preemptively to prevent disease relapse. Moreover, using data from a recent Rituximab Versus Cyclophosphamide for ANCA-associated Vasculitis (RAVE) trial, some association was found between an increase in PR3-ANCA level and risk of subsequent relapse under several specific conditions, although in general an increase in the PR3-ANCA titer was not a very sensitive or specific predictor of subsequent relapse. These conditions include: (1) PR3-ANCA detection methodology (stronger association for direct capture ELISA [hazard ratio 4.57], compared with indirect ELISA); (2) disease with renal vasculitis [HR 7.94], or alveolar hemorrhage [HR 24.19], but not in those with a granulomatous phenotype; and (3) choice of remission induction treatment (stronger association for patients treated with rituximab as opposed to cyclophosphamide/azathioprine). Morgan et al. [19] examined the association between ANCA status and relapse using long-term follow-up data from two previous randomized, controlled clinical trials: CYCLOPS and IMPROVE. They demonstrated that persistent ANCA positivity at the switch from CYC to maintenance therapy is associated with an increased risk of relapse. In multivariable analysis, this association remained significant after adjusting for initial CYC therapy regimen, maintenance therapy, age, and renal function. Moreover, in this study, the increased risk of relapse observed in patients remaining ANCA-positive was true for both MPO-ANCA- and PR3 ANCA-positive patients.

Taken together, persistent ANCA or a rising titer seems insufficient to change treatment decisions. However, in patients with AAV and renal involvement, an increase in ANCA titer could be used as a rationale to follow these patients more carefully, monitoring for other earlier signs of disease flare [11].

ANCA as a biomarker for treatment response

To investigate whether ANCA serotype can be used to predict treatment response, a subsidiary study of the adult RAVE trial was evaluated [20]. PR3-AAV patients treated with rituximab achieved remission at 6 months more frequently than patients treated with cyclophosphamide followed by azathioprine (65% vs 48%; $P=.04$; odds ratio [OR] 2.11, 95% CI 1.04–4.30). Furthermore, PR3-AAV patients with relapsing disease achieved remission more often following rituximab treatment at 6 months (OR 3.57; 95% CI 1.43–8.93), 12 months (OR 4.32; 95% CI 1.53–12.15), and 18 months (OR 3.06; 95% CI 1.05–8.97). There was no association between treatment and remission rates in patients with MPO-AAV, or in groups divided according to AAV diagnosis (i.e., GPA vs MPA) [20]. This study suggested that adult patients with PR3-AAV respond better to rituximab than to conventional induction/remission maintenance treatment with cyclophosphamide and azathioprine, and that ANCA serotype may guide the type of treatment in AAV.

Other biomarkers in AAV

Inflammatory biomarkers

Inflammatory markers, such as erythrocyte sedimentation rate (ESR) and C-reactive protein (CRP), are sensitive but not specific. When they are negative, they are very helpful for ruling out active vasculitis for the patients who never have history of vasculitis. For the patients with prior diagnosis, when they are positive, they are limited by not being able to differentiate from infection [5]. Similar to ANCA directed against PR3 or MPO, these markers are helpful for diagnosis in AAV, in conjunction with appropriate clinical and pathological findings; however, they are less useful when assessing disease activity in patients with an established diagnosis, or in predicting relapses.

Other neutrophil activation biomarkers

Neutrophil activation is thought to be a central event in the initiation of vasculitis caused by ANCAs, though the mechanisms by which ANCAs may contribute to the pathogenesis of ANCA-associated vasculitis are not well understood. One proposed paradigm of ANCA pathogenesis is that after cytokine priming, neutrophils are fully activated by ANCAs, either in the blood or within lesional tissue, and firmly adhere to the vascular endothelium [12]. These neutrophils degranulate and release numerous cytotoxic mediators, provoking endothelial injury and vasculitis.

Increased cellular microparticles (MPs) have been described in AAV, although their pathologic significance in this context is currently unknown. MPs are membrane vesicles released upon activation or apoptosis from various cell types including neutrophils, platelets, and endothelial cells. Loss of phospholipid asymmetry and increased surface expression of phosphatidylserine are crucial events in this process. Hong et al. [21] studied the role of neutrophil microparticles (NMPs) in the pathogenesis of AAV. Primed neutrophils release these membrane vesicles. These microparticles expressed a variety of markers, including the ANCA autoantigens proteinase 3 and myeloperoxidase. They bound endothelial cells via a CD18-mediated mechanism and induced an increase in endothelial intercellular adhesion molecule-1 expression, production of endothelial reactive oxygen species, and release of endothelial IL-6 and IL-8. Removal of the neutrophil microparticles by filtration or inhibition of reactive oxygen species production with antioxidants abolished microparticle-mediated endothelial activation. In addition, these microparticles can bind to and activate endothelium via induction of reactive oxygen species (ROS) and generate thrombin. In vivo, we detected more neutrophil microparticles in the plasma of children with ANCA-associated vasculitis compared with that in healthy controls or those with inactive vasculitis. Taken together, these results support a role for neutrophil microparticles in the pathogenesis of ANCA-associated vasculitis, potentially providing a target for future therapeutics, though currently their clinical utility has not been validated.

Neutrophil extracellular traps (NETs) are auto-antigenic strands of extracellular DNA covered with MPO and PR3 that can be a source for the formation of ANCAs. NETs contain proinflammatory proteins and are thought to contribute to vessel inflammation directly by damaging endothelial cells and by activating the complement system, and indirectly by acting as a link between the innate and adaptive immune system through the generation of PR3- and MPO-ANCA. The presence of NETs was recently demonstrated in renal tissue of patients with AAV. NET formation was enhanced in AAV, suggesting that ANCA could trigger NETs formation, supporting a vicious circle, placing NETs in the center of AAV pathogenesis. Studies [22] have showed that excessive NETs formation was predominantly found during active disease, more so than during remission. Excessive NETs formation was found in patients with AAV hospitalized for disease relapse but not during severe infection. Thus, excessive NETs formation in AAV is independent of ANCA, and an excess of ex vivo NETs formation was related to active clinical disease in patients with AAV and a marker of autoimmunity rather than infection. NETs degradation was also highlighted in another study showing that AAV patients have reduced DNase I activity resulting in less NET degradation. With this in mind, it might be that prolonged exposure to proteins in the NETs due to the overproduction of NETs and/or reduced clearance of NETs is important in AAV. However, not all ANCAs are pathogenic and some might possibly also aid in the clearance of NETs. A dual role for ANCAs in relation to circulating NETs levels has been proposed because a negative correlation was observed between PR3-ANCA and NETs remnants in patients in remission.

Calprotectin has gained widespread interest in studies of acute and chronic inflammation and associated diseases. Calprotectin is a heterodimeric complex of two S100 calcium-binding proteins, myeloïd-related protein (MRP)-8

(S100A8), and MRP-14 (S100A9), expressed in granulocytes and monocytes. Calprotectin is an important proinflammatory factor of innate immunity acting as endogenous damage-associated molecule regulating myeloid cell function via toll-like receptor-4 activation. Its release at sites of inflammation makes calprotectin a potent acute-phase reactant. The serum calprotectin levels were elevated during active AAV and relapse. To validate calprotectin as a biomarker of relapse, samples from the Rituximab in ANCA-associated Vasculitis (RAVE) trial were used [23]. Serum samples were obtained from 182 subjects enrolled in the RAVE trial at baseline (B), 1 month (m1), and 2 months (m2) following enrollment and initiation of treatment. A significantly higher risk of relapse was associated with an increase in calprotectin between baseline and months 2 and 6 for all patients, suggesting that an individual patient's level of calprotectin from the time of diagnosis stratifies the patient's relapse risk. Failure to suppress calprotectin following induction AAV treatment is associated with greater and faster rates of disease relapse. Subgroup analysis demonstrated that patients treated with rituximab and with increased levels of serum calprotectin at baseline had the greatest risk for future relapse. Nevertheless, some patients treated with cyclophosphamide/azathioprine, relapsed despite decreases in serum calprotectin, which requires further investigation. Calprotectin appears to be a promising biomarker predictive of relapse in AAV and should be investigated for use in clinical trials and clinical practice.

Other antibodies

Numbers of additional antibodies, other than ANCAs, could be involved in AAV, such as autoantibodies (aAbs) against plasminogen, lysosome-associated membrane protein-2 (LAMP-2), moesin, and pentraxin-3.

Plasminogen is a key component of the fibrinolytic system. Antiplasminogen antibodies were detected in both PR3-ANCA- and MPO-ANCA-positive patients and found in 24% of United Kingdom and 26% of Dutch patients with AAV compared with 2% in controls [24]. In patients with AAV, antiplasminogen antibodies were associated with reduced renal function and the presence of fibrinoid necrosis and cellular crescents in renal histology. Circulating antiplasminogen antibodies were associated with systemic disease activity and renal disease activity of AAV [24]. The role of plasminogen as a biomarker for AAV disease activity needs to be validated further clinically.

LAMP-2 is a major constituent of the lysosomal membrane, plays a role in autophagy, and is involved in HLA class II antigen presentation. It is a highly glycosylated protein that is found not only lining neutrophil granules but also on the cell surface. These diverse roles link LAMP-2 to the clearance of intracellular pathogens and the immune response, thereby raising the possibility that anti-LAMP-2 antibodies contribute significantly to the pathogenicity of ANCA. Initial findings of high anti-LAMP-2 antibody, with a frequency of 81%, are reported in patients presenting with an AAV, becoming rapidly undetectable after immunosuppressive treatment, rare during remission, and recurring with clinical relapses [25]. However, a much lower prevalence, 21%, was reported by other studies in AAV patients compared with 16% in controls [26]. Furthermore, the testing assays for anti-LAMP-2 have not been standardized. With these conflicting results, the role of anti-LAMP-2 antibodies in the pathogenesis and clinical phenotype of ANCA-associated vasculitis needs additional clarification.

Moesin is a heparin-binding protein and belongs to the ezrin/radixin/moesin family of proteins distributed in the plasma membrane in the cellular cortex. Antimoesin is observed in sera of SCG/Kj mice, which spontaneously develop MPO-ANCA-associated RPGN. In a Japanese study [27], antimoesin autoantibody existed in the serum of patients with MPO-AAV and was associated with the production of inflammatory cytokines/chemokines targeting neutrophils with a cytoplasmic profile, which suggests that the antimoesin autoantibody has the possibility to be a novel autoantibody developing vasculitis via neutrophil and endothelial cell activation. The antimoesin autoantibody was detected in both active disease and during remission, but was more associated with renal damage. However, neither the antimoesin antibody testing on a non-Japanese population nor the utility on assessing AAV disease relapse has been reported.

Pentraxin-3 (PTX-3) is stored in human neutrophil granules and is expressed on apoptotic neutrophil surface. Anti-PTX3 aAbs were detected in 56 of 150 (37.3%) of the AAV patients versus 2 of 227 (5.3%) of healthy controls, and, interestingly, in 7 of 14 (50%) MPO and PR3 ANCA-negative AAV patients [28], indicating a possible role in diagnosing ANCA-negative AAV. Moreover, by indirect immunofluorescence on fixed neutrophils, anti-PTX3 aAbs gave rise to a specific cytoplasmic fluorescence pattern distinct from the classical cytoplasmic (C-ANCA), perinuclear (P-ANCA), and atypical (a-ANCA) pattern. Anti-PTX3 aAbs levels were higher in patients with active AAV compared to patients with inactive disease, but considerable overlap was found [5, 28]. There are no reports regarding the association between anti-PTX3 aAbs and AVV disease relapse.

In summary, these novel antibody biomarkers appear to represent only a portion of the AAV patients and are helpful to identify subsets of disease features, which may be useful for applying individual therapy [5]. Nevertheless, none of the studies has investigated the impact of these antibodies on predicting disease relapse and compared them with ANCA titer change. The significance and utility of these biomarkers are being subjected to further clinical trials [5].

Leukocyte activation biomarkers

The role of B lymphocyte in AAV has been confirmed by B cell targeted therapy by rituximab (RTX), although how particular B cell subsets modulate immunopathogenesis remains unknown. Though their phenotype remains controversial, regulatory B cells (Bregs), play a role in immunological tolerance via interleukin (IL)-10. Patients with active disease had decreased levels of CD5 (+) CD24(hi) CD38(hi) B cells and IL-10(+) B cells compared to patients in remission and healthy controls (HCs) [29]. As IL-10(+) and CD5(+) CD24 (hi) CD38(hi) B cells normalized in remission within an individual, ANCA titers decreased. The CD5(+) subset of CD24(hi) CD38(hi) B cells decreases in active disease and rebounds during remission similarly to IL-10-producing B cells [29]. Moreover, CD5(+) B cells are enriched in the ability to produce IL-10 compared to CD5(neg) B cells. Together these results suggest that CD5 may identify functional IL-10-producing Bregs. The malfunction of Bregs during active disease due to reduced IL-10 expression may thus permit ANCA production. However, the role of Bregs in predicating relapse is still not established.

T lymphocytes are reported to be involved in AAV pathogenesis. Circulating effector T-cell populations are expanded in AAV in a persistent state of activation. Circulating regulatory T-cell subsets are less well characterized, but seem to be impaired in function. Lesional effector T cells are present in granulomas, vasculitic lesions, and nephritis. Lesional T cells usually show pro-inflammatory properties and promote granuloma formation. By transcriptional profiling of purified CD8 T cells, two distinct patient subgroups were identified predicting long-term prognosis in AAV. Gene expression-based biomarkers defining the poor prognostic group were enriched for genes of the IL7R pathway, TCR signaling, and those expressed by memory T cells [30]. Furthermore, the poor prognostic group is associated with an expanded CD8 T cell memory population. Prospective studies are needed to validate these promising findings.

Granulocytes are also involved in AAV pathogenesis. The Granularity Index (GI) measures the difference between the percentage of hypergranular and hypogranular granulocytes. Subset data [31] from the RAVE trial showed that rituximab-treated patients who achieved remission had a significantly higher GI at baseline than those who did not ($P=.0085$), and that this pattern was reversed in cyclophosphamide-treated patients ($P=.037$). Cyclophosphamide was superior to rituximab in inducing remission in patients with GI below −9.25% (67% vs 30%, respectively; $P=.033$), whereas rituximab was superior to cyclophosphamide for patients with GI greater than 47.6% (83% vs 33%, respectively; $P=.0002$). These distinct subsets of granulocytes found at baseline in patients with AAV may help the clinician to choose the best responding treatment.

Complement

The main histological feature in the kidneys of patients with AAV is pauciimmune necrotizing crescentic glomerulonephritis with little immunoglobulin and complement deposition in the glomerular capillary walls. The complement system was not initially thought to be associated with the development of AAV. Despite this comparatively lower level and more localized distribution of vessel wall complement, experimental and clinical observations strongly indicate alternative complement pathway activation as critically important in the pathogenesis of ANCA disease. Experimental data in animal models and in vitro experiments has shown that primed neutrophils are activated by ANCA, which generates C5a, which engages C5a receptors on neutrophils. This attracts and in turn primes more neutrophils for activation by ANCA. In patients with ANCA disease, plasma levels of C3a, C5a, soluble C5b-9, and Bb have been reported to be higher in active disease than in remission, whereas no difference was reported in plasma C4d in active versus ANCA disease remission [31]. Thus, experimental and clinical data support the hypothesis that ANCA-induced neutrophil activation activates the alternative complement pathway and generates C5a. C5a not only recruits additional neutrophils through chemotaxis but also primes neutrophils for activation by ANCA. This creates a self-fueling inflammatory amplification loop that results in the extremely destructive necrotizing vascular injury. However, there are currently no large cohort studies to validate the complement biomarkers in AAV.

Endothelial injury and repair biomarkers

Endothelial injury is central to AAV pathogenesis. The increase of endothelial cell adhesion molecule expression and a switch to a prothrombotic endothelial phenotype both contribute to the vascular pathology of vasculitis.

Two major endothelial injury biomarkers have been studied. The release of endothelial microparticles (EMPs) could be used to track endothelial injury in children with vasculitis [32]. Elevated levels of the detachment of whole circulating endothelial cells (CECs) are associated with vascular injury, including systemic vasculitis in adults. In one pediatric vasculitis study, both CECs and EMPs were suggested to monitor disease activity [33].

Some endothelial repair biomarkers are also involved. The adaptive response to endothelial injury is the migration of endothelial progenitor cells (EPCs) from the bone marrow to sites of endothelial injury under the control of several cytokines and growth factors. A reduced number of EPCs are seen in adults with AAV; however, higher circulating levels are observed in children with active vasculitis [33]. The different response may suggest a compensatory repair process to endothelial injury in pediatric patients. Growth factors such as vascular endothelial growth factor (VEGF) and angiopoietin 2 (Ang-2) promote angiogenesis and vascular genesis and are released in response to vascular injury, possibly representing a repair response. Ang-2, although less well studied, has also been shown to be elevated during active disease in adults with AAV [34].

In addition, Von Willebrand factor (VWF), a plasma glycoprotein synthesized in megakaryocytes and endothelial cells regulating hemostasis, has been described as a biomarker of disease activity in vasculitis, including AAV and HSP. One study of VWF in adults with AAV and renal involvement showed high levels of VWF during the active phase and these persisted in clinical remission phase, casting doubt on the validity of VWF as a robust biomarker of disease activity in AAV [5].

Biomarker panel

Recent studies have tested numerous biomarkers to identify promising predictors of treatment response and relapse in patients with AAV. Twenty-eight markers of inflammation, angiogenesis, tissue damage, and repair were tested in patients enrolled in the RAVE study, before and 6 months after treatment to distinguish active disease from remission [35]. Three promising biomarkers—MMP-3, TIMP-1, and CXCL13 (BCA-1)—were identified particularly to distinguish the activity [35]. However, due to the poor correlation of these markers between each other and lack of follow-up, more studies on clinically relevant endpoints are needed.

Urinary biomarkers

Urinary analysis is easily obtained and noninvasive, which may be helpful in monitoring renal vasculitis. If reliable, it could spare patients from repeat renal biopsy when there is a suspicion of a renal flare of vasculitis. The presence of red blood cell casts is regarded as having high specificity for active glomerulonephritis (GN). However, once kidney damage has occurred due to GN, proteinuria, hematuria, and even red blood cell casts may persist without evidence of progressive kidney disease [5], so the high sensitivity and specificity of different aspects of urinalysis are apparent only during the first episode of GN. Other biomarkers are not specific for AAV. However, two urinary biomarkers appear to be promising in predicting disease activity.

One is urinary monocyte chemoattractant protein 1 (MCP-1), a chemokine that promotes monocyte recruitment to areas of inflammation, which is important in the pathogenesis of glomerulonephritis in AAV. Higher urinary MCP-1 levels were found in patients with active renal vasculitis compared with healthy controls, patients with inactive vasculitis, and patients with extra-renal vasculitis only, with a specificity of 94% and a sensitivity of 89% (AUC 0.93, positive likelihood ratio: 8.5 and negative likelihood ratio: 0.07), and elevated levels were associated with poor prognosis and relapse [36]. Reduction in urinary MCP-1 levels following treatment preceded the improvement of renal function by a median of 2 weeks [36].

Another urinary biomarker is soluble CD163 (sCD163), shed by monocytes and macrophages. O'Reilly et al. [37] observed that the urinary sCD163 level is markedly higher in patients with AAV compared to patients in remission, disease controls, or healthy controls. The authors suggested that a cutoff for urinary sCD163 at 0.3 ng/mmol would detect active renal vasculitis with a sensitivity of 83%, specificity of 96%, and positive likelihood ratio of 20.8. Since sCD163 is highly stable in urine (stable for at least a week at room temperature) and better suited to storage/transport than urinary MCP-1 (degrades quickly), it is thought to be a superior biomarker. However, as some sepsis patients also had high levels, it is advised to use urinary sCD163 to monitor disease activity in renal vasculitis in patients with established diagnosis, but not necessarily for the diagnosis of renal vasculitis.

These two biomarkers could have an interesting use for diagnosing renal flare in AAV and to triage patients with undifferentiated acute kidney injury, when serologic markers or histologic samples are delayed [5].

Anti-GBM disease

Antiglomerular basement membrane (anti-GBM) antibody disease is a rare autoimmune disorder affecting glomerular capillaries, pulmonary capillaries, or both, with GBM deposition of anti-GBM autoantibodies. The disease is a prototype of autoimmune disease, where the patients develop autoantibodies that bind to the basement membranes and activate the classical pathway of the complement system, which start a neutrophil dependent inflammation. The target antigen is noncollagen domain 1 of the human alpha-3 chain of type IV collagen, α3(IV)NC1. Anti-GBM disease that only affects the kidneys is called anti-GBM glomerulonephritis. Anti-GBM disease that causes both kidney disease and lung disease is called Goodpasture's syndrome. Frequencies vary from 0.5 to 1 cases per million inhabitants per year, and there is a strong genetic linkage to HLA-DRB1-1501 and DRB1-1502 [38]. Overt clinical symptoms are most prominent in the glomeruli, where the inflammation usually results in a severe rapidly progressive glomerulonephritis. Despite modern treatment, fewer than one-third of patients survive with a preserved kidney function after 6 months' follow-up. Therapy is effective only at an early stage, making early detection of vital importance.

The diagnosis of anti-GBM disease is made by the demonstration of circulating anti-GBM antibodies in the patient serum and/or by kidney biopsy confirming linear deposition of IgG, or rarely IgA, along the glomerular capillary walls. Data from anti-GBM antibody enzyme-linked immunosorbent assay (ELISA) tests performed at one medical center on 1914 patients showed the anti-GBM serology tests with low sensitivity of 41.2% with specificity of 85.4% [39]. However, immunoassay-based anti-GBM antibodies kits showed a comparable good sensitivity (between 94.7% and 100%), whereas specificity varied considerably (from 90.9% to 100%) [40]. Better performance in terms of sensitivity/specificity was achieved by a fluorescence immunoassay that utilizes a recombinant antigen.

It has been reported that approximately 5% of anti-GBM disease with no circulating anti-GBM antibodies is detectable in serum by well-established or Western blotting techniques [40]. The diagnosis was confirmed by renal biopsy, with linear deposition of immunoglobulin along the GBM and crescentic glomerulonephritis. Though alternative methods of antibody detection using a highly sensitive biosensor system may improve the sensitivity and specificity, those techniques are not routinely used. Therefore, renal biopsy is suggested to confirm the diagnosis, and despite negative serological test results using immunoassay, the diagnosis of anti-GBM disease should still be considered in the correct clinical context.

As many as 25%–30% of patients with anti-GBM disease will also have ANCA present at some time during their disease, most commonly in a perinuclear fluorescent pattern (P-ANCA) with antimyeloperoxidase (MPO) reactivity [41]. It is not clear whether the ANCA-associated glomerulonephritis predisposes to the development of anti-GBM disease or whether ANCA positivity occurs in the course of anti-GBM disease. Anti-GBM antibodies in double-positive patients have been shown to have a broader spectrum of target antigens and lower levels of autoantibodies against α3 (IV) NC1, compared with those without ANCA. There have been variable data on the prognostic significance of ANCA positivity. A retrospective cohort [41] analyzed clinical features and long-term outcomes of 568 contemporary patients with ANCA-associated vasculitis, 41 patients with anti-GBM disease, and 37 double-positive patients with ANCA and anti-GBM disease from four European centers. Double-positive patients shared characteristics of ANCA-associated vasculitis (AAV), such as older age distribution and longer symptom duration before diagnosis, and features of anti-GBM disease, such as severe renal disease and high frequency of lung hemorrhage at presentation [41]. Despite having more evidence of chronic injury on renal biopsy compared to patients with anti-GBM disease, double-positive patients had a greater tendency to recover from being dialysis-dependent after treatment and had intermediate long-term renal survival compared to the single-positive patients. However, overall patient survival was similar in all three groups. Predictors of poor patient survival included advanced age, severe renal failure, and lung hemorrhage at presentation. No single-positive anti-GBM patients experienced disease relapse, whereas approximately half of surviving patients with AAV and double-positive patients had recurrent disease during a median follow-up of 4.8 years [41]. Thus, double-positive patients may have a truly hybrid disease phenotype, requiring aggressive early treatment for anti-GBM disease, careful long-term follow-up, and consideration for maintenance immunosuppression for AAV [41]. Since double-positivity appears to be common, further work is required to define the underlying mechanisms of this association and define optimum treatment strategies. The anti-GBM and MPO-ANCA coexistence also necessitates testing for anti-GBM whenever an acute test for ANCA is ordered in patients with renal disease.

Oligoanuria is the strongest predictor of patient and renal survival while the percentage of glomerular crescents is the only pathologic parameter associated with poor renal outcome in anti-GBM disease [42]. Kidney biopsy may not be necessary in oligoanuric patients without pulmonary hemorrhage [42].

CD^{8+} T cells were found to be significantly accumulated around the glomeruli with cellular crescents without IgG deposits compared to those with IgG deposits. The prevalence of infiltrating CD^{8+} T cells was correlated with the percentage of ruptured Bowman's capsules, which suggested that cellular immunity might play a crucial role in the inflammatory kidney injury in anti-GBM patients [43]. The periglomerular infiltration of T cells, especially CD^{8+} T cells, may participate in the pathogenic mechanism of glomerular damage.

In addition, experimental data [44] have shown urinary cystatin C, β2-microglobulin, clusterin, GST-α, GST-μ, KIM-1, and NGAL could be useful biomarkers of renal damage in anti-GBM glomerulonephritis rats.

Some molecules, including serum amyloid P (SAP), PG D synthase, superoxide dismutase, renin, and total protease, were validated by urinary proteome assay to be elevated in the urine and kidneys of mice during anti-GBM disease. Among these, urinary protease was the only marker that appeared to be exclusively renal in origin, whereas the others were partly serum-derived. However, the clinical use of these biomarkers needs to be investigated further.

IgA vasculitis/Henoch-Schönlein purpura

IgA vasculitis (IgAV)/Henoch-Schönlein purpura, one of the most common forms of vasculitis in childhood, is a systemic vasculitis characterized by IgA deposits, which target the skin, joints, and kidneys, among other organs. Most cases are self-limiting but nephritis occurs in 20%–54% of children, leading to complications. The natural history of nephritis ranges from persistent asymptomatic microscopic hematuria to progressive kidney failure, with an estimated long-term risk of end-stage renal failure of up to 15% in children. Noninvasive biomarkers of nephritis are specifically required for children to avoid kidney biopsies.

Etiology of IgAV remains elusive. Susceptibility may be conferred by major histocompatibility complex. There are limited reports on the association of human leucocyte antigens (HLA) and IgAV. The current studies mostly focus on the association of HLA class II antigen with IgAV. It has been reported that the frequency of HLA-B35 was increased in patients who developed nephritis [45]. The incidence of HLA-DRB1*01 was also increased compared with matched controls, and HLADRB1*07 was decreased [44]. A study of unselected children with IgAV from Turkey showed that HLA A2, A11, and B35 antigens were associated with a significantly increased risk of IgAV, whereas HLA A1, B49, and B50 antigens were associated with a decreased risk of the disease [46]. However, there was no association with HLA class 1 alleles and renal involvement [46].

Though many questions remain unanswered in pathogenesis, increased serum levels of galactose-deficient IgA1 (Gd-IgA1) remain the most consistent finding in patients with IgAV nephritis and IgA nephropathy, and these almost certainly predispose to the formation of IgA immune complexes with resulting vasculitis.

To identify biomarkers that are able to distinguish patients of HSP initially with or without nephritis, a prospective study [47] was performed including 50 children at the time of IgAV rash between 2010 and 2015, and 21 controls. At the time of inclusion (Day 1), the serum levels of Gd-IgA1 and urinary concentrations of IgA, IgG, IgM, neutrophil gelatinase-associated lipocalin (NGAL), interleukin (IL)-1β, IL-6, IL-8, IL-10, IgA-IgG, and IgA-sCD89 complexes were higher in the IgAV patients with nephritis than in the IgAV patients without nephritis. After follow-up (1 year), 22 patients showed a poor outcome. Among the tested markers, urine IgA at disease onset adequately reclassified the risk of poor outcome over conventional clinical factors, including estimated glomerular filtration rate, proteinuria, and age in IgAV patients.

Berthelot et al [48]. applied comparative analysis of serum proteomes to obtain an insight about IgAV pathogenesis. This study has utilized high sensitivity nanoscale ultra-performance liquid chromatography-mass spectrometry (nanoLC-MS/MS) to investigate the alterations in serum proteomic profiles in patients with IgAV, patients with IgAVN, and healthy subjects. There were 107 differentially expressed proteins among three different groups, and functional analysis suggested that, in addition to earlier reported pathways, such as acute phase response, immune response, complement, and blood coagulation pathways, hemostasis and Wnt signaling pathway were probably involved in the pathogenesis of IgAV. A few differentially abundant proteins identified, such as C4a, serum amyloid A, angiotensinogen, and kininogen 1, were further validated by ELISA. Additionally, angiotensinogen concentration is correlated with IgAVN and could be used as a potential marker for the progression of IgAV.

Another study [49] was performed to explore a panel of serum biomarkers for laboratory diagnosis of pediatric HSP. Compared with healthy individuals, serum levels of WBC, CRP, IL-6, SAA, IgA, and IgM were significantly increased in patients with HSP. Biomarker index = (SAA (serum levels of serum amyloid A) + IgA/4000 + IgM/4000) × 0.4 CRP (mean value)/CRP. The biomarker index in HSP patients was significantly higher than that of the healthy controls, and the biomarker index in septicemia patients was significantly lower than the control.

Neutrophil-to-lymphocyte ratio (NLR) is an inflammatory marker. Some studies have shown that it could be a potential indicator for prognosticating systemic involvement in adult IgAV [50]. In addition, Makay et al. [51] suggested that an elevated circulating neutrophil-to-lymphocyte ratio (NLR) might be a predictor of gastrointestinal hemorrhage in children with IgAV. However, this finding was not replicated in a separate study that excluded those already bleeding, which is of particular relevance as acute hemorrhage is associated with neutrophilia. Thus, the clinical utility of the NLR in IgAV is questionable.

The presence of significant proteinuria is an important marker for staging and prognosis in Henoch-Schönlein purpura, and is often used to make decisions regarding treatment.

Cryoglobinemic vasculitis

Cryoglobulinemic vasculitis (CV) is a systemic small-to-medium vessel vasculitis due to vascular deposition of cold-precipitable serum proteins called cryoglobulins. There are three types of cryoglobulins. Type I are monoclonal immunoglobulins, generally associated with underlying premalignant or malignant disorders, such as plasma cell dyscrasia, Waldenström macroglobulinemia, and chronic lymphocytic leukemia. Type II are the most common type, consisting of monoclonal IgM plus polyclonal IgG with a rheumatoid factor (RF) activity, associated with infectious diseases, particularly hepatitis C (HCV), HIV, or HCV and HIV coinfection, hematological diseases, particularly B cell disorders, and autoimmune diseases. Type III cryoglobulins comprise polyclonal IgG and IgM with RF activity, associated with autoimmune diseases, particularly Sjögren's syndrome, less commonly SLE, and rheumatoid arthritis, as well as infectious diseases such as HCV infection. Type II and type III cryoglobulinemic diseases are often grouped together and referred to as mixed cryoglobulinemia or mixed cryoglobulinemic disease. While the glomerulonephritis occurring in mixed disease appears to be due to inflammatory vasculitis, the glomerulonephritis occurring in type I disease appears to be due to the interruption of blood flow [52].

A positive cryoglobulin test has high sensitivity for cryogobulinemic disease [53]. However, testing for cryoglobulins is complicated by lack of reference range, standards, and stringency in maintaining testing temperature conditions [53]. Identification of cryoprecipitate can be critical for patient care; therefore, correct testing conditions are crucial for reliable cryoglobulin testing. The patient's blood sample should be kept at 37°C initially to avoid premature precipitation of cryoglobulins and thereby decreasing the yield for subsequent identification, which could cause a false negative result. After warm centrifugation or warm cell precipitation, the clear serum is observed at 4°C for formation of cryoprecipitate. The cryoprecipitate is then washed in cold buffer, and the resulting precipitate is warmed to 37°C and subjected to further analysis by immunodiffusion and immunofixation. Cryoglobulin testing has been neglected in routine clinical laboratories and by clinicians due to several factors, such as the lengthy time for serum cryoglobulin analysis and failure to appreciate that low levels of cryoglobulin can be associated with severe symptoms. In a series of 194 serum samples that gave positive tests for cryoglobulin, the majority contained low cryoglobulin concentrations (65% of the samples were type II with a mean of 372 mg/L and 39% of type III with a mean of 216 mg/L; reference range 0–60 mg/L) [54]. Case studies are presented to illustrate the importance of such low levels of cryoglobulin. There is a need for more rapid and more reliable methods for quantification and phenotyping of low concentrations of serum cryoglobulin [54].

Routine antibody tests such as rheumatoid factor may be performed, which are positive in 80%–90% of cases. Individuals with cryoglobulinemia often have low levels of complement (hypocomplementemia), especially low C4 levels. Blood tests can reveal characteristic low levels of complement, which can aid in a diagnosis of cryoglobulinemia. Additional tests may also be performed to detect underlying disorders associated with cryoglobulinemia such as liver function tests and hepatitis C virus infection testing. Practitioners should be alert to the possibility of HCV infection in patients presenting with palpable purpura, livedo reticularis, or urticaria, in which the underlying histologic features are those of leukocytoclastic vasculitis or necrotizing pan-arteritis. Positive serologic test results for HCV antibody and rheumatoid factor in such patients virtually confirm the diagnosis of HCV-induced mixed cryoglobulinemia [52].

Concentrations of cryoglobulins and the C4 component of complement are also used as secondary endpoints in treatment trials of cryoglobulinemic vasculitis, but the ability of these markers to distinguish the clinical states of complete remission, partial remission, and ongoing active vasculitis is modest [55]. Levels of various inflammatory cytokines and chemokines are higher in patients with cryoglobulinemic vasculitis than in controls, but it is unclear whether these markers will be of value for clinically relevant endpoints.

In patients with HCV-CV, eradication of the virus is a very favorable prognostic sign for long-term remission, but cases of recurrent vasculitis without evidence of recurrent viral infection have been reported [52].

A recent study [56] further subgrouped cryoglobulinemic vasculitis associated with hepatitis C virus infection (HCV-CV) according to the presence or absence of IgG3 (potent pro-inflammatory antibodies). Higher IgM-RF was detected in the IgG3-negative group. IgG3-positive group showed higher IgG-RF compared to the IgG3-negative group. IgG3-negative/monoclonal-IgM patients had higher cryocrit compared to IgG3-negative/polyclonal-IgM patients. These findings led to a new hypothesis of clonal selection with IgG3 being involved in the initiation of early stages of cryoglobulinemia as viral infections usually lead to the onset of an IgG1 and IgG3 response. At the onset of HCV infection, the IgG1 recognizes the virus and forms a complex that activates the IgG3 RF. Then, the persistence of antigenic stimulus activates the production of polyclonal IgM-RF, giving rise to cryoglobulins (oligoclonal IgG and polyclonal IgM). Through clonal selection, the next step is the transformation of polyclonal IgM into oligoclonal ones with the simultaneous presence of an IgG3-RF characterized by oligoclonal cryoglobulins. The last stage involves the formation of a monoclonal IgM-FR clone, and, after a period of time, the end of IgG3 production [55, 56].

Anti-C1q vasculitis

Hypocomplementemic urticarial vasculitis (HUV) is an uncommon immune complex-mediated small vessel vasculitis characterized by urticaria and hypocomplementemia (low C1q with or without low C3 and C4), and usually associated with circulating anti-C1q autoantibodies. Arthritis, pulmonary disease, ocular inflammation, and glomerulonephritis are common systemic manifestations. Pediatric cases are reported to have more frequent and more severe renal involvement.

C1q is specifically bound to early apoptotic blebs of keratinocytes and vascular endothelial cells, which subsequently leads to activation of complement system, thereby facilitating clearance of apoptotic bodies [57]. Anti-C1q antibodies (Anti-C1q Abs) are IgG autoantibodies bound to collagen like C1q. Other than HUV, anti-C1q Abs are also seen in connective tissue diseases, most commonly systemic lupus erythematosus (SLE), other vasculitic disorders, such as polyarteritis nodosa (PAN), giant cell arteritis, vascular Behçet's disease, vasculitis of infectious etiologies, which include hepatitis C virus-related vasculitis, and toxocariasis. The presence of anti-C1q Ab does not differ between patients with primary and secondary vasculitis. Anti-C1q Ab titers were correlated with younger age, high ESR, and low C3 in patients with vasculitis [57]. However, the role of anti-C1q Ab in pathogenesis of vasculitis is not clear.

In the largest series of HUV tests reported in the literature [58], including 57 patients, slightly more than half of the patients had anti-C1q antibodies. In contrast, the vast majority of patients exhibited decreased levels of C1q, indicating that low C1q level in the setting of HUV could be a more sensitive marker than anti-C1q antibodies for diagnosis of HUV. Anti-C1q antibodies are also not specific enough since these were previously detected in other systemic diseases, mainly SLE. When patients did have positive anti-C1q antibodies, they were more likely to have systemic HUV, angioedema, livedo reticularis, ocular, musculoskeletal, or kidney involvement, and less likely to have pulmonary or gastrointestinal involvement than patients without anti-C1q antibodies. After first-line therapy, time to treatment failure (TTF) was 7.9 months. The presence or absence of anti-C1q antibodies did not affect treatment response or TTF, as investigators observed.

References

[1] Jennette JC, Falk RJ, Bacon PA, et al. 2012 revised International Chapel Hill Consensus Conference Nomenclature of Vasculitides. Arthritis Rheum 2013;65(1):1–11.

[2] John R, Herzenberg AM. Vasculitis affecting the kidney. Semin Diagn Pathol 2009;26(2):89–102.

[3] Jennete JC, Thomas DB. Heptinstall's pathology of the kidney. 7th ed. 2015.

[4] Vizjak A, Rott T, Koselj-Kajtna M, et al. Histologic and immunohistologic study and clinical presentation of ANCA-associated glomerulonephritis with correlation to ANCA antigen specificity. Am J Kidney Dis 2003;41:539.

[5] Draibe JB, Fulladosa X, Cruzado JM, et al. Current and novel biomarkers in anti-neutrophil cytoplasm-associated vasculitis. Clin Kidney J 2016;9(4):547–51.

[6] Jennette JC, Nachman PH. ANCA glomerulonephritis and vasculitis. Clin J Am Soc Nephrol 2017;12:1–12.

[7] Villacorta J, Diaz-Crespo F, Acevedo M, et al. Antineutrophil cytoplasmic antibody negative pauci-immune extracapillary glomerulonephritis. Nephrol Ther 2016;21:301–7.

[8] Cornec D, Cornec-Le Gall E, Fervenza FC, et al. ANCA-associated vasculitis—clinical utility of using ANCA specificity to classify patients. Nat Rev Rheumatol 2016;12:570–9.

[9] Judy S, Wendy P, Michelle T. What do antineutrophil cytoplasmic antibodies (ANCA) tell us? Clin Rheumatol 2005;19(2):263–76.

[10] Bossuyt X, Cohen Tervaert JW, Arimura Y, et al. Position paper: revised 2017 international consensus on testing of ANCAs in granulomatosis with polyangiitis and microscopic polyangiitis. Nat Rev Rheumatol 2017;13(11):683–92.

[11] Brogan P, Eleftheriou D. Vasculitis update: pathogenesis and biomarkers. Pediatr Nephrol 2018;33:187–98.

[12] Pendergraft III WF, Nachman PH. Recent pathogenetic advances in ANCA-associated vasculitis. Presse Med 2015;44:e223–9.

[13] Lyons PA, Rayner TF, Trivedi S, et al. Genetically distinct subsets within ANCA-associated vasculitis. N Engl J Med 2012;367:214–23.

[14] Tsuchiya N. Genetics of ANCA-associated vasculitis in Japan: a role for HLA-DRB1*09:01 haplotype. Clin Exp Nephrol 2013;17(5):628–30.

[15] Csernok E, Ahlquist D, Ullrich S, Gross WL. A critical evaluation of commercial immunoassays for antineutrophil cytoplasmic antibodies directed against proteinase 3 and myeloperoxidase in Wegener's granulomatosis and microscopic polyangiitis. Rheumatology 2002;41:1313–7.

[16] Damoiseaux J, Csernok E, Rasmussen N, et al. Detection of antineutrophil cytoplasmic antibodies (ANCAs): a multicentre European Vasculitis Study Group (EUVAS) evaluation of the value of indirect immunofluorescence (IIF) versus antigen-specific immunoassays. Ann Rheum Dis 2017;76(4):647–53.

[17] Brogan P, Eleftheriou D, Dillon MJ. Small vessel vasculitis. Pediatr Nephrol 2010;25:1025–35.

[18] Kemna MJ, Damoiseaux J, Austen J, et al. ANCA as a predictor of relapse: useful in patients with renal involvement but not in patients with nonrenal disease. J Am Soc Nephrol 2015;26(3):537–42.

[19] Morgan MD, Szeto M, Walsh M, et al. Negative anti-neutrophil cytoplasm antibody at switch to maintenance therapy is associated with a reduced risk of relapse. Arthritis Res Ther 2017;19.

[20] Unizony S, Villarreal M, Miloslavsky EM, et al. Clinical outcomes of treatment of anti-neutrophil cytoplasmic antibody (ANCA)-associated vasculitis based on ANCA type. Ann Rheum Dis 2016;75(6):1166–9.

[21] Hong Y, Eleftheriou D, Hussain AA, et al. Anti-neutrophil cytoplasmic antibodies stimulate release of neutrophil microparticles. J Am Soc Nephrol 2012;23(1):49–62.

[22] Kraaij T, Kamerling SWA, van Dam LS, et al. Excessive neutrophil extracellular trap formation in ANCA-associated vasculitis is independent of ANCA. Kidney Int 2018;94(1):139–49.

[23] Pepper RJ, Draibe JB, Caplin B, et al. Association of serum calprotectin (S100A8/A9) level with disease relapse in proteinase 3-antineutrophil cytoplasmic antibody-associated vasculitis. Arthritis Rheum 2017;69:185–93.

[24] Hao J, Wang C, Gou SJ, et al. The association between anti-plasminogen antibodies and disease activity in ANCA-associated vasculitis. Rheumatology (Oxford) 2014;53(2):300–6.

[25] Kain R, Tadema H, McKinney EF, et al. High prevalence of autoantibodies to hLAMP-2 in anti-neutrophil cytoplasmic antibody-associated vasculitis. J Am Soc Nephrol 2012;23:556–66.

[26] Roth AJ, Brown MC, Smith RN, et al. Anti-LAMP-2 antibodies are not prevalent in patients with antineutrophil cytoplasmic autoantibody glomerulonephritis. J Am Soc Nephrol 2012;23:545–55.

[27] Suzuki K, Nagao T, Itabashi M, et al. A novel autoantibody against moesin in the serum of patients with MPO-ANCA-associated vasculitis. Nephrol Dial Transplant 2014;29(6):1168–77.

[28] Simon A, Subra J-F, Guilpain P, et al. Detection of anti-pentraxin-3 autoantibodies in ANCA-associated vasculitis. PLoS ONE 2016;11(1).

[29] Aybar LT, McGregor JG, Hogan SL, et al. Reduced CD5(+) CD24(hi) CD38(hi) and interleukin-10(+) regulatory B cells in active anti-neutrophil cytoplasmic autoantibody-associated vasculitis permit increased circulating autoantibodies. Clin Exp Immunol 2015;180(2):178–88.

[30] McKinney EF, Lyons PA, Carr EJ, et al. A CD8 T cell transcription signature predicts prognosis in autoimmune disease. Nat Med 2010;16(5):586–91.

[31] Nasrallah M, Pouliot Y, Hartmann B, et al. Reanalysis of the Rituximab in ANCA-Associated Vasculitis trial identifies granulocyte subsets as a novel early marker of successful treatment. Arthritis Res Ther 2015;17:262.

[32] Erdbruegger U, Grossheim M, Hertel B, et al. Diagnostic role of endothelial microparticles in vasculitis. Rheumatology (Oxford) 2008;47(12):1820–5.

[33] Clarke LA, Hong Y, Eleftheriou D, et al. Endothelial injury and repair in systemic vasculitis of the young. Arthritis Rheum 2010;62(6):1770–80.

[34] Kumpers P, Hellpap J, David S, et al. Circulating angiopoietin-2 is a marker and potential mediator of endothelial cell detachment in ANCA-associated vasculitis with renal involvement. Nephrol Dial Transplant 2009;24:1845–50.

[35] Monach PA, Warner RL, Tomasson G, et al. Serum proteins reflecting inflammation, injury and repair as biomarkers of disease activity in ANCA-associated vasculitis. Ann Rheum Dis 2013;72:1342–50.

[36] Ohlsson S, Bakoush O, Tencer J, et al. Monocyte chemoattractant protein 1 is a prognostic marker in ANCA-associated small vessel vasculitis. Mediat Inflamm 2009;1–7.

[37] O'Reilly VP, Wong L, Kennedy C, et al. Urinary soluble CD163 in active renal vasculitis. J Am Soc Nephrol 2016;27(9):2906–16.

[38] Huey B, McCormick K, Capper J, et al. Associations of HLA-DR and HLA-DQ types with anti-GBM nephritis by sequence-specific oligonucleotide probe hybridization. Kidney Int 1993;44 (2):307–12.

[39] Watad A, Luigi Bragazzi N, Sharif K. Anti-glomerular basement membrane antibody diagnostics in a large cohort tertiary center: should we trust serological findings? Isr Med Assoc J 2017;19(7):424–8.

[40] Sinico RA, Radice A, Corace C, Sabadini E, Bollini B. Anti-glomerular basement membrane antibodies in the diagnosis of Goodpasture syndrome: a comparison of different assays. Nephrol Dial Transplant 2006;21(2):397–401.

[41] McAdoo SP, Tanna A, Hrušková Z, et al. Patients double-seropositive for ANCA and anti-GBM antibodies have varied renal survival, frequency of relapse, and outcomes compared to single-seropositive patients. Kidney Int 2017;92(3):693–702.

[42] Alchi B, Griffiths M, Sivalingam M, Jayne D, Farrington K. Predictors of renal and patient outcomes in anti-GBM disease: clinicopathologic analysis of a two-centre cohort. Nephrol Dial Transplant 2015;30(5):814–21.

[43] Hu SY, Jia XY, Li JN, et al. T cell infiltration is associated with kidney injury in patients with anti-glomerular basement membrane disease. Sci China Life Sci 2016;59(12):1282–9.

[44] Togashi Y, Imura N, Miyamoto Y. Urinary cystatin C as a renal biomarker and its immunohistochemical localization in anti-GBM glomerulonephritis rats. Exp Toxicol Pathol 2013;65(7–8):1137–43.

[45] Soylemezoglu O, Peru H, Gonen S, et al. HLA-DRB1 alleles and Henoch-Schönlein purpura: susceptibility and severity of disease. J Rheumatol 2008;35(6):1165–8.

[46] Peru H, Soylemezoglu O, Gonen S, et al. HLA class 1 associations in Henoch Schonlein purpura: increased and decreased frequencies. Clin Rheumatol 2008;27(1):5–10.

[47] Pillebout E, Jamin A, Ayari H, et al. Biomarkers of IgA vasculitis nephritis in children. PLoS ONE 2017;12(11).

[48] Berthelot L, Jamin A, Viglietti D, Chemouny JM, et al. Value of biomarkers for predicting immunoglobulin A vasculitis nephritis outcome in an adult prospective cohort. Nephrol Dial Transplant 2018;33(9):1579–90.

[49] Purevdorj N, Mu Y, Gu Y, et al. Clinical significance of the serum biomarker index detection in children with Henoch-Schonlein purpura. Clin Biochem 2018;52:167–70.

[50] Gayret OB, Erol M, Nacaroglu HT. The relationship of neutrophil-lymphocyte ratio and platelet-lymphocyte ratio with gastrointestinal bleeding in Henoch-Schonlein purpura. Iran J Pediatr 2016;26(5):e8191.

[51] Tanoglu A, Karagoz E. Predictive role of neutrophil to lymphocyte ratio in Henoch-Schonlein purpura related gastrointestinal bleeding. Rheumatol Int 2014;34(9):1331–2.

[52] Monach PA. Biomarkers in vasculitis. Curr Opin Rheumatol 2014;26(1):24–30.

[53] Motyckova G, Murali M. Laboratory testing for cryoglobulins. Am J Hematol 2011;14:500–2.

[54] Shihabi ZK. Cryoglobulins: an important but neglected clinical test. Ann Clin Lab Sci 2006;36 (4).

[55] Elefante E, Bond M, Monti S, et al. One year in review 2018: systemic vasculitis. Clin Exp Rheumatol 2018;36, N2. Suppl. 111(2):12–32.

[56] Basile U, Gulli F, Gragnani L, et al. IgG3 subclass: a possible trigger of mixed cryoglobulin cascade in hepatitis C virus chronic infection. Dig Liver Dis 2017;49:1233–9.

[57] Jayakanthan K, Gupta N, Mathew J, et al. Clinical utility of anti-C1q antibody in primary and secondary vasculitic conditions. Int J Health Sci (Qassim) 2017;11(5):3–6.

[58] Marie J, Beatrice F, Alban D, et al. The clinical spectrum and therapeutic management of hypocomplementemic urticarial vasculitis. Arthritis Rheum 2015;67(2):527–34.

CHAPTER 6

Biomarkers in renal cancer

Brian Castillo
Houston Methodist Hospital, Houston, TX, United States

Introduction

Renal cell carcinoma (RCC) is the ninth most common malignancy worldwide with the highest incidence observed in the Czech Republic and North America [1, 2]. In the United States alone, annually, there are approximately 74,000 newly diagnosed cases of RCC with approximately 15,000 deaths per year [3]. Classically, RCC presents with a triad of flank pain, palpable abdominal mass, and gross hematuria. However, in clinical practice, the diagnosis of RCC is often discovered incidentally on imaging evaluation for nongenitourinary-related symptoms [4]. The majority of cases are not identified until advanced stages in the disease because of the asymptomatic presentation of RCC [5].

The primary treatment for RCC includes either partial or radical nephrectomy [6]. Individuals that undergo surgical intervention for localized disease have a 5-year survival rate of 70%–80% [7]. However, in advanced stages of the disease with metastasis, RCC has a poor response to adjunct therapies such as chemotherapy, radiation therapy, or hormonal therapy [8].

Imaging techniques cannot definitively distinguish between benign and malignant lesions [9]. As a result, histologic classification has remained the standard in differentiating between benign and malignant lesions as well as classifying subtypes of RCC. According to the World Health Organization (WHO), RCC is considered a heterogenous disease which has the following recognized histopathologic subtypes. These include clear cell RCC (cRCC), papillary RCC (pRCC), chromophobe RCC (chRCC), and collecting duct RCC (cdRCC). The majority of these cases are represented by cRCC which accounts for approximately 70% of cases followed by pRCC 10%–15% of cases, chRCC 4%–6% of cases, and very rarely cdRCC representing less than 1% of cases [10].

The histologic appearance of cRCC comes from the clear cytoplasm seen in the malignant cells that are encased in an endothelial architecture. This subtype of

Seema S. Ahuja and Brian Castillo: Kidney Biomarkers. https://doi.org/10.1016/B978-0-12-815923-1.00007-9

RCC arises from the proximal tubules of the kidney, and its development is often attributed to deletion of chromosome 3p [11]. It has also been associated with the loss of several other genes to include von Hippel Lindau (VHL) gene. Loss of function of the VHL gene is associated with increased tumor angiogenesis through the dysregulation of hypoxia-inducible factor (HIF) and vascular endothelial growth factor (VEGF) [12]. It is the identification of molecular pathways, like VEGF, that has led to the development of targeted therapies [13].

pRCC has been classified into two types, type 1 and type 2, based on the clinical presentation as well as genetic abnormalities. Type 2 shows a less favorable prognosis and tends to be more often associated with gene mutation than sporadic incidences [14]. Histologically, pRCC is characterized by papillary architecture with basophilic cytoplasm and foamy histocytes [15].

ChRCC arises from intercalated cells of the collecting system. Histologically, it is characterized by sheets of cells with empty cytoplasm that are darker than cRCC and occasionally have perinuclear clearing [16]. The histologic appearance is due to the lack of copious amounts of lipid and glycogen within the cells. ChRCC is also histologically characterized by a low mitotic rate and clinically, compared to cRCC, has a lower risk for disease progression and mortality [17]. However, sarcomatoid differentiation is associated with a more aggressive tumor and poorer survival [18].

Although rare, cdRCC is more frequently seen in African Americans. It more commonly presents with gross hematuria and seen in individuals at a younger age [19]. This tumor arises from cells of the distal collecting ducts of Bellini and histologically characterized by irregular, infiltrating tubules with high-grade features. Of all the subtypes, chRCC has the worst prognosis.

Since RCC is an inconspicuous lesion with a heterogenous presentation, it often presents in advanced stages of the disease. It is paramount to be able to reliably identify and classify these lesions and provide appropriate treatment regimens to improve the overall care of patients with RCC. Biomarkers are a potential option that could prove to be a valuable tool in the management of RCC.

According to the World Health Organization "A biomarker is any substance, structure, or process that can be measured in the body or its products and influence or predict the incidence of outcome or disease" [20]. A clinically useful biomarker in cancer would be a biomarker able to "measure the risk of developing cancer," or "measure risk of cancer progression or potential response to therapy" [21]. As a result, cancer biomarker can be classified as either diagnostic, prognostic, or predictive biomarkers.

A *diagnostic biomarker* is a biomarker that is used to determine if a patient has a certain disease state. A *prognostic biomarker* provides information regarding the course of a disease state and provides the treating physician with information regarding clinical outcomes. A *predictive marker* is a biomarker that predicts the response of patients with the same disease process to a specific therapeutic intervention [21]. There are numerous biomarkers that have emerged which may aid in the management of patients with RCC. These biomarkers can be categorized based on their characteristics such as imaging biomarkers, urine biomarkers, serum biomarkers, and tissue biomarkers. In this chapter, we will provide an overview of various noninvasive biomarkers in renal cancer.

Imaging biomarkers in renal cancer

One of the advantages of imaging biomarkers is that they provide a noninvasive method for evaluating renal lesions that would otherwise require biopsy or surgical excision of the renal mass for histological classification; a feature that is required for tissue biomarkers. Alternatively, imaging techniques may serve as a possible diagnostic marker by providing information similar to histologic classification. In addition, imaging has the potential of being a predictive biomarker by identifying features in the renal mass that would predict response to therapy as well as prognostic implications. It is important to note that imaging biomarkers alone would be inadequate for evaluation and monitoring of renal masses but in conjunction with other biomarkers would provide improved personalized medicine [22].

Positron emission tomography

Positron emission tomography (PET) imaging is a type of nuclear medicine imaging that uses radioactive material, called radiotracers, to evaluate diseases. In the evaluation of RCC, numerous radiotracers have been identified that have the potential as either prognostic or predictive biomarkers. In addition, select subtypes of RCC can be evaluated by PET imaging with the potential to identify these subtypes of RCC as well as differentiate between benign and malignant renal lesions.

^{18}F-fluro-deoxy-glucose (FDG) is one of the most well-described radiotracers used in the evaluation of RCC. FDG functions as a glucose analog. As a result, FDG serves a biomarker for tissue uptake of glucose. Currently, FDG-PET cannot be used as a diagnostic biomarker of RCC. This is due to the inconsistent uptake of the radiotracer into RCC as well as the high expression of FDG in normal renal tissue. As a result, there is limited utility of FDG as it has a reported sensitivity of 22% for the diagnosis of localized RCC [23]. Although FDG has limited value as a diagnostic biomarker for localized lesions, it has demonstrated significant value in the evaluation of metastatic disease with a 94% detection rate of metastasis [23].

FDG does show promise as a prognostic biomarker. In total 101 patients with advanced stages of RCC, primarily metastatic disease, were evaluated by standardized uptake values (SUV) before the initiation of systemic therapies. Patients with max SUV_{max} (the highest SUV recorded in each patient) had a poorer prognosis. In addition, analysis demonstrated max SUV_{max} independently predicted overall survival [24]. A similar study evaluating FDG as a prognostic marker in patients with metastatic cRCC prior to treatment also found that baseline SUV_{max} correlated with overall survival [25]. FDG may also have a role as a predictive marker in patients with metastatic cRCC undergoing treatment with sunitinib. In the same study previously mentioned, patients treated with sunitinib showed a correlation with progression-free survival and overall survival at 16 weeks but not 4 weeks when evaluated by FDG-PET [25]. However, a more recent systematic review of FDG as a predictive biomarker in patients with metastatic RCC treated with tyrosine-kinase inhibitors revealed that the value of FDG as a predictive biomarker continues to be uncertain due to heterogenous nature of the studies to date [26].

Carbonic anhydrase IX (CA-IX) is another protein that shows utility as an imaging biomarker in RCC. CA-IX is an enzyme that has several biological functions and is found in low levels in normal renal tissue. However, mutations in the VHL pathway lead to overexpression of CA-IX particularly in cRCC. Iodine-124-cG250 (I-cG250) is a radiotracer that can be used to evaluate CA-IX. I-cG250 functions as a monoclonal antibody that binds to CA-IX. One of the major disadvantages of I-cG250-PET is the long half-life of the radiotracer thus requiring long intervals before proper evaluation of the tumor can take place. One of the major studies examining I-cG250 was the REDECT trial. This study evaluated the utility of I-cG250 as a diagnostic marker. This trial was a multicenter study with patients with renal masses scheduled for surgical intervention that were evaluated by I-cG250-PET prior to resection. This study found the average sensitivity and specificity for cRCC to be 86.2% and 85.9%, respectively, and was the first study to validate the clinical utility of I-cG250 as a diagnostic imaging biomarker [27]. However, critics of the REDECT trail question the utility of I-cG250 as the REDECT trial includes renal masses as large as 22 cm. These critics advocate that the true value of a diagnostic biomarker is in its ability to accurately evaluate small renal masses [28]. In a subgroup analysis of T1 tumors, found to be 2 cm or less in diameter, it was determined that the sensitivity of I-cG250 was 70.8% [29]. As mentioned previously, one of the major disadvantages of I-cG250 is the time delay required before evaluating a renal mass. In the REDECT trial, imaging was delayed 2–6 days after intravenous administration of I-cG250. ^{18}F-VM4-037 is an alternative radiotracer to I-cG250 that also binds to CA-IX but has a shortest half-life and thus clinically more practical. A small phase II pilot study of 11 patients with renal masses were evaluated by ^{18}F-VM4-037 prior to surgical intervention. The majority of patients, 9 of 11, had localized disease while the remainder had evidence of metastatic disease. This study demonstrated difficulty in discerning the localized tumor as the radiotracer showed high uptake in the normal renal parenchyma while metastatic lesions were well visualized by ^{18}F-VM4-037-PET [30]. This study shows that while ^{18}F-VM4-037 is advantageous in allowing same-day imaging assessment of CA-IX, it comes with challenges in being able to evaluate primary cRCC and is more efficacious in evaluating metastatic disease.

Prostate-specific membrane antigen (PSMA) targeted ^{18}F-DCFPyL is another potential radiotracer that has been studied to determine its reliability as a biomarker in RCC. PSMA is a membrane protein found in numerous tissues to include the proximal tubules of the kidney and shows increased expression in vascular tumors like RCC. ^{18}F-DCFPyL-PET has been used to evaluate metastatic RCC. In a small study, ^{18}F-DCFPyL-PET demonstrated a sensitivity of 94.7% compared to a sensitivity of 78.9% for conventional imaging [31]. Therefore, ^{18}F-DCFPyL-PET appears to be a useful biomarker in the staging of metastatic RCC. However, this finding is limited by the small sample size and lack of histological confirmation of imaging findings. Larger studies would be required to validate the utility of ^{18}F-DCFPyL-PET as a biomarker for metastatic RCC.

Magnetic resonance imaging

Magnetic resonance imaging (MRI) is another modality that has potential as either a diagnostic or predictive imaging biomarker. Arterial spin labeling (ASL) MRI is a technique that does not require intravenous contrast but instead enhances the capability of MRI to magnetically label arterial blood flow and thus quantify perfusion. A prospective study of ASL MRI has been evaluated in 42 patients with renal masses. In this study, perfusion levels were determined and later correlated with histopathologic findings after surgical intervention. This study found that ASL MRI can differentiate between benign and malignant lesions, oncocytoma versus RCC, as well as differentiate malignant lesions into papillary RCC versus other subtypes of RCC [32]. In addition, there has been a systematic review of diffusion-weighted MRI (DWI) as a diagnostic imaging biomarker for RCC. Overall, DWI shows moderate accuracy in diagnosing benign and malignant lesions with an estimated sensitivity of 86% and specificity of 78% [33]. As a predictive marker, ASL MRI has been studied in patients with metastatic RCC following antiangiogenic therapy [34]. In this study, 10 patients with metastatic RCC were evaluated by ASL MRI at 1 and 4 months. Overall, the results of this study suggest that decreased tumor blood flow seen as early as 1 month may predict a favorable outcome and promise as a predictive imaging biomarker.

Urine biomarkers in renal cancer

Urine biomarkers provide a convenient way to monitor RCC through tumor substances excreted in the urine. Although the data is limited in evaluating urine as a biomarker in RCC, the available literature shows it to be a potential diagnostic and prognostic biomarker in RCC.

Kidney Injury Molecule-1 (KIM-1) is a protein that is expressed on the surface of tubular epithelial cells in the kidney and is undetectable in normal kidneys. However, under certain conditions, such as ischemia seen in acute kidney injury (AKI), the level of KIM-1 increases. While KIM-1 has been evaluated as a possible biomarker in AKI, it has also been assessed as a possible diagnostic biomarker in RCC. A prospective study evaluating the diagnostic sensitivity and specificity of KIM-1 in cRR and pRCC showed significant excretion of KIM-1 compared to controls. Additionally, the study also found the level of KIM-1 excreted to correlate with tumor size. Although KIM-1 shows diagnostic sensitivity for RCC, its use as a diagnostic biomarker is limited by its poor specificity due to elevated levels of KIM-1 in noncancerous kidney conditions [35].

Neutrophil Gelatinase-Associated Lipocalin (NGAL) is a protein involved in innate immunity and regulation of bacterial infections. It is primarily expressed in neutrophils but also expressed at low levels by epithelial cells in the kidney. Similar to KIM-1, NGAL can be secreted at high levels in noncancer-related conditions of the kidney. A study evaluating NGAL has found it to have a poor sensitivity and specificity for renal cancers [35]. A separate study also had a similar conclusion in that NGAL is unable to distinguish between benign and malignant renal tumors [36]. Therefore, based on the current literature, NGAL has little utility as a diagnostic biomarker in RCC.

Aquaporin-1 (AQP-1) and perilipin 2 (PLIN2) are two proteins that have been found to be increased in a variety of cancer types. As a result, both of these markers have been studied in RCC. Initial studies examining AQP-1 and perilipin 2 found that patients with either cRCC or pRCC had significantly higher levels of AQP-1 (23-fold increase) and PLIN2 (fourfold increase) when compared to control patients [37, 38]. These studies illustrate the sensitivity of AQP-1 and PLIN2 in the identification of cRCC and pRCC and their potential as a diagnostic biomarker in these subtypes. A follow-up study also demonstrated that these urinary biomarkers, AQP-1 and PLIN2, have a respectable specificity for renal cancers [38]. This study looked at 36 patients with either cRCC or pRCC and found that nonmalignant conditions of the kidney (urinary tract infection, diabetic nephropathy, or glomerulonephritis) did not significantly increase AQP-1 and PLIN2 levels. Therefore, both AQP-1 and PLIN2 show promise as a diagnostic biomarker in RCC.

Nuclear Matrix Proteins-22 (NMP-22) is a well-recognized biomarker and is Federal Drug Administration (FDA) approved for the screening of bladder cancer [39]. There has also been interest in the use of NMP-22 as a biomarker in RCC. A couple of studies have found NMP-22 to be significantly higher in patients with RCC when compared to controls which included patients with a variety of benign conditions such as kidney stones, simple renal cysts, and two patients with oncocytomas [40, 41]. These studies demonstrate that NMP-22 may aid in the diagnosis of RCC and promise as a diagnostic biomarker.

Glycosaminoglycans (GAGs) are a ubiquitous membrane protein with heparin sulfate proteoglycan being the major type. In the kidney, they play a major role as a selective filter for the glomerular basement membrane. GAGs have been investigated as a potential prognostic biomarker in RCC. A study of 31 patients diagnosed with metastatic cRCC were prospectively followed and urine biomarkers scores were correlated with progression-free survival and overall survival. The study found that GAGs significantly independently predict both progression-free survival and overall survival. This was the first study to evaluate GAG scores and demonstrate a role for urine GAGs as a prognostic marker for cRCC [42].

Serum biomarkers in renal cancer

Similar to urine, serum biomarkers are a noninvasive test that could be used to personalize the care of patients with RCC. There are several studies that have looked at a variety of serum substances in RCC with identification of select serum proteins that may provide useful information as a serum biomarker in RCC.

Heat shock protein 27 (Hsp 27) is a protein that promotes cell survival under stressful conditions. Overexpression of Hsp 27 has been linked to the development of some cancers [43]. Hsp 27 has also been studied as a potential diagnostic biomarker in RCC. A study of 199 patients with cRCC were evaluated by quantitative proteomics analysis. The study identified 39 dysregulated proteins that could serve as a serum biomarker. Of these serum proteins, five of the most likely candidates were further validated as potential biomarkers for cRCC. The study revealed Hsp 27 as a potential diagnostic biomarker in cRCC as a correlation between elevated levels of Hsp 27 and high-grade tumors was demonstrated [44].

Neutrophil-lymphocyte ratio (NRL) is a biomarker of inflammation and found to be an independent prognostic factor in many types of cancer [45]. In these cancers, NRL has shown to be a poor prognostic factor of overall survival and progression-free survival. A literature review of NRL as a prognostic biomarker in RCC found that in localized RCC, a low NRL (<3) was predictive of a reduced risk of recurrence. Furthermore, a low NRL in metastatic or locally advanced disease predicted a better overall survival and progression-free survival. Overall, this study illustrated that a high NRL is associated with a worse prognosis for patients with RCC [46]. Additionally, a study evaluating the use of NRL as a prognostic biomarker in patients with metastatic RCC undergoing cytoreductive nephrectomy found that NRL independently predicted overall survival in this patient population [47].

C-Reactive protein (CRP) is another marker of inflammation that like NRL has shown potential as a biomarker in RCC. CRP is an acute-phase reactant and elevated in a variety of inflammatory states. In addition to being a nonspecific marker of inflammation, it has also been found to be elevated in malignant conditions [48]. As a result, CRP has been evaluated as a possible prognostic marker in RCC. A study involving 88 patients treated with metastasectomy for metastatic renal cell carcinoma evaluated the prognostic ability of CRP. This study found that CRP levels greater than 5 mg/dL independently predicted overall survival in this population [49]. Furthermore, a later study evaluating the prognostic ability of CRP in a large cohort of patients with cRCC found that CRP was an independent predictor of overall survival, cancer-specific survival, and metastasis-free survival [50].

Carbonic anhydrase IX (CA-IX), as mentioned earlier, is an enzyme that is found in low levels in normal renal tissue that shows overexpression typically in cRCC due to mutations in the VHL gene. Although CA-IX has utility as a diagnostic imaging biomarker, its role as a serum biomarker in RCC is unclear. Serum CA-IX was analyzed in 361 patients for prognostic information and compared to histopathologic findings. While CA-IX demonstrated higher levels in cRCC compared to other subtypes, it did not correlate with survival [51]. However, analysis of a subset of patients enrolled in the Treatment Approaches in Renal cancer Global Evaluation Trial (TARGET) study looking at potential prognostic biomarkers identified CA-IX as a prognostic biomarker for survival. Moreover, the other evaluated markers: vascular endothelial growth factor (VEGF), tissue inhibitor of metalloproteinase 1 (TIMP-1) and Ras p21 also showed significance as prognostic biomarkers for survival with TIMP-1 being independently significant [52]. Contrary, a later study examining CA-IX in patients receiving sorafenib or placebo found that CA-IX does not provide prognostic information or have a predictive role in patients with metastatic cRCC treated with sorafenib [53].

Vascular endothelial growth factor (VEGF) is a protein involved in the formation of new blood vessels. Normally, it is produced under conditions of hypoxia via increased production of HIF which stimulates the increased release of VEGF. While VEGF has a vital role in angiogenesis, its overexpression, such as in mutations of VHL, is a major factor contributing to RCC. As a result, the VEGF pathway is a major focus for targeted therapies and there have been several studies examining VEGF role as a predictive biomarker as well as a prognostic biomarker in RCC. One major study was the TARGET study, which included over 900 patients with cRCC randomized to either treatment with sorafenib or placebo. VEFG was analyzed for overall survival and progression-free survival. The study concluded that VEGF is prognostic for overall survival and progression-free survival in RCC [54].

Lactate dehydrogenase (LDH) is an enzyme involved in anaerobic glycolysis and regulated by the mammalian target of rapamycin (mTOR) pathway. It is expressed in numerous tissues, such as the heart and muscle, and widely expressed during times of tissue injury. LDH is a well-studied marker and has prognostic significance in many malignancies such as lymphoma, prostate, melanoma, and RCC [55]. Furthermore, LDH is included in the Memorial Sloan-Kettering Cancer Center (MSKCC) score for metastatic renal cell carcinoma. In a study evaluating LDH as a prognostic and predictive biomarker for patients with metastatic RCC treated with an mTOR inhibitor, it was determined that LDH is both a statistically significant prognostic factor and predictive of overall survival in patients treated with temsirolimus [55].

Composite Biomarkers are a consolidation of select serum biomarkers in an attempt to improve the performance of these individual biomarkers in clinical application. The composite index is often reported as a score. The Renal Cell Cancer Treatment with Oral RAD001 given Daily (RECORD) study was an open-label, randomized, phase II study that compared everolimus versus sunitinib in patients with untreated metastatic RCC [56]. Plasma samples were collected for biomarker analysis. Five biomarkers were identified as having the strongest potential as predictive markers of progression-free survival and were included in a composite biomarker score (CBS). These biomarkers included CSF1, ICAM1, IL-18BP, KIM1, and TNFRII. Nearly two-thirds of patients were classified as having a low CBS (score 0–3) with the remainder of patients having a high CBS (score 4–5). The results showed a significant association between progression-free survival for patients treated with everolimus but not for patients treated with sunitinib when evaluated by the CBS. Overall, this study highlights the usefulness of composite biomarkers in directing therapy and the prospect alternative combinations of composite biomarkers may provide in managing patients with RCC.

Summary

RCC remains a prominent malignancy with some of the highest incidences being observed in the United States. It is an inconspicuous tumor, typically remaining asymptomatic and often discovered incidentally or not until advanced stages of the disease when it is symptomatic. Furthermore, RCC has a poor response to adjunct therapies such as chemotherapy, radiation therapy, or hormonal therapy. In light of the challenges associated with RCC, biomarkers have been studied to determine their utility in renal cancer as they could prove to be an easy noninvasive tool to aid in the clinical approach to RCC. There are numerous studies evaluating biomarkers of RCC for diagnostic, prognostic, and predictive response to therapy. While many of these studies have demonstrated promising results, they are often limited by sample size, retrospective in design, or have yet to be replicated. Although a definitive biomarker has yet to be determined, their adjunct to the current management of RCC has shown to be beneficial in patients with renal cancer.

References

[1] Chow WH, Dong LM, Devesa SS. Epidemiology and risk factors for kidney cancer. Nat Rev Urol 2010;7(5):245–57.

[2] Lipworth L, Tarone RE, McLaughlin JK. The epidemiology of renal cell carcinoma. J Urol 2006;176(6):2353–8.

[3] Siegel RL, Miller KD, Jemal A. Cancer statistics 2019. CA Cancer J Clin 2019;69:7–34.

[4] Pastore AL, Palleschi G, Silvestri L, Moschese D, et al. Serum and urine biomarkers for human renal cell carcinoma. Dis Markers 2015;2015.

[5] Lee CT, Katz J, Fearn PA, Russo P. Mode of presentation of renal cell carcinoma provides prognostic information. Urol Oncol 2002;7(4):135–40.

[6] MacLennan D, Imamura M, Lapitan MC, Omar MI, et al. Systematic review of oncological outcomes following surgical management of localized renal cancer. Eur Urol 2012;61(5):972–93.

[7] Lasseigne BN, Burwell TC, Patil MA, Absher DM, et al. DNA methylation profiling reveals novel diagnostic biomarkers in renal cell carcinoma. BMC Med 2014;12:235.

[8] Yang SF, Hsu HL, Chao TK, Hsiao CJ, et al. Annexin A2 in renal cell carcinoma: expression, function, and prognostic significance. Urol Oncol 2015;33:11–21.

[9] Schloer B, Figenshau RS, Yan Y, Venkatesh R, et al. Pathologic features of renal neoplasms classified by size and symptomatology. J Urol 2006;176:1317–20.

[10] Prasad SR, Humphrey PA, Catena JR, Narra VR, et al. Common and uncommon histologic subtypes of renal cell carcinoma: imaging spectrum with pathologic correlation. Radiographics 2006;26(6):1795–806.

[11] Presti JC, Rao PH, Chen Q, Reuter VE, et al. Histopathological, cytogenetic, and molecular characterization of renal cortical tumors. Cancer Res 1991;51:1544–52.

[12] Keegan KA, Schupp CW, Chamie K, Hellenthal NJ, et al. Histopathology of surgically treated renal cell carcinoma: survival difference by subtype and stage. J Urol 2012;188(2):391–7.

[13] Di Pietro G, Luu HN, Spiess PE, Sexton W, et al. Biomarkers and new therapeutic targets in renal cell carcinoma. Eur Rev Med Pharmacol Sci 2018;22:5874–91.

[14] Cancer Genome Atlas Research Network, Linehan WM, Spellman PT, Ricketts CJ, et al. Comprehensive molecular characterization of papillary renal cell carcinoma. N Engl J Med 2016;374(2):135–45.

[15] Sukov WR, Lohse CM, Leibovich BC, Thompson RH, et al. Clinical and pathological features associated with prognosis in patients with papillary renal cell carcinoma. J Urol 2012; 187(1):54–9.

[16] Paner GP, Amin MB, Alvarado-Cabrero I, Young AN, et al. A novel tumor grading scheme for chromophobe renal cell carcinoma: prognostic utility and comparison with Fuhrman nuclear grade. Am J Surg Pathol 2010;34(9):1233–40.

[17] Volpe A, Novara G, Antonelli A, Bertini R, et al. Chromophobe renal cell carcinoma (RCC): oncological outcomes and prognostic factors in a large multicenter series. BJU Int 2012; 110(1):76–83.

[18] Kutikov A, Uzzo RG. The R.E.N.A.L. nephrometry score: a comprehensive standardized system for quantitating renal tumor size, location, and depth. J Urol 2009;182(3):844–53.

[19] Tokuda N, Naito S, Matsuzaki O, Nagashima Y, et al. Collecting duct (Bellini duct) renal cell carcinoma: a nationwide survey in Japan. J Urol 2006;176:40–3.

[20] World Health Organization. Biomarkers in risk assessment: Validity and validation. WHO; 2001.

[21] Goosens N, Nakagawa S, Sun X, Hoshida Y. Cancer biomarker discovery and validation. Transl Cancer Res 2015;4(3):256–69.

[22] Modi PK, Farber NJ, Singer EA. Precision oncology: identifying predictive biomarkers for the treatment of metastatic renal cell carcinoma. Transl Cancer Res 2016;5:76–80.

[23] Gofrit ON, Orevi M. Diagnostic challenges of kidney cancer: a systemic review of the role of positron emission tomography computerized tomography. J Urol 2016;196:648–57.

[24] Nakaigawa N, Kondo K, Tateishi U, Minamimoto R, et al. FDG PET/CT as a prognostic biomarker in the era of molecular-targeting therapies: max SUVmax predicts survival of patients with advanced renal cell carcinoma. BMC Cancer 2016;16:67.

[25] Kayani I, Avril N, Bomanji J, Chowdhury S, et al. Sequential FDG-PET/CT as a biomarker of response to sunitinib in metastatic clear cell renal cancer. Clin Cancer Res 2011; 17(18):6021–8.

[26] Caldarella C, Muoio B, Isgro MA, Porfiri E, et al. The role of fluorine-18-fluorodeoxyglucose positron emission tomography in evaluating the response to tyrosine kinase inhibitors in patients with metastatic primary renal cell carcinoma. Radiol Oncol 2014;48(3):219–27.

[27] Divgi CR, Uzzo RG, Gatsonis C, Bartz R, et al. Positron emission tomography/computer tomography identification of clear cell carcinoma: results from the REDECT trial. J Clin Oncol 2013;31(2):187–94.

[28] Farber NJ, Kim CJ, Modi PK, Hon JD, et al. Renal cell carcinoma: the search for a reliable biomarker. Transl Cancer Res 2017;6(3):620–32.

[29] Farber NJ, Wu Y, Zou L, Belani P, et al. Challenges in RCC imaging: renal insufficiency, post-operative surveillance, and the role of radiomics. Kidney Cancer J 2015;13(4):84–90.

[30] Turkbey B, Lindenberg ML, Adler S, Kurdziel KA, et al. PET/CT imaging of renal cell carcinoma with (18)F-VM4-037: a phase II pilot study. Abdom Radiol 2016;41(1):109–18.

[31] Rowe SP, Gorin MA, Hammers HJ, Som Javadi M, et al. Imaging of metastatic clear cell renal cell carcinoma with PSMA-targeted (18)F-DCFPyL PET/CT. Ann Nucl Med 2015; 29(10):877–82.

[32] Lanzman RS, Robson PM, Sun MR, Patel AD, et al. Arterial spin-labeling MR imaging of renal masses: correlation with histopathologic findings. Radiology 2012;265(3):799–808.

[33] Kang SK, Zhang A, Pandharipande PV, Chandarana H, et al. DWI for renal mass characterization: systematic review and meta-analysis of diagnostic test performance. AJR Am J Roentgenol 2015;205(2):317–24.

[34] de Bazelaire C, Alsop DC, George D, Pedrosa I, et al. Magnetic resonance imaging-measured blood flow change after antiangiogenic therapy with PTK787/ZK 222584 correlates with clinical outcome in metastatic renal cell carcinoma. Clin Cancer Res 2008;14(17):5548–54.

[35] Morrissey JJ, London AN, Lambert MC, Kharasch ED. Sensitivity and specificity of urinary neutrophil gelatinase-associated lipocalin and kidney injury molecule-1 for the diagnosis of renal cell carcinoma. Am J Nephrol 2011;34(5):391–8.

[36] Di Carlo A. Evaluation of neutrophil gelatinase-associated lipocalin (NGAL), matrix metalloproteinase-9 (MMP-9) and their complex MMP-9/NGAL in sera and urine of patients with kidney tumors. Oncol Lett 2013;6(5):1677–81.

[37] Morrissey JJ, London AN, Luo J, Kharasch ED. Urinary biomarkers for the early diagnosis of kidney cancer. Mayo Clin Proc 2010;85(5):413–21.

[38] Morrissey JJ, Kharasch ED. The specificity of urinary aquaporin 1 and perilipin 2 to screen for renal cell carcinoma. J Urol 2013;189(5):1913–20.

[39] Rybotycha Z, Dlugosz A. Diagnostic significance of protein NMP22 in bladder cancer. Pol Merkur Lekarski 2015;38(228):309–14.

[40] Kaya K, Ayan S, Gokce G, Kilcarslan H, et al. Urinary nuclear matrix protein 22 for diagnosis of renal cell carcinoma. Scand J Urol Nephrol 2005;39(1):25–9.

[41] Huang S, Rhee E, Patel H, Park E, et al. Urinary NMP22 and renal cell carcinoma. Urology 2000;55(2):227–30.

[42] Gatto F, Maruzzo M, Magro C, Basso U, et al. Prognostic value of plasma and urine glycosaminoglycan scores in clear cell renal cell carcinoma. Front Oncol 2016;24(6):253.

[43] Katsogiannou M, Andrieu C, Rocchi P. Heat shock protein 27 phosphorylation state is associated with cancer progression. Front Genet 2014;5:346.

[44] White NMA, Masui O, DeSouza LV, Krakovska-Yutz O, et al. Quantitative proteomic analysis reveals potential diagnostic markers and pathways involved in pathogenesis of renal cell carcinoma. Oncotarget 2014;5(2):506–18.

[45] Templeton AJ, McNamara MG, Seruga B, Vera-Badillo FE, et al. Prognostic role of neutrophil to lymphocyte ratio in solid tumors: a systematic review and meta-analysis. J Natl Cancer Inst 2014;106(6):124.

[46] Boissier R, Campagna J, Branger N, Karsenty G, et al. The prognostic value of the neutrophil lymphocyte ratio in renal oncology: a review. Urol Oncol 2017;35(4):135–41.

[47] Ohno Y, Nakashima J, Ohori M, Tanaka A, et al. Clinical variables for predicting metastatic renal cell carcinoma patients who might not benefit from cytoreductive nephrectomy: neutrophil to lymphocyte ratio and performance status. Int J Clin Oncol 2014;19(1):139–45.

[48] Wang CS, Sun CF. C-reactive protein and malignancy: clinic-pathologic association and therapeutic implication. Chang Gung Med J 2009;32(5):471–82.

[49] Rausch S, Kruck S, Walter K, Stenzl A, et al. Metastasectomy for metastatic renal cell carcinoma in the era of modern systemic treatment: C-reactive protein is an independent predictor of overall survival. Int J Urol 2016;23(11):916–21.

[50] Dalpiaz O, Luef T, Seles M, Stotz M, et al. Critical evaluation of the potential prognostic value of the pretreatment-derived neutrophil-lymphocyte ratio under consideration of C-reactive protein levels in clear cell renal cell carcinoma. Br J Cancer 2017;116(1):85–90.

[51] Papworth K, Sandllund J, Grankvist K, Ljungberg B, et al. Soluble carbonic anhydrase IX is not an independent prognostic factor in human renal cell carcinoma. Anticancer Res 2010; 30(7):2953–7.

[52] Pena C, Lathia C, Shan M, Escudier B, et al. Biomarkers predicting outcome in patients with advanced renal cell carcinoma: results from sorafenib phase III treatment approaches in renal cancer global evaluation trial. Clin Cancer Res 2010;16(19):4853–63.

[53] Choueiri TK, Cheng S, Qu AQ, Pastorek J, et al. Carbonic anhydrase IX as a potential biomarker of efficacy in metastatic clear-cell renal cell carcinoma patients receiving sorafenib or placebo: analysis from the treatment approaches in renal cancer global evaluation trial (TARGET). Urol Oncol 2013;31(8):1788–93.

[54] Escudier B, Eisen T, Stadler WM, Szczylik C, et al. Sorafenib for treatment of renal cell carcinoma: final efficacy and safety results of the phase III treatment approaches in renal cancer global evaluation trial. J Clin Oncol 2009;27(20):3312–8.

[55] Armstrong AJ, George DJ, Halabi S. Serum lactate dehydrogenase predicts for overall survival benefit in patients with metastatic renal cell carcinoma treated with inhibition of mammalian target of rapamycin. J Clin Oncol 2012;30(27):3402–7.

[56] Voss MH, Chen D, Marker M, Hakimi AA, et al. Circulating biomarkers and outcome from a randomized phase II trial of sunitinib vs everolimus for patients with metastatic renal cell carcinoma. Br J Cancer 2016;114(6):642–9.

CHAPTER 7

Biomarkers in essential hypertension

Smita Pattanaik
Additional Professor of Clinical Pharmacology, Post Graduate Institute of Medical Education and Research, Chandigarh, India

Introduction

Hypertension is the single most important cause of cardiovascular mortality and morbidity worldwide [1]. The diagnosis of hypertension is often made based on repeated measurements of elevated blood pressure, reading of systolic blood pressure (SBP) more than equal to 130 mmHg, and diastolic blood pressure (DBP) more than equal to 90 mmHg [2]. About 90% of all cases of hypertension have no detectable cause for persistently elevated blood pressure and are labeled as essential hypertension (EH). The diagnosis is often that of exclusion after ruling out the secondary causes (secondary hypertension). EH is an age-related phenomenon and the incidence rates rise with increasing age, unlike secondary hypertension, which presents irrespective of age [2]. The estimated prevalence of EH in the general adult population is about 30%–45%. The incidence is about 7% in younger individuals (18–39 years) which increases to about 65% in individuals more than 59 years old [3, 4]. Hypertension leads to damage of several other organ systems in the body, and this damage is often irreversible. Hypertension is the most important age-related disease causing disability and death. An estimated 9.4 million deaths every year, which constitute about 16% of all deaths annually, are attributed to hypertension. Nearly 45% of all cardiovascular deaths and 51% of deaths due to stroke are due to hypertension [1]. Despite the availability of several classes of antihypertensive medications, which are efficacious and well tolerated, the prevalence of hypertension is alarmingly high [5]. Patients even with controlled hypertension appear to be at higher risk of cardiovascular and cerebrovascular death compared to those of normotensive individuals. The rate of mortality is particularly high in individuals with uncontrolled or difficult-to-control hypertension [6]. Approximately 50% of hypertensive patients develop end organ damage if blood pressure is untreated or not optimally treated [7].

Seema S Ahuja and Brian Castillo: Kidney Biomarkers. https://doi.org/10.1016/B978-0-12-815923-1.00008-0

In conventional practice, blood pressure measurement is the parameter most relied upon for diagnosis and treatment of EH. Adequate treatment with antihypertensive drugs results in lowering of blood pressure, causing delay in the progression of end organ damage and decreasing the rate of mortality. It has been demonstrated in several clinical studies that treatment with antihypertensive drugs could reduce the incidence of heart failure (50%), stroke (35%–40%), and myocardial infarction (20%–25%). A reduction of 10 mm of Hg in systolic blood pressure led to 22% reduction in coronary events and 41% reduction in stroke, with a higher degree of mortality benefit in men compared to women (46% vs 41%) [8–15]. The beneficial effects of antihypertensive therapy are mostly independent of the individual pathophysiological mechanism contributing to its genesis. Even in the secondary forms of hypertension, nonspecific blood pressure lowering agents have been found to confer survival benefits compared to no treatment [16, 17].

The NIH biomarker definition group defines a biomarker as "a characteristic that is objectively measured and evaluated as an indicator of normal biological processes, pathogenic processes, or pharmacologic responses to a therapeutic intervention" [18]. Blood pressure is the universal biomarker for diagnosis as well as the target for monitoring therapeutic interventions in hypertension. "Blood pressure" is the phenotype that is treated with a universal approach irrespective of the pathophysiologic mechanism at an individual level contributing to elevated blood pressure. The reduction in morbidity and mortality is related to reduction in blood pressure. However, it does not answer many questions that may be important in the management of hypertension and reduction of the adverse outcomes further [3]. These include: (1) predicting the risk of developing hypertension so that the primordial and primary preventive measures could be employed to delay the onset; (2) the predominant etiopathogenic mechanism and the extent of derangement in a particular patient; (3) the effect of treatment on the particular pathophysiology so as to individualize antihypertensive treatment; and (4), preexisting or imminent target organ damage as a consequence of hypertension.

Even with our progressive understanding of the pathophysiology of hypertension, advent of efficacious antihypertensive medications, and increasing awareness of target organ damage due to hypertension, approximately 40% of patients remain underdiagnosed according to the consecutive NHANES (National Health and Nutrition Examination Survey) reports in the United States [19–21], and this figure could be similar or higher in other countries. Of the 60% diagnosed and treated, the treatment goal (BP reduction to <140/90 mm of Hg) is achieved only in 30%–40% of patients [22, 23]. Therefore, there is a huge unmet need to improve the diagnosis and management of this common disease. It is highly desirable to find certain genetic, molecular, biochemical, or imaging biomarkers, which would enable better understanding of the pathophysiology, improve the diagnosis, objectively quantify the prognosis, and improve the therapeutic efficacy in hypertension. This chapter intends to focus on the proposed biomarkers so far and their current status in addressing diverse facets of the disease process and management of essential hypertension. These biomarkers can be divided into two broad categories: (1) circulating biochemical markers in the blood, and (2) noninvasively measured imaging parameters. Both types of biomarkers have been arranged into different sections as per their intended use.

Biomarkers of pathophysiology

To understand the origin, rationale for use, and utility of a particular biomarker, it is important to understand some of the important mechanisms contributing to the pathogenesis.

Pathophysiologic mechanisms

Blood pressure is the product of cardiac output and peripheral resistance. Hypertension, therefore, results due to increase in the cardiac output and/or peripheral resistance. Cardiac output is a product of stroke volume (amount of blood pumped by the heart in each beat) and the heart rate [24]. Stroke volume is dependent on the myocardial contraction and size of the vascular compartment. Therefore, any change that augments stroke volume, heart rate, or peripheral vascular resistance can increase blood pressure [25]. Hypertension reflects the state of the vascular system of an individual. Though the exact cause of essential hypertension remains obscure, the development is multifactorial. Genetic predisposition, deranged physiological systems, and environmental influence play important roles. The kidneys play a crucial role in development of essential hypertension. Inability of the kidneys to handle excess salt load, rightward shifting of the pressure natriuresis curve, and activation of the renin angiotensin aldosterone systems (RAAS) as well as the sympathetic nervous system (SNS) lead to increase in the blood pressure [26].

It has been proposed that hypertension develops in two phases. In the first phase, neurohumoral stimuli lead to changes in pressure natriuresis and autoregulatory responses are activated leading to episodic renal vasoconstriction, induced by SNS and RAAS. During the second phase, tubulointerstitial inflammation sets in, which is associated with local angiotensin-II (Ang-II) formation, and salt and fluid retention [27]. An ever-increasing body of evidence also suggests that increase in blood pressure is due to vascular endothelial dysfunction (impaired release of NO) and subsequent remodeling is induced due to a state of chronic inflammation in the vessels, which leads to cytokine production, oxidative stress, and generation of ROS [28]. The functional and structural changes in the resistance vessels mainly contribute to increased peripheral resistance. Hypertension in its earlier phase of onset is usually devoid of clinical symptoms. However, long-term effects on the target organs (heart, brain, kidney, eye) are irreversible and devastating. Therefore, ideal management of hypertension is to prevent these complications.

Biomarkers for genetic susceptibility

Genomic components of hypertension have been extensively studied. Since the genome of an individual is believed to be stable throughout life, the identification of genomic biomarkers offers a potential avenue for detecting individuals at risk early, before development of the disease. It is now increasingly evident that essential hypertension exists as a polygenic trait, with no single gene exerting a predominant effect on the blood pressure, and these follow a complex non-Mendelian mode of inheritance [29]. There are at least 25 rare mutations and 53 SNPs detected that are proposed to cause hypertension. The total number of genomic variants extrapolated to affect blood pressure thus far is about 116. Large-scale genome-wide association studies (GWAS) have established that the overall effect size of each of these genetic variations is modest (to the tune of 1 mm of Hg for SBP and 0.5 mm of Hg for DBP) and are significantly confounded by environmental effects [30]. The collective effect of 29 of these variants taken together in an experiment accounted for only 1%–2% of SBP and DBP variance [31]. Hence it was proposed that we have only deciphered a minuscule subset of the SNP responsible for essential hypertension and it is expected that more loci are yet to be discovered that may include variants with larger effect sizes.

Since a very small portion of the heritability of essential hypertension is currently explained by the GWAS studies, it has been labeled as a case of missing heritability. The major challenge here is to transform the SNP discoveries from the GWAS studies to causal molecular mechanisms contributing to the pathophysiology [32, 33]. The large GWAS study concluded that the overall phenotypic variance of hypertension might not be adequately predicted by taking into consideration even a large subset of genetic variants. Therefore, it may not be a useful biomarker for risk prediction even with powerful genomic analysis techniques like next generation sequencing (NGS) [31]. It is now a well-acknowledged fact that the relationship between genotype and phenotype are modified throughout life due to by external factors like the environment [33]. This provides a rationale to study the epigenetic regulations and gene expressions as biomarkers for hypertension. Transcriptomic, proteomic, and metabolomics biomarkers are of special interest for exploring the difference between normotensive and hypertensive individuals. Recent studies have indicated that there is a difference between the expression of gene coding renin and the circulating levels of microRNA miR-181a in hypertensive compared to the normotensive individuals [34–36]. Metabolomic markers, like hexadecanedioate, have been found to be dysregulated in hypertensive individuals. Several studies are underway that are expected to provide a wealth of omics markers [37, 38]. However, as of now, these biomarkers are still investigational.

Several rare forms of familial syndromes have been recognized that are due to monogenic hypertension genes and are inherited as autosomal-dominant or recessive disorders [39]. A list of 12 genes has been identified linked to the renal sodium pathway and steroid hormone metabolism pathway, which further strengthens the role of the kidneys in blood pressure regulation [30, 40]. The effect size of these genetic abnormalities is high, frequently leading to hypertensive urgencies. The syndromic cases usually have an early presentation with progressive treatment resistant form of hypertension and high cardiovascular morbidity and mortality. These phenotypic extremes could be candidates for NGS studies so as to enable better understanding of the pathophysiology and according treatment [30].

Biomarkers for kidney, RAAS, and SNS dysfunction

Renal function could serve as a biomarker for hypertension as it is the primary site for the pathogenesis and affected by changes in blood pressure. Blood pressure and kidney are connected in a cyclic manner. Decline in renal function impairs the auto regulatory mechanisms that control the blood pressure. This leads to persistently elevated blood pressure that further damages the kidneys [41–43]. Therefore, biomarkers related to renal function can fall into several categories such as predicting the development of HTN, monitoring the treatment efficacy and progression of the disease, or the prevention of renal damage.

Glomerular filtration rate, serum creatinine, cystatin-C

Serum creatinine, estimated glomerular filtration rate (eGFR), and urinary albumin creatinine ratio are considered early biomarkers of renal dysfunction [44]. Serum creatinine correlates well with the eGFR and is the most commonly measured biomarker. However, serum cystatin-C has recently emerged as a more robust marker. Creatinine is a by-product of muscle metabolism. It is freely filtered by the proximal tubules and also partly secreted into the tubular system. Since the muscle mass varies with age, sex, race, and level of fitness, the serum creatinine may differ among individuals with similar renal function. However, this variation seems to be minimal for cystatin-C. It is a cysteine protease inhibitor produced by all nucleated cells in the body and a major part of cystatin-C undergoes metabolism in the proximal tubules excreting only a smaller fraction. Thus, under normal health conditions, cystatin-C does not enter the final excreted urine to any significant degree, and the serum concentration is unaffected by infections, body mass, diet, drugs, or inflammatory or neoplastic states. Studies have shown that serum cystatin-C varies less compared to serum creatinine between individuals and remains constant in people with the same eGFR [45–47]. The correlation of 1/cystatin-C with GFR ($r = 0.90$) was significantly superior to 1/creatinine ($r = 0.82$, $P < 0.05$) with GFR in a wide spectrum of individuals especially with normal renal function and with renal dysfunction [48]. In the Heart and Soul study, the investigators found that there is a stronger relationship between cystatin-C and SBP (1.19 ± 0.55 mm of Hg increase per 0.4 mg/L cystatin-C, $P = 0.03$) across all ranges of kidney function, but creatinine clearance with SBP was significantly associated with kidney function only in subjects with CrCl <60 mL/min (6.4 ± 2.13 mmHg increase per 28 mL/min, $P = 0.003$) but not >60 mL/min [46]. In this study, it was also shown that DBP was not associated with cystatin-C or CrCl. However, PP was linearly associated with both cystatin-C (1.28 ± 0.55 mmHg per 0.4 mg/L cystatin-C, $P = 0.02$) and CrCl <60 mL/min (7.27 ± 2.16 mmHg per 28 mL/min, $P = 0.001$).

The cystatin-C estimation methods are now well established. In addition, the availability of an international reference standard (ERM-DA471/IFCC) has led to an increasing number of laboratories performing the tests. However, confounding by nonspecific chronic inflammation should be kept in mind. Therefore, the use as a prognostic marker in hypertension could be challenging, though it may be able to predict the risk well. The eGFR estimated by cystatin-C is usually higher than by creatinine and it may have advantages over creatinine eGFR in certain patient groups in whom muscle mass is abnormally high or low (e.g., quadriplegics, the elderly, or malnourished individuals). Blood levels of cystatin-C also equilibrate more quickly than creatinine, and therefore serum cystatin-C may be more accurate than serum creatinine when kidney function is rapidly changing (e.g., hospitalized individuals). Some clinicians prefer assessing the risk in patients by using both the methods [49]. eGFR, irrespective of its method of estimation, has been found to be a good predictor of onset of hypertension in the general population, hence reinforcing the role of the kidneys [50].

Urinary albumin creatinine ratio

Urinary albumin to creatinine ratio (UACR in mg/g Cr) has been proposed as a possible predictor of the onset of hypertension as well as for progression of the disease [50–52]. Both microalbuminuria ($300 > \text{UACR} \geq 30$ mg/g Cr) and proteinuria ($\text{UACR} \geq 300$ mg/g Cr) have been found to indicate the progressive stages of the disease. Microalbuminuria also reflects a state of vascular endothelial dysfunction that precedes the development of hypertension [52] and proteinuria indicates renal damage due to hypertension. It has also been found that higher quartiles of normal UACR (<30 mg/g Cr) are independently associated with risk of development of hypertension [52]. A large Japanese study involving more than 6000 participants reiterated the same findings that any level of increased urinary albumin was closely associated with the risk for developing hypertension, and UACR is a significant predictor of future increases in blood pressure [53]. Studies from other parts of the world have also confirmed these findings [54, 55]. The close association between UACR and hypertension suggests that it can be considered as a continuous predictor variable for risk of development of hypertension. Longitudinal increase in UACR has been found to correlate well with increase in the SBP. Detection of a longitudinal increase could be important from a clinical standpoint so as to introduce primary prevention measure like lifestyle modifications early in the course, which could prevent onset of hypertension [53].

Interstitial tissue sodium content

Renal salt handling is known to be important in the pathogenesis of hypertension, but a recent study has indicated not only that intravascular salt is important but also how the interstitial sodium content interacts with the immune system plays an important role in development of hypertension [56] and has been found to increase hypertension [57]. It has been proposed that interstitial sodium content is a new biomarker that can be measured by magnetic resonance imaging (MRI) noninvasively [58]. This can be used as a biomarker for primary prevention as it can be modified by physical exercise [59]. Moreover, it can also be used as a treatment target, as suggested by the study on end-stage renal disease patients [60].

Other biomarkers

Other biomarkers related to the SNS and RAAS activity have been evaluated. These include 24-h urine catecholamine, plasma renin concentration (PRC), angiotensin A, plasma aldosterone concentration (PAC), and natriuretic peptides (BNP and N-terminal of pro ANP). However, they were found to be of little value individually [61]. The aldosterone-to-renin ratio (ARR=PRC/PAC) has been demonstrated to represent the relative aldosterone excess and found to correlate significantly with median SBP as well as DBP ($P<0.001$ for both) after multivariate adjustment for age, sex, BMI, renal function, and antihypertensive medication use [62].

Certain novel investigational biomarkers like measurement of noradrenalin spill over and muscle sympathetic nerve activity have been evaluated, but have not yet been established [63]. Other non-RAS peptide biomarkers include vasoconstriction-inhibiting factor, vasopressin, and its inert copeptin [64–66]. Adrenomedullin is another investigational biomarker, which is a peptide derived from the brain, and which has multiple biological action like inhibiting the secretion of renin, aldosterone, inhibition of vascular smooth muscle migration and proliferation, and promoting vasodilatation and natriuresis [67, 68]. It has been found to be associated with aging as well as increased BMI. High serum adrenomedullin concentration was found to precede the development of hypertension in normotensive individuals [69].

Biomarkers for vascular dysfunction and damage

Vascular dysfunction is a well-established mechanism for the development of EH [44]. The relationship is cyclical, like that of the kidneys and hypertension. Alteration in the normal vascular endothelial function leads to impaired vasodilator response. This subsequently causes structural damage to the vasculature, leading to remodeling and stiffening [70]. This is a continuum, applicable to both small and large vessels [71, 72]. Markers of vascular dysfunction help not only in understanding the pathogenesis but also in assessing the disease progression, efficacy of antihypertensive drugs, and development of hypertension-associated complications [3, 44]. Several circulating biomarkers as well as imaging biomarkers have been explored for assessing endothelial dysfunction [73].

Biomarkers of endothelial dysfunction

A strong body of evidence suggests that brachial artery endothelial dysfunction is an important indicator of systemic endothelial dysfunction [73]. This can be assessed by invasive as well as noninvasive methods. Invasive techniques use vasoactive agents like acetylcholine via an intraarterial route, which activates the endothelial cells to release nitric oxide (NO), resulting in vasodilatation and hyperemia. This response is measured by high-resolution ultrasound or strain gauge plethysmography. In individuals with endothelial dysfunction, this response is impaired [74]. However, this procedure involves risk, as in some individuals there is paradoxical vasoconstriction leading to myocardial ischemia [75]. Therefore, the utility of such an invasive test is limited. Its status is currently only investigational.

Brachial artery flow-mediated dilatation

Among noninvasive techniques, brachial artery flow-mediated dilatation (brachial-FMD) is most commonly employed [76]. Brachial-FMD uses an arterial occlusion cuff that is inflated to occlude the arteries of the distal forearm followed by deflation. This leads to a vasodilatation response increasing the flow in the proximal segment of the vessel, i.e., the brachial artery, by releasing NO. This is measured by high-resolution ultrasound and is standardized [77, 78]. It has been observed that there is a proportional relationship between the magnitude of dilation and endothelial dysfunction [76]. Impaired FMD is one of the earliest manifestations of vascular dysfunction even in individuals who are asymptomatic [79]. It is known to predict vascular complication considered as a forerunner of cardiovascular events. FMD measurement has been used to stratify individuals at low, moderate, and high risk of future cardiovascular events [80]. Impaired FMD is thought to be potentially reversible with lifestyle modification and other primary prevention measures [80]. Hence it is an important biomarker in the initial phases of disease pathogenesis. However, despite all the available evidence in its favor, FMD has not gained enough attention as a screening biomarker due to issues like observer dependence, high biological variability, and lack of consensus for age-specific cutoff thresholds for intervention [74]. FMD has also been evaluated in patients with late stage hypertension and found to correlate well with target organ damage [81]. Nevertheless, the utility of this marker in advanced stages of the disease appears to be limited [82].

Peripheral artery tonometry

Peripheral artery tonometry (PAT) measures the reactive vasodilatations in the small digital arteries [98]. Though large studies like the Framingham heart study found that it may be associated with increased cardiovascular risk and PAT correlates well with FMD [99, 100], studies in Korean hypertensive individuals did not support the use of PAT as a biomarker in this population [101]. As a result, there is not enough evidence currently to consider PAT as a biomarker for the management of hypertension. Other noninvasive modalities for the assessment of vascular endothelial dysfunction, like laser Doppler flowmetry (LDF) which measures blood flow in the skin, and venous occlusion plethysmography (VOP), which measures blood flow to various organs, have not been found to be useful biomarkers for hypertension.

Serum vitamin D

Vitamin D has been increasingly linked to vascular health. The physiologically active moiety 25-hydroxy vitamin D3 is known to be essential in the homeostasis of calcium and phosphate ions in the body. The important micronutrients affected are iron, magnesium, phosphate, and zinc. The epidemiological association of low vitamin D3 with hypertension was established about 20 years ago and several observational studies have subsequently supported the inverse association with blood pressure [102]. It has been demonstrated that several cells in the body such as vascular endothelial cells, renal tubules, and T and B cells cause endothelial dysfunction and vascular damage. This leads to stiffness, RAAS activation, and modulation of immune and inflammatory responses through toll-like receptors. The resulting effects on organ systems have led to the proposed mechanism for the development of hypertension from low vitamin D [103]. However, vitamin D supplementation studies in hypertension have resulted in equivocal results [104], with some of them showing positive [105, 106] and some negative results [107–110]. Despite the presumably strong epidemiological evidence, several randomized controlled clinical trials failed to show benefit of treatment with vitamin D even up to 1 year. This could be because of selection bias, information bias, or confounding during the observational study. The role of vitamin D3 is also not convincing in prevention with regard to long-term outcomes of hypertension [111–115]. Therefore, the current scientific literature does not support the use of vitamin D3 as either a diagnostic or prognostic biomarker for hypertension.

Markers of vascular damage and remodeling

Endothelial microparticles

Endothelial microparticles (EMP) are small vesicles formed of membrane fragments of the endothelial cells and released into the circulation in response to cell activation, stress, damage, or apoptosis [83–85]. They serve as biomarkers and transmitters of signal between organ systems [84]. The EMPs express adhesion molecules, enzymes, receptors, and constitutive antigens of the cells of origin. The characteristics of these cells are similar to endothelial progenitor cells and of special interest for hypertension-induced endothelial damage [85]. The Framingham heart study found that there is a difference in the number and composition of EMPs between normotensive and hypertensive individuals [86]. The EMPs in EH have been found to be $CD144^+$ and $CD32^+$ [86]. Other hypertensive conditions, such as preeclampsia, have been shown to have increased numbers of circulating EMPs, providing evidence for hypertension-induced systemic endothelial damage [76, 87]. A correlation has also been established between the circulatory biomarkers for endothelial dysfunction and inflammation-oxidative stress in EH [88]. In addition, it has been demonstrated in studies that antihypertensive therapy (calcium channel blockers, beta blockers, angiotensin receptor blockers) and lipid-lowering therapies impact EMPs and change their characteristics [89]. EMPs are examples of novel and promising biomarkers that could explain the pathophysiology of hypertension and vascular dysfunction. However, they are not yet ready for clinical use. The estimations are typically done by flow cytometry using different centrifugation techniques to separate the platelet-free plasma followed by tagging the surface antigens on the microparticles with antibodies. This requires larger sample volumes, specialized handling, and special sample preparation. There is ongoing effort to define EMPs precisely and standardize the protocol to overcome the technical difficulties [44]. Therefore, additional research is needed before EMPs could be used as a biomarker in the management of hypertension.

Pulse wave velocity

Pulse wave velocity (PWV) is now widely accepted as the gold standard measure of arterial stiffness, which reflects vascular damage as well as vascular structural remodeling [90, 91]. PWV is a simple measure of the speed (meters per second) at which blood pressure pulse travels from the heart to a peripheral artery after systole, e.g., carotid to femoral (cf-PWV), or brachial artery to ankle (ba-PWV) [92]. Due to its intrinsic relationship to the mechanical properties of the artery by the Meons-Kortweg formula (PWV= $\sqrt{(Eh/2)}\ R\rho$, where E = Young's modulus for the artery, h = wall thickness, R = end distolic radius, and ρ = density of the blood), PWV reflects the hemodynamic status and its consequences [93]. Several studies have demonstrated PWV as a valuable biomarker for prediction and progression of hypertension. The Baltimore study, which included 449 normotensive or untreated hypertensives with a median follow-up of a 4.3-year period, found that PWV is an independent predictor of longitudinal increase in SBP and incident hypertension after adjusting for covariates with a significant interaction term for time (P = 0.003). The hazard ratio for development of hypertension was 1.1 (95% CI 1.0–1.3) [94]. A recently published Finnish study, which included 1149 individuals from 2007 to 2011, concluded that the PWV measured was an independent predictor of incident hypertension (OR 1.96 per 1 SD increase; 95% CI 1.51–2.57). This study also found that by adding PWV as a covariate to the prediction model beyond the traditional cardiovascular risk factors, the area under receiver operative characteristic (ROC) curve improved to 0.833 from 0.809 with a continuous reclassification improvement of 59.5% [95]. This study reiterates the importance of PWV as a prediction tool for hypertension in young adults. Despite the clear association of PWV as a valuable tool, routine use of this biomarker is limited in the management of hypertension, and it continues to be used in research studies only.

A few other measures have also been proposed which depict the arterial stiffness, like central pulse pressure (PP), aortic pressure augmentation (ΔP), aortic augmentation index (ΔP/PP), and analysis of the arterial pulse wave form. However, some experts find PWV to be a better indicator of arterial stiffness compared to the other measures [96]. Though there has been no consensus on the cutoff for PWV, a value of 13 m/s was proposed as a cutoff for predicting the risk of incident hypertension and cardiovascular events in normotensive individuals in the Baltimore study [94]. However, aging and mean arterial blood pressure (MAP) remain the most important confounders in evaluation of the desirable cutoff values [97]. Nevertheless, evidence from these studies suggests that PWV as a biomarker could allow identification of individuals who warrant intensive follow-up.

Biomarkers of inflammation and oxidative stress

The growing body of evidence suggests the development and progression of hypertension is due to a state of chronic inflammation of the vessels [116]. The process of endothelial dysfunction, extracellular matrix deposition, and vascular remodeling, along with systemic inflammation and oxidative stress, exist as a complex process in EH [44]. Several biomarkers indicative of inflammation and oxidative stress have been studied in isolation or together as a group. Some of the promising ones are listed here along with the evidence and potential clinical applicability.

C-reactive protein

C-reactive protein (CRP) is an acute phase reactant that has been extensively studied in EH in both cross-sectional and longitudinal studies. CRP estimations performed by a technically optimized sensitive method with a detection limit as low as 0.3 mg/L are labeled as high sensitive CRP (hs-CRP). The relationship between hs-CRP and essential hypertension is well established as CRP is not only a marker of atherothrombotic process but a mediator as well [117]. In 2003, the American Heart Association published cutoff points for hsCRP based on cohort studies in healthy populations stratifying 10-year cardiovascular risk. The established cutoff points are low risk (<1.0 mg/L), average risk (1.0–3.0 mg/L), and high risk (>3 mg/L) [118]. Several studies in hypertension have referred to these values. A large cohort of 15,000 healthy women was studied for 8 years and it was found that elevated CRP levels (≥ 3 mg/L) along with increasing BP were independent predictors of future cardiovascular events. Hence, CRP had incremental prognostic value at all BP levels [119]. Another study by *Sesso et al.*, evaluating 20,000 individuals for 8 years, identified 5000 individuals who developed hypertension. It was concluded that women in the highest CRP quartile had a twofold increased risk of developing hypertension compared to those in the lowest quartile. Additionally, CRP marginally improved the prediction of incident hypertension [120].

Two subsequent studies demonstrated similar findings [121, 122]. However, in recently conducted nested case control studies, it was found that the association between baseline CRP levels [123] or increasing CRP levels [124] lost significance in the risk of developing hypertension after adjustment for body mass index (BMI). Subsequently, two more studies demonstrated that hs-CRP is independently associated with the risk of hypertension even after adjustment for abdominal obesity. Association of hs-CRP to EH in the context of level of physical activity was explored in 2475 normotensive men, of whom 226 developed hypertension. This study concluded that unfit men with high hs-CRP are at 1.81 times greater risk of developing EH compared to fit men with low CRP [125]. Analyzing the current evidence, it is evident that hs-CRP is an important biomarker despite confounders like BMI and physical fitness. Irrespective of the trigger and the concomitant co-mediators (genetic, dietary, metabolic, lifestyle factors like inactivity) for the development of EH, hs-CRP could be used as a diagnostic biomarker to stratify the risk in an individual patient. Current evidence is unequivocal to support the prognostic role of hs-CRP as a biomarker [126–128]. Conversely, therapeutic intervention studies have supported its use in hyperlipidemia during the JUPITOR trial for rosuvastatin [129]. There is need for more research to understand the prognostic impact of hs-CRP in hypertension to characterize the effect of therapeutic as well as lifestyle interventions on long-term outcomes of EH.

Serum uric acid

Serum uric acid (SUA) was first associated with EH in 1874. Its role in this condition remains unclear and it was thought to be a secondary response to hypertension [130, 131]. Several experimental and clinical studies suggest that it is a potential mediator for inflammation and oxidative stress in hypertension [132, 133]. The epidemiological association of SUA concentrations and hypertension as well as cardiovascular complication due to hypertension is strong [134, 135]. The association of increased uric acid is stronger for individuals younger than 60 years compared to older individuals [136]. A Japanese study suggests that middle-aged individuals (>40 years) are more prone when SUA levels are >6.3 mg/dL [137]. SUA between 2.40 and 5.70 mg/dL in females and 3.40 and 7.00 mg/dL in males are considered the reference range for normal [138]. SUA has also been demonstrated to have a strong relationship with PWV, showing a linear trend in women and a J-shaped association in men [139]. The longitudinal data from the Baltimore aging study also supports the relationship [140]. The role of SUA has been strengthened by at least two studies where therapeutic intervention to reduce uric acid lowered the blood pressure. Feig et al. found that administration of allopurinol to obese adolescents who were newly diagnosed stage-1 EH with serum uric acid 6 mg/dL or higher resulted in −6.3 mm of Hg reduction in SBP and −4.6 mm of Hg reduction in DBP [141]. In an observational study, the exposure to allopurinol was found to reduce rates of stroke and cardiac events in adults in a propensity-score matched design [142]. The role of serum uric acid as a biomarker to predict risk of development of hypertension is intriguing and should be systematically evaluated to obtain age and gender-specific cutoff values.

Homocysteine

Homocysteine is a sulfur-containing amino acid of methionine, and hyperhomocystinemia (>10 μmol/L) has been proposed to cause hypertension and cardiovascular disease by generating endothelial dysfunction and vascular damage [143]. Larger observational studies like NHANES III and SHEP (systolic hypertension examination survey) demonstrated that there is an increased prevalence of hypertension in individuals with higher homocysteine concentration compared to lower [144, 145]. However, recent studies have failed to demonstrate similar findings [146, 147]. Moreover, an interventional clinical trial that used vitamin B12 and folate therapy to reduce plasma homocysteine levels demonstrated no significant reduction in blood pressure despite reductions in homocysteine concentrations at 2 years [148]. The role of homocysteine in the pathogenesis of hypertension has been called into question and it is currently believed that there is a complex interplay between advancing age and altered homocysteine metabolism. As a result, homocysteine may not be as useful a biomarker for hypertension.

Multibiomarker approach

Other biomarkers that have been studied are related to: (1) the coagulation system, specifically plasminogen activator inhibitor-1 and D-dimer [120]; (2) inflammatory cytokines such as IL-1, IL-6, TNF-alfa [123, 149, 150]; (3) peptide biomarkers like BNP, angiotensin-II, serum aldosterone, and plasma renin; and (4) biochemical markers like homocysteine, uric acid, fibrinogen, and 25-hydroxy vitamin D3. Wang et al. studied 3500 participants using a multimodal biomarker predictive model approach and estimated the above listed markers. A set of three biomarkers, CRP, PAI-1, and UACR, proved to have strong association with development of EH. This study emphasizes that none of the markers of inflammation can be used as a standalone modality to predict risk [61]. Even as a group, their value in predicting risk is limited due to practical constraints.

Novel biomarkers of oxidative stress

Nicotinamide adenine dinucleotide phosphate reduced oxidase 5

Inflammation leads to reactive oxygen species (ROS) and reactive nitrogen species (RNS), which lead to oxidative stress. Various signaling pathways are triggered that result in functional (endothelial dysfunction) as well as structural damage to the vessel wall. These processes are critically involved in the development and progression of hypertension [44]. Beyond the vasculature, the ROS-induced damage also occurs in the target organs. Nicotinamide adenine dinucleotide phosphate reduced oxidase 5 (NoX5) is a kidney-specific ROS that has recently been described in EH [151, 152].

Oxidized LDL

Oxidized low density lipoproteins (ox-LDL) are reported to be increased significantly in individuals with EH [153]. Ox-LDL has proinflammatory, prothrombotic, and proapoptotic properties. It causes stimulation of the monocyte migration and infiltration into the vessel wall as well as proliferation of the vascular smooth muscle cells causing atheromatous plaques. Therefore, this might indicate progression of the disease.

8-Isoprostaglandin F2a

The lipid peroxidation process is also enhanced in EH. There is significant increase in the plasma levels of plasma 8-isoprostaglandin F2a (8-ISO PGF2a), which is a by-product of esterified arachidonic acid in the walls of cell membranes. It is a potent renal vasoconstrictor and also stimulates release of endothelin-1 from the endothelial cells lining the aorta [154]. 8-ISO PGF2a is a highly stable isoprostane and excellent biomarker for oxidative stress *in-vivo*. Other biomarkers for oxidative stress are breakdown products of arginine like asymmetric dimethylarginine (ADMA) and symmetric dimethylarginine (SDMA) [151]. There has been considerable progress in our understanding of the interplay between these pathogenic mechanisms in the past two decades. However, given the complex nature of oxidative stress and the nonspecific nature of biomarkers for oxidative stress, it is unlikely that any single biomarkers would be clearly useful [3]. None of the biomarkers has proven to be qualified for diagnosis, prognosis, or management of hypertension. At this time, their application remains investigational.

Neutrophil-to-lymphocyte ratio

A few hematological biomarkers have gained attention recently for their association with cardiac and noncardiac diseases. An elevated neutrophil-to-lymphocyte ratio (NLR) was found to correlate significantly with the increased risk of development of hypertension in a Chinese cohort of 28,850 normotensive individuals who were followed for 6 years [155]. Apart from being a biomarker for predicting the risk, it has also been found to be a good prognostic biomarker. Hypertensive patients with higher NLR quartiles have been found to have higher all-cause mortality and the ROC area under the curve was 0.714 (95% CI 0.629–0.798) with a sensitivity of 92.6%, specificity of 52.5%, and critical value of 2.97 [156]. Similarly, NLR has been found to be associated with resistant hypertension [157], diastolic dysfunction [158], and BP variability [159].

Red cell distribution width

Red cell distribution width (RDW) has also gained considerable popularity as a marker of inflammation and found to be increased in several diseases including hypertension [160]. It is a part of the complete blood count profile (CBC) that is easily measurable and reflects the index of heterogeneity of red blood cells (RBC) size in the circulation. It was found that RDW is increased in individuals with prehypertension and hypertension compared to healthy individuals [161]. Hypertensive patients with increased RDW were found to have more carotid intima media thickness (IMT) and plaques [162]. Though RDW is a simple and readily available test that may be helpful in identifying patients at greater risk of development of hypertension, it is still not widely used in clinical settings.

Biomarkers predictive of development of hypertension

Peripheral blood pressure

Blood pressure has been the sole biomarker to diagnose and manage hypertension. Blood pressure above the cutoff value of 140 mm of Hg for systolic blood pressure (SBP) and 90 mm of Hg for diastolic blood pressure (DBP) is labeled as hypertension in an adult (<60 years) [163]. Prehypertension was proposed by JNC-7 and retained as a class by JNC-8 [22, 163]. This defines SBP 120–139 mm of Hg and DBP 80–9 mm of Hg, which lies between normal blood pressure (SBP <120, DBP <80) and hypertension. High normal blood pressure (HNBP; SBP 13–139, DBP 85–89) proposed by JNC-6 was embedded in the prehypertension subsequently. Established hypertension is preceded by abnormalities in several pathophysiological pathways and hence transition from normotension to prehypertension could foretell the increased risk of hypertension. Various longitudinal studies suggest that individuals with HNBP were two to three times more likely to develop hypertension compared to the normotensives [164]. In the TROPHY (TRial Of Preventing HYpertension) trial, 52% of the subjects developed hypertension in the placebo group over a period of 4 years [165]. Apart from predicting the risk of development, HNBP and prehypertension also predicted the risk of cardiovascular events even after adjusting for the risk of progression (hazard ratio ranged from 1.42 to 2.33 in a meta-analysis of eight studies) [164]. Detection within the window period provides an opportunity to screen for other cardiovascular risk factors like obesity, hyperlipidemia, impaired glucose tolerance, and smoking, and enables introducing necessary lifestyle modifications as well as drug therapy as primary prevention measures. Therefore, blood pressure is a satisfactory biomarker for diagnosis as well as predicting risk and prognosis.

Markers of renal function

As described in the above sections in the process of pathogenesis, eGFR and UACR are important renal function parameters that could indicate an impending risk of development of hypertension.

Biochemical markers

Several circulating biomarkers as described in the pathogenesis section have been found to be associated with development of hypertension. The biomarkers that look promising for risk stratification after reviewing the available data are CRP (>3 mg/L as cutoff) and serum uric acid (>5.7 mg/dL for women and >7.0 mg/dL for men). However, their use as a sole biomarker may not be free of prejudice, hence the overall all risk evaluation should take into consideration other traditional risk factors like hyperlipidemia, hyperglycemia, insulin resistance, physical inactivity, high BMI, etc.

The other circulating biomarkers that seem to be of value are adrenomodulin, ARR, PAI-1, fibrinogen, renin, aldosterone, BNP, N-terminal proatrial natriuretic peptide, parathyroid hormone, cardiac troponin, insulin sensitivity, and lipoprotein. The list of biomarkers appears long and clearly these associations are of research interest for further evaluation. A multimarker approach as described by *Wang et al.* could be a useful strategy, but identifying the confounding factors during interpretation remains a challenge. Nevertheless, understanding the relative importance of these biomarkers would aid understanding of the predominant mechanism responsible for causing elevated blood pressure in an individual so that specific system-targeted prevention therapies could be instituted instead of a general approach.

Biomarkers predicting treatment response and monitoring therapy

Therapeutic response to a particular antihypertensive drug is typically difficult to predict. Though about 5–10 mm of Hg reduction is expected with any antihypertensive drug, it may vary from patient to patient. Most of the time the treatment is chosen on a trial and error basis until the target blood pressure is attained. Therefore, cycling of the drugs is the usual clinical practice in the early part of therapy. The currently available drugs cause either vasodilatation, reducing peripheral vascular resistance, or volume contraction, leading to reduced cardiac output. These drugs are used either alone or in combination to attain the blood pressure goal. If there are biomarkers that could help predict the response to a specific antihypertensive agent, these may possibly curb the predominant pathophysiology and provide a useful means of monitoring therapy.

The UK's National Institute of Clinical Excellence (NICE) guidelines propose age and ethnicity as important surrogates to choose the first-line antihypertensive. Calcium channel blockers (CCB) are the first-line drugs to be started in individuals over 55 years of age and for those at any age who are of African or Caribbean origin, but an angiotensin inhibitor or angiotensin receptor blocker for individuals who are less than 55 years and of any other ethnic origin. This is based on the assessment of RAAS activity in these populations [166]. This strategy has recently been validated by the PATHWAY-2 clinical trial, where spironolactone, the aldosterone antagonist, was found to provide the maximum blood pressure reduction in cases of resistant hypertension that had high levels of plasma renin [167]. The veteran affairs study suggests that age and ethnicity were better predictors of response compared to the plasma renin activity

[168]. Other studies found that plasma renin activity is an independent predictor of response to atenolol and hydrochlorothiazide other than age and race [169].

However, age, race, and plasma renin activity have been proposed to explain only 50% of the variability in blood pressure response to single drug therapy [170], and use of combination treatment makes the situation more complex. Several genetic and nongenetic studies have attempted to identify the predictors of response to various antihypertensive drugs, but these findings correlate only modestly and many of them were not reproducible [171–177].

Blood pressure and its several dimensions

The multimodal inheritance, complex pathophysiologic mechanisms, and significant environmental influence prevent simplification of treatment approach based on biomarkers. Therefore, the reductionist approach in clinical management is still prevalent, which treats the phenotype peripheral blood pressure and relies on the SBP/DBP as the most important biomarker to monitor response to treatment.

Cardiovascular complications still occur in hypertensive patients despite attaining the recommended treatment goals and maintaining blood pressure at optimal levels. Prediction of risk in well-controlled hypertension has been challenging. Several characteristics related to blood pressure measurements have been investigated to determine if any of these parameters could be used as a biomarker.

Blood pressure variability

Blood pressure is a dynamic variable with minute-to-minute variability. Short and long-term fluctuations are perceptible. These fluctuations are due to a complex interplay between several cardiovascular regulatory mechanisms, varied types of physical activity, environmental influences, and circadian rhythms. The extent of fluctuations, which is termed "blood pressure variability" (BPV), varies from person to person. It has been classified into two main types: short term (over 24-h) and long term (over days, weeks, and months). Several factors have been proposed to underlie episodes of BPV like impaired autonomic reflexes, renal dysfunction, increased arterial stiffness, etc. It has also been observed that the higher the blood pressure, the higher the BPV [178]. Therefore, it is reasonable to expect that high BPV would be a good prognostic biomarker to predict risk of complications of hypertension, which could be used to monitor therapy antihypertensive treatment. The reference value for adults for normal CBP is 110/80, and >130/90 is considered as hypertension [179]. BPV has been extensively studied in the last two decades. Unfortunately, the prognostic significance of BPV remains controversial. Studies have reported it to be an important biomarker for cardiovascular events [180–182], target for organ damage [183–185], and mortality [186]. However, several large observational studies have failed to demonstrate similar findings [187]. A recently published observational study has also reported discordant findings. This study found that there was a significant correlation between BPV and cardiovascular mortality. However, the addition of BPV into the risk model did not improve the performance of cardiovascular risk score. Therefore, despite the plethora of studies evaluating BPV as a biomarker for targeting the antihypertensive treatment, clinical use of BPV as a biomarker is limited.

Central blood pressure

Peripheral blood pressure (PBP) is a blood pressure reading taken in the arm from the brachial artery. The SBP varies throughout the arterial tree. The pressure in the aorta is known as the central blood pressure (CBP) and can be measured invasively or noninvasively. The CBP is usually lower than the PBP and varies with age, posture, heart rate, and BP itself [188]. The difference between CBP and PBP is more evident in hypertension because of the stiffening of the peripheral arteries. The difference is highly variable between individuals, and antihypertensive drugs may have a differential effect on the CBP and PBP [189–191]. Differences between CBP and PBP may be clinically important since CPB directly reflects the left ventricular load [192]. In recent years, great emphasis has been placed on the role of central aortic blood pressure, and it has been observed that individuals with similar PBP may have different CBP. CBP has been found to predict the risk of development of hypertension [193]. Many studies also suggested that CBP may be a superior target compared to brachial pressure in monitoring response to antihypertensive treatment [192, 194]. In addition, CBP was found to be an independent predictor and demonstrated a stronger relationship with target organ damage in several studies [179, 195–197]. Contrary to this, other studies did not find a better predictive value, though it was found to be an independent risk factor [198–200]. To resolve these conflicting results, there is a need to develop better noninvasive measurement methods that must be standardized and validated. Moreover, well-designed clinical trials are needed to assess the outcomes and prove the robustness of CBP as a biomarker to monitor treatment response and risk prediction [189, 201, 202].

Pulse pressure

Pulse pressure (PP) refers to the difference between SBP and DBP. Like SBP and CBP, PP is also suggested to be an independent predictor of cardiovascular complications in hypertension. A wide pulse pressure is usually indicative of a noncompliant stiff aorta with a reduced ability to distend and recoil, hence the SBP increase and DPB reduces, leading to increased PP. The principle is the same as in PWV, indicating stiffening of the arteries. Repeated analysis of data from large observational studies, like the Framingham heart study, showed that PP could be a better indicator of cardiovascular risk [203–205]. An association between PP and cardiovascular outcomes has also been seen in the antihypertensive drugs treatment trials. In the Systolic Hypertension in Elderly Program (SHEP) study, a 10 mm Hg rise in pulse pressure was associated with a 32% increase in the risk of heart failure and a 24% increase in the risk of stroke even after controlling for SBP. However, this increase was not seen in patients assigned to placebo [206]. In the Systolic Hypertension in Europe (Sys-Eur) study, every 10 mmHg increase in pulse pressure was associated with a hazard ratio ranging from 1.25 to 1.68 in the placebo arm but not significant in the treatment arm [207]. The reduction in pulse pressure with antihypertensive drugs may not be similar to the degree of reduction in SBP [208, 209]. In the Losartan Intervention for Endpoints (LIFE) trial, the extent of reduction of PP was better with losartan compared to atenolol. There was also a significant increase in cardiovascular death seen with increasing quartile of PP for atenolol. This was not seen for losartan, although there was an upward trend [210]. Similarly, in the Conduit Artery Functional Evaluation (CAFE) study, atenolol was found to decrease PP less compared to calcium channel blockers for similar reduction in SBP [211]. However, other studies suggested that SBP is a better predictor of cardiovascular risk reduction than PP [212, 213]. As a result, the current scientific literature is divided on the prognostic role of PP as a target for treatment and monitoring [214, 215]. The current treatment guidelines do not consider it as a target for treatment. However, clinicians may consider increased PP in the elderly as a biomarker to upscale the risk estimate in conjunction with SBP, which is the primary treatment goal. Nevertheless, PP is a potential tool for research to understand the pharmacodynamic effect of the antihypertensive drugs, and further research is needed to understand its role as a prognostic biomarker [216].

Biomarkers in disease progression and target organ damage

There is an established causal relationship between the blood pressure and cardiovascular events [44, 217]. Vascular damage leads to damage of the target organs, which results in cardiovascular events. Though there is a relationship between the extent of blood pressure elevations and cardiovascular events, this does not equate strongly when it comes to target organ damage [44]. For example, the degree of left ventricular hypertrophy is less closely related to raised SBP compared to the risk of congestive cardiac failure or myocardial ischemic events and raised SBP [218]. Therefore, there is an opportunity to identify the individuals at intermediate risk of events who are yet to develop full-blown target organ damage, providing scope for prevention.

Renal damage

Serum creatinine, eGFR, and UACR are the recommended biomarkers to assess the excretory function of the kidneys [219]. Therefore, any decrease in eGFR to less than 60 mL/min and an increase in UACR above 30 mg/g of creatinine is an indicator of onset of renal damage. These are well-recognized screening tools but the balance between sensitivity and specificity is suboptimal for both. Neither are biomarkers specific for hypertension-induced renal damage, nor are they specific enough to detect very early stages of renal disease. Therefore, patients with uncomplicated hypertension need more avenues of detection of risk [44]. A large number of other biomarkers have been investigated to predict nephropathy, such as adrenomodulin, natriuretic peptides, plasma renin, and serum aldosterone [220]. But neither solely or as a group do they merit replacing the three biomarkers used clinically.

Novel biomarkers are being investigated to overcome the challenges posed by conventional circulating biomarkers. Advanced imaging modalities have become very popular. Blood oxygenation level dependent (BOLD) MRI and MRI-based arterial spin labeling (ASL) provide insight into renal perfusion and oxygenation [221–223]. These noninvasive imaging-based biomarkers have the potential to improve diagnostic accuracies for early stage renal damage [44]. Future studies need to examine whether they would be successful biomarkers in the management of hypertension.

Several proteomics studies have been conducted to look for biomarkers of renal damage in the plasma as well as urine [224, 225]. They have been encouraging, but it would be desirable if they could be classified into early signals of damage and disease progression so as to enable monitoring of antihypertensive therapy [226].

Cardiac damage

The heart is cyclically associated with blood pressure. Physiologically, blood pressure is dependent on cardiac output and elevated blood pressure affects the heart pathologically [227]. Hypertensive heart disease is a result of diverse perturbations in the changes of cardiac function as well as structure. Diastolic dysfunction, such as impaired diastolic filling, increases left ventricular filling pressure, leading to concentric hypertrophy of the left ventricle, and subsequently left arterial enlargement. Systolic dysfunction sets in at later stages, leading to reduction in the ejection fraction (<50%). A concomitant right ventricular hypertrophy (RVH) is also noted in 20%–80% of individuals, and its severity and prevalence correlates with left ventricular hypertrophy (LVH) [228]. Studies evaluating antihypertensive drugs have reported significant reduction in LVH [229–232]. However, the magnitude of reduction correlates only modestly with the extent of decline in BP [229, 230]. Improvement in diastolic function with antihypertensive therapy has not been consistent in clinical studies. Some studies reported that diastolic dysfunction improves with therapy [233–235]. However, two large RCTs with longitudinal serial echocardiographic evaluation have failed to demonstrate the impact of antihypertensive treatment on diastolic function despite adequate blood pressure control and significant regression of LV mass [230, 231]. Therefore, there seems to be a dissociation between the functional and structural biomarkers of cardiac damage in hypertension. This has been attributed to the multifactorial origin of diastolic dysfunction [228]. As a result, LVH is considered an independent risk for cardiovascular events and as a standalone treatment target. ECG based screening methods are usually recommended for LVH followed by echocardiography or MRI depending on the availability and expertise [44].

Microalbuminuria has been considered an important predictor of left ventricular function and cardiovascular events. This is commonly relied upon in clinical practice for risk stratification of patients in conjunction with blood pressure [236]. A novel biomarker, Galectin-3, was found to be increased in hypertension with a strong association with LVH and may be helpful in detecting subclinical cardiac damage in hypertensive patients [237]. Similarly, soluble angiotensin converting enzyme 2 (sACE2) activity has been found to correlate with imminent heart failure [238]. These biomarkers are still investigational and future studies are required to find their applicability. Several other indicators like circulatory markers of collagen turnover, brain natriuretic peptides, and cardiac troponins have been reported to be altered in hypertension. However, their incremental value over the currently used traditional biomarkers, blood pressure, and ECG appear too small to justify their use in routine clinical practice [44].

PRA has been linked to risk of adverse cardiovascular outcomes. As the cardiac remodeling and fibrosis is due to the activated RAAS system, the PRA is a logical biomarker for assessing the risk of end organ damage. Higher PRA levels demonstrated increased risk for ischemic heart events and congestive heart failure and a trend toward higher mortality among individuals with SBP $\geq$140 mmHg. However, it was not seen among those with SBP $<$140 mmHg [239]. PRA has been a suggested marker for more intense antihypertensive therapy as it could be a surrogate of cardiac damage [240]. Higher PRA was associated with greater likelihood for prevalent IHD and CHF but not CED [241]. However, other studies do not support the use of PRA to predict future risk of cardiovascular events in established CHD. Therefore, use of PRA as a biomarker for cardiac damage is limited in clinical settings [242].

Multiparameter biomarkers based on omics techniques appear to have the potential to overcome the limitations of a single biomarker. Several studies have evaluated the proteomic signatures for early functional cardiac damage as well as markers predictive of late cardiac dysfunction and failure [243, 244]. At the moment, such biomarkers are not yet ready for clinical use, but with rapidly developing technologies and computational biology, omics-based biomarkers are likely to become more precise and less expensive [38]. More exploratory research and clinical investigations are required in this field.

Cerebral damage

Several studies have established a strong correlation between cerebrovascular events and acute cerebrovascular events like hemorrhagic stroke [217]. Intracranial bleeds are significantly related to the severity of rise of blood pressure. The odds of bleeding are 3.68 for SBP >160 mm of Hg and 2.2 for 140–159 mm of Hg [230, 245]. The hypertension-induced damage to the cerebral blood vessels is responsible for spontaneous bleeds. The damage is due to cerebral arterial remodeling, vascular endothelial activation, inflammation, and impairment of cerebral autoregulation. The vessels that are more prone to bleeding are the perforating arteries that supply the deeper part of the brain like the basal ganglia, thalamus, and brain stem [246]. Apart from these devastating consequences, the slow damage to small as well as large cerebral vessels is responsible for the progressive cognitive decline in hypertensive individuals.

Cerebral microvascular damage is considered to be the surrogate of vascular dementia in hypertension [247]. Certain features observed in the MRI scans characterize the structural damage to the vessels [248]. White matter lesions are usually located around ventricles closer to the ventricular horns. They are hyperintensities on fluid-attenuated inversion recovery (FLAIR) and proton density/T2-weighted images, without prominent hypointensity on T1-weighted images [249–252]. Enlarged perivascular spaces (EPVS) or the Virchow Robin spaces are another marker of hypertension induced cerebral small vessel damage (CSVD). EPVS are identified in T2-weighted MRI and are characterized by punctate or linear signal intensities [251, 252]. These changes have been demonstrated in basal ganglia as well as white matter. There is no consensus about the most common site of EPVS, as higher SBP has been demonstrated to be associated with basal ganglia CSVD [252, 253]. Recently, a study by *Zhang et al.* found that white matter lesions are more commonly associated with high SBP than basal ganglia [254]. Lacunar infarcts (LI) are 3–15 mm in diameter and typically located in the internal capsule, basal ganglia, corona radiata, thalamus, or brain stem. They are caused by the occlusion of a perforating artery [255]. They are detected as hypodense lesions on T1, hyperintense on T2/FLAIR with restricted diffusion, and T1C+ may show enhancement. Chronic lesions are isointense to CSF but may demonstrate a hyperintense rim of marginal gliosis in T2/FLAIR lesions. LI is usually considered benign, but appropriate secondary preventive measures may decrease the onset of cognitive decline and stroke recurrence [256].

The Ohasama study with 7-year follow-up revealed that LI is a better indicator of declines in motivation, interest, and reaction to the environment compared to white matter lesion [256]. These imaging biomarkers are promising in determining the most effective antihypertensive regimen to delay cerebral microvascular damage and dementia. It appears plausible that appropriate blood pressure control can help reduce the damage to blood vessels and curtail incidence of dementia [257]. However, tight blood pressure control, especially in the elderly, may result in cerebral hypoperfusion and increase the incidence of falls [258]. Therefore, MRI-based biomarkers are generally used to assist diagnosis, but they are of little utility in clinical settings as most of the damage has already occurred [259]. PET-based biomarkers have also been proposed but have not been extensively evaluated [260]. PWV, a marker of aortic stiffness, has also been related to cognitive decline [261].

Many circulatory biomarkers have been evaluated but only a few novel markers merit mention. Ubiquitin C-terminal hydrolase L1 (UCH-L1) has been found to be associated with cerebral white matter lesion [262]. Vascular cell adhesion molecule-1 (VCAM-1) has been shown to predict decreased cerebral blood flow and could identify individuals at increased risk of falling. Neuron-specific enolase (NSE) looks promising in terms of cost as a biomarker [263]. At this point, these biomarkers are still investigational. With increasing life expectancy and prevalence of vascular dementia, more research is needed to discover biomarkers of cognition that could contribute to better therapeutic decisions and improve outcomes.

Retinal damage

Hypertension affects the retinal blood vessels. The changes seen are choroidopathy, retinopathy, and optic neuropathy [264]. The incidence of retinopathy is related to the degree of severity and duration of hypertension [265]. Substantial evidence suggests that hypertensive retinopathy could be a predictor of cardiovascular [266, 267], cerebrovascular [268], and renal [269] morbidity and mortality due to hypertension. Blood pressure is the most important biomarker to predict development of retinopathy, and studies have shown that treatment of hypertension also leads to regression of retinal changes [270]. In a large multicentric cohort, a 10 mm of Hg reduction in blood pressure reduced the risk of retinopathy by 10% [271].

There is a group of genetic biomarkers for risk of hypertensive retinopathy. Deletion of the allele of the angiotensin-converting enzyme has a higher risk associated with the development of hypertensive retinopathy [271, 272]. In a genome-wide association study (GWAS), it was found that a novel loci 12q24 is associated with narrowing of retinal vessels, hypertension, and cardiovascular disease [273]. In another GWAS, a locus of 80 kb on chromosome 5 was found to be associated with renal arteriolar caliber, though it did not correlate significantly with the retinal endpoints [274]. In a large discovery cohort of 24,000 participants of multiple ethnicities and a validation cohort of 5000 European subjects, single-nucleotide polymorphism rs201255422 was associated with both systolic and diastolic BPs [275]. Several micro RNA have also been proposed as biomarkers for retinal damage [276]. Despite the number of proposed genetic markers, their role has yet been clearly affirmed, as hypertensive end organ damage is multifactorial with a combination of genetic susceptibilities and environmental factors. Future clinical trials are needed to prove their role as diagnostic or prognostic biomarkers in hypertensive retinal damage.

Several fundus imaging methods and analysis techniques are being used to assess retinal damage in hypertensive patients. Optical coherence tomography (OCT) provides objective measurements of retinal vessel lumen diameters and wall thicknesses [41]. Schuster et al. demonstrated a relationship between mean arterial BP and OCT-based arterial-venous ratio [42]. A decrease in choroidal thickness and accumulation of subretinal fluid has been found to be associated with poor visual outcomes for patients with hypertension [44]. OCT is useful in assessing retinal damage so that appropriate antihypertensive medications may be used to reduce the blood pressure and minimize long-term damage [277]. It has been demonstrated that retinal vessel caliber is an independent predictor of cardiovascular outcomes [278]. Adaptive optics techniques allow vascular phenotyping in vivo at a near histology level and could be considered as an excellent tool to assess retinal damage [279]. An increase in the wall-to-lumen ratio (WLR) is the hallmark of retinal microangiopathy and predictive of hypertensive end organ damage [280]. It has also been demonstrated that WLR normalization could be achieved in treated and hypertension-controlled patients [279]. Therefore, adaptive retinal imaging could be considered as an important biomarker for stratification of end organ damage [278]. Scanning laser Doppler flowmetry is another valuable tool that measures retinal blood flow in real time. The flow is dependent on the WLR, which in turn depends upon the blood pressure [281]. Retinal imaging analysis provides an excellent opportunity to study disease pathophysiology and risk stratification [282]. Nevertheless, standardized measurement protocols evaluating their clinical use and risk prediction have yet to be established.

Biomarkers to individualize antihypertensive treatment

The epidemiological figures suggest that only 66% of patients treated with antihypertensive drugs attain blood pressure goals. A "one-size-fits-all" approach is the usual norm in choosing a drug, and most patients undergo a tedious titration process. This results in poor compliance and a sizable mortality despite the availability of a myriad class of antihypertensive drugs. The modern approaches should use diagnostic and screening tools to characterize the distinctive traits or predominant pathological types that may differentiate the response to individual drugs.

As discussed in the previous section, the NICE guidelines suggest that age and ethnicity are the two most important considerations for choosing drugs. Since there are no competing claims, one might believe that it is almost universal. However, data are lacking for several other ethnicities like Asians of Mongoloid origin, Southeast Asians, and several others.

Stratifying the patients to the physiological phenotype, high PRA or salt-sensitive (low PRA) was probably the starting point of personalizing antihypertensive treatments about two decades ago. The recommended drug for high PRA is ACEI/ARB and for low PRA it is CCB or diuretics. Limited evidence suggested that the PRA-informed treatments had better blood pressure control than those not informed. However, PRA never gained clinical utility due to the complex nature of the disease and associated confounders.

Several pharmacogenomics studies suggest a range of biomarkers as elaborated in the previous section, but their utility in clinical practice is negligible for treating EH. However, the cases of secondary hypertension and other syndromic conditions with resistant hypertension are of great use. Targeted screening for these conditions certainly helps the clinician to choose the best therapeutic regimen and combination of drugs.

The future of personalized medicine in managing hypertension is a systems biology approach with multiomics data to build a network of integrated models with input from genes, RNA, protein, and small molecules. Computer-assisted bioinformatic tools and advanced mathematical modeling, such as machine learning techniques, are being used to understand the complex nature of the disease and find models for risk stratification, treatment individualization, and monitoring. This holds great promise and it is currently a priority area of research worldwide. Blood pressure as a phenotype is expected to be dissected into meaningful model phenotypes that would enable practice of precision medicine in management of hypertension.

Conclusion

Blood pressure has been the eternal biomarker in the management of hypertension despite its limitations. Despite a considerable amount of research, none of the other biomarkers has proven to be as robust as blood pressure in clinical practice. Ancillary use of other biomarkers like PWV and imaging techniques may add to better risk stratification. The goals for future research are to elucidate key biomarkers of hypertension management and individualize therapy, thereby allowing a reduction in the end organ damage.

References

[1] WHO. A global brief on hypertension. World Health Organization; 2013.

[2] Ovbiagele B, Smith SC, Spencer CC, Stafford RS, Taler SJ, Thomas RJ, Williams KA, Williamson JD, Wright JT. 2017 ACC/AHA/AAPA/ABC/ACPM/AGS/APhA/ASH/ASPC/NMA/PCNA guideline for the prevention, detection, evaluation, and management of high blood pressure in adults: a report of the American College of Cardiology/American Heart Association task force on clinical practice guidelines. Hypertension 2018;71(6):e13–e115.

[3] Shere A, Eletta O, Goyal H. Circulating blood biomarkers in essential hypertension: a literature review. J Lab Precis Med 2017;2(12)https://doi.org/10.21037/jlpm.2017.12.06.

[4] Hajjar I, Kotchen TA. Trends in prevalence, awareness, treatment, and control of hypertension in the United States, 1988–2000. JAMA 2003;290:199–206.

[5] Chow CK, Teo KK, Rangarajan S, Islam S, Gupta R, Avezum A, et al. Prevalence, awareness, treatment, and control of hypertension in rural and urban communities in high-, middle-, and low-income countries. JAMA 2013;310(9):959–68.

[6] Wang TJ, Vasan RS. Epidemiology of uncontrolled hypertension in the United States. Circulation 2005;112:1651–62.

[7] Fields LE, Burt VL, Cutler JA, Hughes J, Roccella EJ, Sorlie P. The burden of adult hypertension in the United States 1999 to 2000: a rising tide. Hypertension 2004;44:398–404.

[8] Staessen JA, Fagard R, Thijs L, Celis H, Arabidze GG, Birkenhager WH, et al. Randomised double-blind comparison of placebo and active treatment for older patients with isolated systolic hypertension. The Systolic Hypertension in Europe (Syst-Eur) Trial Investigators. Lancet 1997;350(9080):757–64.

[9] Beckett NS, Peters R, Fletcher AE, Staessen JA, Liu L, Dumitrascu D, et al. Treatment of hypertension in patients 80 years of age or older. N Engl J Med 2008;358(18):1887–98.

[10] Rodgers A, Ezzati M, Vander Hoorn S, Lopez AD, Lin RB, Murray CJ. Distribution of major health risks: findings from the Global Burden of Disease study. PLoS Med 2004;1(1):e27.

[11] D'Agostino Sr RB, Vasan RS, Pencina MJ, Wolf PA, Cobain M, Massaro JM, et al. General cardiovascular risk profile for use in primary care: the Framingham Heart Study. Circulation 2008;117(6):743–53.

[12] Conroy RM, Pyorala K, Fitzgerald AP, Sans S, Menotti A, De Backer G, et al. Estimation of ten-year risk of fatal cardiovascular disease in Europe: the SCORE project. Eur Heart J 2003;24(11):987–1003.

[13] Woodward M, Brindle P, Tunstall-Pedoe H. Adding social deprivation and family history to cardiovascular risk assessment: the ASSIGN score from the Scottish Heart Health Extended Cohort (SHHEC). Heart 2007;93(2):172–6.

[14] Hippisley-Cox J, Coupland C, Vinogradova Y, Robson J, Minhas R, Sheikh A, et al. Predicting cardiovascular risk in England and Wales: prospective derivation and validation of QRISK2. BMJ 2008;336(7659):1475–82.

[15] Assmann G, Cullen P, Schulte H. Simple scoring scheme for calculating the risk of acute coronary events based on the 10-year follow-up of the prospective cardiovascular Munster (PROCAM) study. Circulation 2002;105(3):310–5.

[16] Thomopoulos C, Parati G, Zanchetti A. Effects of blood pressure lowering on outcome incidence in hypertension. 1. Overview, meta-analyses, and meta-regression analyses of randomized trials. J Hypertens 2014;32(12):2285–95.

[17] Xie X, Atkins E, Lv J, Bennett A, Neal B, Ninomiya T, et al. Effects of intensive blood pressure lowering on cardiovascular and renal outcomes: updated systematic review and meta-analysis. Lancet 2016;387(10017):435–43.

[18] Biomarkers Definitions Working Group. Biomarkers and surrogate endpoints: preferred definitions and conceptual framework. Clin Pharmacol Ther 2001;69(3):89–95.

[19] https://www.cdc.gov/features/undiagnosed-hypertension/index.html.

[20] Banerjee D, Chung S, Wong EC, Wang EJ, Stafford RS, Palaniappan LP. Underdiagnosis of hypertension using electronic health records. Am J Hypertens 2012;25(1):97–102.

[21] Mozaffarian D, Benjamin EJ, Go AS, Arnett DK, Blaha MJ, Cushman M, et al. Heart disease and stroke statistics—2016 update. Circulation 2016;133(4):e38–e360.

[22] Chobanian AV. Shattuck lecture. The hypertension paradox—more uncontrolled disease despite improved therapy. N Engl J Med 2009;361(9):878–87.

[23] Tocci G, Rosei EA, Ambrosioni E, Borghi C, Ferri C, Ferrucci A, et al. Blood pressure control in Italy: analysis of clinical data from 2005–2011 surveys on hypertension. J Hypertens 2012; 30(6):1065–74.

[24] Saxena T, Ali AO, Saxena M. Pathophysiology of essential hypertension: an update. Expert Rev Cardiovasc Ther 2018;16(12):879–87.

[25] Johnson RJ, Rodriguez-Iturbe B, Kang DH, Feig DI, Herrera-Acosta J. A unifying pathway for essential hypertension. Am J Hypertens 2005;18:431–40.

[26] Cain AE, Khalil RA. Pathophysiology of essential hypertension: role of the pump, the vessel, and the kidney. Semin Nephrol 2002;22(1):3–16.

[27] Johnson RJ, Feig DI, Nakagawa T, Sanchez-Lozada LG, Rodriguez-Iturbe B. Pathogenesis of essential hypertension: historical paradigms and modern insights. J Hyperten 2008;26:381–91.

[28] Androulakis ES, Tousoulis D, Papageorgiou N, Tsioufis C, Kallikazaros I, Stefanadis C. Essential hypertension: is there a role for inflammatory mechanisms? Cardiol Rev 2009;17:216–21.

[29] Padmanabhan S, Caulfield M, Dominiczak AF. Genetic and molecular aspects of hypertension. Circ Res 2015;116(6):937–59.

[30] Dodoo SN, Benjamin IJ. Genomic approaches to hypertension. Cardiol Clin 2017;35:185–96.

[31] Ehret GB, Munroe PB, Rice KM, Bochud M, Johnson AD, Chasman DI, et al. Genetic variants in novel pathways influence blood pressure and cardiovascular disease risk. Nature 2011; 478(7367):103–9.

[32] Manolio TA, Collins FS, Cox NJ, Goldstein DB, Hindorff LA, Hunter DJ, et al. Finding the missing heritability of complex diseases. Nature 2009;461(7265):747–53.

[33] Padmanabhan S, Melander O, Hastie C, Menni C, Delles C, Connell JM, et al. Hypertension and genome-wide association studies: combining high fidelity phenotyping and hypercontrols. J Hypertens 2008;26(7):1275–81.

[34] Marques FZ, Campain AE, Tomaszewski M, Zukowska-Szczechowska E, Yang YH, Charchar FJ, et al. Gene expression profiling reveals renin mRNA overexpression in human hypertensive kidneys and a role for microRNAs. Hypertension 2011;58(6):1093–8.

[35] Marques FZ, Romaine SP, Denniff M, Eales J, Dormer J, Garrelds IM, et al. Signatures of miR-181a on the renal transcriptome and blood pressure. Mol Med (Cambridge, MA) 2015; 21(1):739–48.

[36] Wang G, Wu L, Chen Z, Sun J. Identification of crucial miRNAs and the targets in renal cortex of hypertensive patients by expression profiles. Ren Fail 2017;39(1):92–9.

[37] Delles C, Carrick E, Graham D, Nicklin SA. Utilizing proteomics to understand and define hypertension: where are we and where do we go? Expert Rev Proteomics 2018;15(7):581–92.

[38] Lindsey ML, Mayr M, Gomes AV, Delles C, Arrell DK, Murphy AM, et al. Transformative impact of proteomics on cardiovascular health and disease: a scientific statement from the American Heart Association. Circulation 2015;132(9):852–72.

[39] Lifton RP, Gharavi AG, Geller DS. Molecular mechanisms of human hypertension. Cell 2001;104:545–56.

[40] Lifton RP. Genetic dissection of human blood pressure variation: common pathways from rare phenotypes. Harvey Lect 2004;100:71–101.

[41] Cowley Jr AW, Roman RJ. The role of the kidney in hypertension. JAMA 1996;275 (20):1581–9.

[42] Fukuda M, Munemura M, Usami T, Nakao N, Takeuchi O, Kamiya Y, et al. Nocturnal blood pressure is elevated with natriuresis and proteinuria as renal function deteriorates in nephropathy. Kidney Int 2004;65(2):621–5.

[43] Kimura G, Dohi Y, Fukuda M. Salt sensitivity and circadian rhythm of blood pressure: the keys to connect CKD with cardiovascular events. Hypertens Res 2010;33(6):515–20.

[44] Currie G, Delles C. Use of biomarkers in the evaluation and treatment of hypertensive patients. Curr Hypertens Rep 2016;18(7):54.

[45] Cabarkapa V, Ilincic B, Deric M, Vucaj Cirilovic V, Kresoja M, Zeravica R, et al. Cystatin C, vascular biomarkers and measured glomerular filtration rate in patients with unresponsive hypertensive phenotype: a pilot study. Ren Fail 2017;39(1):203–10.

[46] Peralta CA, Whooley MA, Ix JH, Shlipak MG. Kidney function and systolic blood pressure new insights from cystatin C: data from the Heart and Soul Study. Am J Hypertens 2006; 19(9):939–46.

[47] Shankar A, Teppala S. Relationship between serum cystatin C and hypertension among US adults without clinically recognized chronic kidney disease. J Am Soc Hypertens 2011; 5(5):378–84.

[48] Inker LA, Schmid CH, Tighiouart H, Eckfeldt JH, Feldman HI, Greene T, et al. Estimating glomerular filtration rate from serum creatinine and cystatin C. N Engl J Med 2012;367(1):20–9.

[49] https://www.aacc.org/publications/cln/articles/2016/april/cystatin-c-and-creatinine-complementary-markers-of-gfr-expert-john-c-lieske-md.

[50] Takase H, Dohi Y, Toriyama T, Okado T, Tanaka S, Sonoda H, et al. Evaluation of risk for incident hypertension using glomerular filtration rate in the normotensive general population. J Hypertens 2012;30(3):505–12.

[51] Wang TJ, Evans JC, Meigs JB, Rifai N, Fox CS, D'Agostino RB, et al. Low-grade albuminuria and the risks of hypertension and blood pressure progression. Circulation 2005;111 (11):1370–6.

[52] Forman JP, Fisher ND, Schopick EL, Curhan GC. Higher levels of albuminuria within the normal range predict incident hypertension. J Am Soc Nephrol 2008;19(10):1983–8.

[53] Takase H, Sugiura T, Ohte N, Dohi Y. Urinary albumin as a marker of future blood pressure and hypertension in the general population. Medicine 2015;94(6):e511.

[54] Park SK, Moon SY, Oh CM, Ryoo JH, Park MS. High normal urine albumin-to-creatinine ratio predicts development of hypertension in Korean men. Circ J 2014;78(3):656–61.

[55] Sung KC, Ryu S, Lee JY, Lee SH, Cheong E, Hyun YY, et al. Urine albumin/creatinine ratio below 30 mg/g is a predictor of incident hypertension and cardiovascular mortality. J Am Heart Assoc 2016;5(9)https://doi.org/10.1161/JAHA.116.003245.

[56] Titze J. Sodium is not just a renal affair. Curr Opin Nephrol Hypertens 2014;23:101–5.

[57] Kopp C, Linz P, Dahlmann A, Hammon M, Jantsch J, Muller DN, et al. 23Na magnetic resonance imaging-determined tissue sodium in healthy subjects and hypertensive patients. Hypertension 2013;61(3):635–40.

[58] Linz P, Santoro D, Renz W, Rieger J, Ruehle A, Ruff J, Deimling M, Rakova N, Muller DN, Luft FC, Titze J, Niendorf T. Skin sodium measured with 23Na MRI at 7.0 T. NMR Biomed 2014;28:54–62.

[59] Hammon M, Grossmann S, Linz P, Kopp C, Dahlmann A, Janka R, et al. 3 Tesla (23)Na magnetic resonance imaging during aerobic and anaerobic exercise. Acad Radiol 2015; 22(9):1181–90.

[60] Dahlmann A, Dorfelt K, Eicher F, Linz P, Kopp C, Mossinger I, et al. Magnetic resonance-determined sodium removal from tissue stores in hemodialysis patients. Kidney Int 2015;87(2):434–41.

[61] Wang TJ, Gona P, Larson MG, Levy D, Benjamin EJ, Tofler GH, et al. Multiple biomarkers and the risk of incident hypertension. Hypertension 2007;49(3):432–8.

[62] Tomaschitz A, Maerz W, Pilz S, Ritz E, Scharnagl H, Renner W, et al. Aldosterone/renin ratio determines peripheral and central blood pressure values over a broad range. J Am Coll Cardiol 2010;55(19):2171–80.

[63] Grassi G, Mark A, Esler M. The sympathetic nervous system alterations in human hypertension. Circ Res 2015;116:976–90.

[64] Salem S, Jankowski V, Asare Y, Liehn E, Welker P, Raya-Bermudez A, et al. Identification of the vasoconstriction-inhibiting factor (VIF), a potent endogenous cofactor of angiotensin II acting on the angiotensin II type 2 receptor. Circulation 2015;131(16):1426–34.

[65] Tenderenda-Banasiuk E, Wasilewska A, Filonowicz R, Jakubowska U, Waszkiewicz-Stojda M. Serum copeptin levels in adolescents with primary hypertension. Pediatr Nephrol 2014; 29(3):423–9.

[66] Afsar B. Pathophysiology of copeptin in kidney disease and hypertension. Clin Hypertens 2017;23:13.

[67] Kato JKK, Eto T. Roles of adrenomedullin in hypertension and hypertensive organ damage. Curr Hypertens Rev 2006;2:283–95.

[68] Wong HK, Cheung TT, Cheung BM. Adrenomedullin and cardiovascular diseases. JRSM Cardiovasc Dis 2012;1(5)https://doi.org/10.1258/cvd.2012.012003.

[69] Kato J, Kitamura K, Eto T. Plasma adrenomedullin level and development of hypertension. J Hum Hypertens 2006;20:566.

[70] Cortese F, Scicchitano P, Gesualdo M, Ciccone MM. The correlation between arterial hypertension and endothelial function. Arch Clin Hypertens 2016;2:001–3.

[71] Feihl F, Liaudet L, Levy BI, Waeber B. Hypertension and microvascular remodelling. Cardiovasc Res 2008;78(2):274–85.

[72] Laurent S, Boutouyrie P. The structural factor of hypertension: large and small artery alterations. Circ Res 2015;116(6):1007–21.

[73] Storch AS, de Mattos JD, Alves R, Galdino ID, Miguens Rocha HN. Methods of endothelial function assessment: description and applications. Int J Cardiovasc Sci 2017;30:262–73.

[74] https://www.escardio.org/Journals/E-Journal-of-Cardiology-Practice/Volume-10/How-to-assess-endothelial-function-for-detection-of-pre-clinical-atherosclerosis.

[75] Ludmer PL, Selwyn AP, Shook TL, Wayne RR, Mudge GH, Alexander RW, Ganz P. Paradoxical vasoconstriction induced by acetylcholine in atherosclerotic coronary arteries. N Engl J Med 1986;315:1046–51.

[76] Campello E, Spiezia L, Radu CM, Dhima S, Visentin S, Valle FD, et al. Circulating microparticles in umbilical cord blood in normal pregnancy and pregnancy with preeclampsia. Thromb Res 2015;136(2):427–31.

[77] Sorensen KE, Celermajer DS, Spiegelhalter DJ, Georgakopoulos D, Robinson J, Thomas O, et al. Non-invasive measurement of human endothelium dependent arterial responses: accuracy and reproducibility. Br Heart J 1995;74(3):247–53.

[78] Patel S, Celermajer DS. Assessment of vascular disease using arterial flow mediated dilatation. Pharmacol Rep 2006;58(Suppl):3–7.

[79] Fathi R, Haluska B, Isbel N, Short L, Marwick TH. The relative importance of vascular structure and function in predicting cardiovascular events. J Am Coll Cardiol 2004;43(4):616–23.

[80] Yeboah J, Folsom AR, Burke GL, Johnson C, Polak JF, Post W, et al. Predictive value of brachial flow-mediated dilation for incident cardiovascular events in a population-based study: the multi-ethnic study of atherosclerosis. Circulation 2009;120(6):502–9.

[81] Xu JZ, Wu SY, Yan YQ, Xie YS, Ren YR, Yin ZF, et al. Left atrial diameter, flow-mediated dilation of brachial artery and target organ damage in Chinese patients with hypertension. J Hum Hypertens 2012;26(1):41–7.

[82] Yang Y, Xu JZ, Wang Y, Tang XF, Gao PJ. Brachial flow-mediated dilation predicts subclinical target organ damage progression in essential hypertensive patients: a 3-year follow-up study. J Hypertens 2014;32(12):2393–400 [discussion 400].

[83] Szmitko PE, Wang CH, Weisel RD, de Almeida JR, Anderson TJ, Verma S. New markers of inflammation and endothelial cell activation: Part I. Circulation 2003;108(16):1917–23.

[84] Meziani F, Tesse A, Andriantsitohaina R. Microparticles are vectors of paradoxical information in vascular cells including the endothelium: role in health and diseases. Pharmacol Rep 2008;60(1):75–84.

[85] Burger D, Schock S, Thompson CS, Montezano AC, Hakim AM, Touyz RM. Microparticles: biomarkers and beyond. Clin Sci (Lond) 2013;124(7):423–41.

[86] Amabile N, Cheng S, Renard JM, Larson MG, Ghorbani A, McCabe E, et al. Association of circulating endothelial microparticles with cardiometabolic risk factors in the Framingham Heart Study. Eur Heart J 2014;35(42):2972–9.

[87] Salem M, Kamal S, El Sherbiny W, Abdel Aal AA. Flow cytometric assessment of endothelial and platelet microparticles in preeclampsia and their relation to disease severity and Doppler parameters. Hematology 2015;20(3):154–9.

[88] de la Sierra A, Larrousse M. Endothelial dysfunction is associated with increased levels of biomarkers in essential hypertension. J Hum Hypertens 2010;24(6):373–9.

[89] Zu LRC, Pan B, Zhou B, Zhou E, Niu C, Wang X, Zhao M, Gao W, Guo L, Zheng L. Endothelial microparticles after antihypertensive and lipid-lowering therapy inhibit the adhesion of monocytes to endothelial cells. Int J Cardiol 2016;202:756–9.

[90] Laurent S, Cockcroft J, Van Bortel L, Boutouyrie P, Giannattasio C, Hayoz D, et al. Expert consensus document on arterial stiffness: methodological issues and clinical applications. Eur Heart J 2006;27(21):2588–605.

[91] Cecelja M, Chowienczyk P. Role of arterial stiffness in cardiovascular disease. JRSM Cardiovasc Dis 2012;1(4)https://doi.org/10.1258/cvd.2012.012016.

[92] Payne RA, Wilkinson IB, Webb DJ. Arterial stiffness and hypertension: emerging concepts. Hypertension 2010;55(1):9–14.

[93] Shahmirzadi D, Li RX, Konofagou EE. Pulse-wave propagation in straight-geometry vessels for stiffness estimation: theory, simulations, phantoms and in vitro findings. J Biomech Eng 2012;134(11):114502.

[94] Najjar SS, Scuteri A, Shetty V, Wright JG, Muller DC, Fleg JL, et al. Pulse wave velocity is an independent predictor of the longitudinal increase in systolic blood pressure and of incident hypertension in the Baltimore Longitudinal Study of Aging. J Am Coll Cardiol 2008;51(14):1377–83.

[95] Koivistoinen T, Lyytikainen LP, Aatola H, Luukkaala T, Juonala M, Viikari J, et al. Pulse wave velocity predicts the progression of blood pressure and development of hypertension in young adults. Hypertension 2018;71(3):451–6.

[96] Lim HS, Lip GYH. Arterial stiffness: beyond pulse wave velocity and its measurement. J Hum Hypertens 2008;22:656–8.

[97] Reference Values for Arterial Stiffness' Collaboration. Determinants of pulse wave velocity in healthy people and in the presence of cardiovascular risk factors: 'establishing normal and reference values'. Eur Heart J 2010;31(19):2338–50.

[98] Kuvin JT, Patel AR, Sliney KA, Pandian NG, Sheffy J, Schnall RP, Karas RH, Udelson JE. Assessment of peripheral vascular endothelial function with finger arterial pulse wave amplitude. Am Heart J 2003;146:168–74.

[99] Tsao CW, Vasan RS. Cohort profile: the Framingham Heart Study (FHS): overview of milestones in cardiovascular epidemiology. Int J Epidemiol 2015;44(6):1800–13.

[100] Heffernan KS, Karas RH, Patvardhan EA, Jafri H, Kuvin JT. Peripheral arterial tonometry for risk stratification in men with coronary artery disease. Clin Cardiol 2010;33(2):94–8.

[101] Yang WI, Park S, Youn JC, Son NH, Lee SH, Kang SM, et al. Augmentation index association with reactive hyperemia as assessed by peripheral arterial tonometry in hypertension. Am J Hypertens 2011;24(11):1234–8.

[102] Al Mheid I, Patel R, Murrow J, Morris A, Rahman A, Fike L, et al. Vitamin D status is associated with arterial stiffness and vascular dysfunction in healthy humans. J Am Coll Cardiol 2011;58(2):186–92.

[103] Jeong HY, Park KM, Lee MJ, Yang DH, Kim SH, Lee SY. Vitamin D and hypertension. Electrolyte Blood Press 2017;15(1):1–11.

[104] Geleijnse JM. Vitamin D and the prevention of hypertension and cardiovascular diseases: a review of the current evidence. Am J Hypertens 2011;24(3):253–62.

[105] Carrara D, Bernini M, Bacca A, Rugani I, Duranti E, Virdis A, et al. Cholecalciferol administration blunts the systemic renin-angiotensin system in essential hypertensives with hypovitaminosis D. J Renin Angiotensin Aldosterone Syst 2014;15(1):82–7.

[106] Larsen T, Mose FH, Bech JN, Hansen AB, Pedersen EB. Effect of cholecalciferol supplementation during winter months in patients with hypertension: a randomized, placebo-controlled trial. Am J Hypertens 2012;25(11):1215–22.

[107] Schleithoff SS, Zittermann A, Tenderich G, Berthold HK, Stehle P, Koerfer R. Vitamin D supplementation improves cytokine profiles in patients with congestive heart failure: a double-blind, randomized, placebo-controlled trial. Am J Clin Nutr 2006;83(4):754–9.

[108] Margolis KL, Ray RM, Van Horn L, Manson JE, Allison MA, Black HR, et al. Effect of calcium and vitamin D supplementation on blood pressure: the Women's Health Initiative randomized trial. Hypertension 2008;52(5):847–55.

[109] Zittermann A, Frisch S, Berthold HK, Gotting C, Kuhn J, Kleesiek K, et al. Vitamin D supplementation enhances the beneficial effects of weight loss on cardiovascular disease risk markers. Am J Clin Nutr 2009;89(5):1321–7.

[110] Jorde R, Sneve M, Torjesen P, Figenschau Y. No improvement in cardiovascular risk factors in overweight and obese subjects after supplementation with vitamin D3 for 1 year. J Intern Med 2010;267(5):462–72.

[111] Wang L, Manson JE, Song Y, Sesso HD. Systematic review: vitamin D and calcium supplementation in prevention of cardiovascular events. Ann Intern Med 2010;152(5):315–23.

[112] Trivedi DP, Doll R, Khaw KT. Effect of four monthly oral vitamin D3 (cholecalciferol) supplementation on fractures and mortality in men and women living in the community: randomised double blind controlled trial. BMJ 2003;326(7387):469.

[113] Prince RL, Austin N, Devine A, Dick IM, Bruce D, Zhu K. Effects of ergocalciferol added to calcium on the risk of falls in elderly high-risk women. Arch Intern Med 2008;168(1):103–8.

[114] Hsia J, Heiss G, Ren H, Allison M, Dolan NC, Greenland P, et al. Calcium/vitamin D supplementation and cardiovascular events. Circulation 2007;115(7):846–54.

[115] Sanders KM, Stuart AL, Williamson EJ, Simpson JA, Kotowicz MA, Young D, et al. Annual high-dose oral vitamin D and falls and fractures in older women: a randomized controlled trial. JAMA 2010;303(18):1815–22.

[116] Savoia C, Schiffrin EL. Inflammation in hypertension. Curr Opin Nephrol Hypertens 2006;15(2):152–8.

[117] Cortez AF, Muxfeldt ES. The role of C-reactive protein in the cardiovascular risk and its association with hypertension. JSM Atheroscler 2016;1:1015–21.

[118] Pearson TA, Mensah GA, Alexander RW, Anderson JL, Cannon 3rd RO, Criqui M, et al. Markers of inflammation and cardiovascular disease: application to clinical and public health practice: a statement for healthcare professionals from the Centers for Disease Control and Prevention and the American Heart Association. Circulation 2003;107(3):499–511.

[119] Blake GJ, Rifai N, Buring JE, Ridker PM. Blood pressure, C-reactive protein, and risk of future cardiovascular events. Circulation 2003;108:2993–9.

[120] Sesso HD, Buring JE, Rifai N, Blake GJ, Gaziano JM, Ridker PM. C-reactive protein and the risk of developing hypertension. JAMA 2003;290:2945–51.

[121] Lakoski SG, Cushman M, Palmas W, Blumenthal R, D'Agostino Jr. RB, Herrington DM. The relationship between blood pressure and C-reactive protein in the Multi-Ethnic Study of Atherosclerosis (MESA). J Am Coll Cardiol 2005;46(10):1869–74.

[122] Davey Smith G, Lawlor DA, Harbord R, Timpson N, Rumley A, Lowe GD, et al. Association of C-reactive protein with blood pressure and hypertension: life course confounding and mendelian randomization tests of causality. Arterioscler Thromb Vasc Biol 2005;25(5):1051–6.

[123] Sesso HD, Jimenez MC, Wang L, Ridker PM, Buring JE, Gaziano JM. Plasma inflammatory markers and the risk of developing hypertension in men. J Am Heart Assoc 2015;4(9): e001802.

[124] Wang L, Manson JE, Gaziano JM, Liu S, Cochrane B, Cook NR, et al. Circulating inflammatory and endothelial markers and risk of hypertension in white and black postmenopausal women. Clin Chem 2011;57(5):729–36.

[125] Jae SY, Kurl S, Laukkanen JA, Lee CD, Choi YH, Fernhall B, et al. Relation of C-reactive protein, fibrinogen, and cardiorespiratory fitness to risk of systemic hypertension in men. Am J Cardiol 2015;115(12):1714–9.

[126] Wang A, Xu T, Xu T, Zhang M, Li H, Tong W, et al. Hypertension and elevated C-reactive protein: future risk of ischemic stroke in a prospective cohort study among inner Mongolians in China. Int J Cardiol 2014;174(2):455–6.

[127] Tanaka F, Makita S, Onoda T, Tanno K, Ohsawa M, Itai K, et al. Prehypertension subtype with elevated C-reactive protein: risk of ischemic stroke in a general Japanese population. Am J Hypertens 2010;23(10):1108–13.

[128] Iwashima Y, Horio T, Kamide K, Rakugi H, Ogihara T, Kawano Y. C-reactive protein, left ventricular mass index, and risk of cardiovascular disease in essential hypertension. Hypertens Res 2007;30(12):1177–85.

[129] Ridker PM, Danielson E, Fonseca FA, Genest J, Gotto Jr. AM, Kastelein JJ, et al. Rosuvastatin to prevent vascular events in men and women with elevated C-reactive protein. N Engl J Med 2008;359(21):2195–207.

[130] Mazzali M, Kanbay M, Segal MS, Shafiu M, Jalal D, Feig DI, et al. Uric acid and hypertension: cause or effect? Curr Rheumatol Rep 2010;12(2):108–17.

[131] Kuwabara M. Hyperuricemia, cardiovascular disease, and hypertension. Pulse 2016;3 (3–4):242–52.

[132] Scheepers LEJM, Boonen A, Dagnelie PC, Schram MT, van der Kallen CJH, Henry RMA, et al. Uric acid and blood pressure: exploring the role of uric acid production in The Maastricht Study. J Hypertens 2017;35(10):1968–75.

[133] Schmitz B, Brand S-M. Uric acid and essential hypertension: the endothelial connection. J Hypertens 2016;34(11):2138–9.

[134] Verdecchia P, Schillaci G, Reboldi G, Santeusanio F, Porcellati C, Brunetti P. Relation between serum uric acid and risk of cardiovascular disease in essential hypertension. Hypertension 2000;36(6):1072–8.

[135] Viazzi F, Garneri D, Leoncini G, Gonnella A, Muiesan ML, Ambrosioni E, et al. Serum uric acid and its relationship with metabolic syndrome and cardiovascular risk profile in patients with hypertension: insights from the I-DEMAND study. Nutr Metab Cardiovasc Dis 2014;24 (8):921–7.

[136] Lee JJ, Ahn J, Hwang J, Han SW, Lee KN, Kim JB, et al. Relationship between uric acid and blood pressure in different age groups. Clin Hypertens 2015;21:14.

[137] Buzas R, Tautu O-F, Dorobantu M, Ivan V, Lighezan D. Serum uric acid and arterial hypertension—data from Sephar III survey. PLoS One 2018;13(7):e0199865.

[138] Yokoi Y, Kondo T, Okumura N, Shimokata K, Osugi S, Maeda K, et al. Serum uric acid as a predictor of future hypertension: stratified analysis based on body mass index and age. Prev Med 2016;90:201–6.

[139] Hwang J, Hwang JH, Chung SM, Kwon MJ, Ahn JK. Association between serum uric acid and arterial stiffness in a low-risk, middle-aged, large Korean population: a cross-sectional study. Medicine 2018;97(36):e12086.

[140] Canepa M, Viazzi F, Strait JB, Ameri P, Pontremoli R, Brunelli C, et al. Longitudinal association between serum uric acid and arterial stiffness: results from the Baltimore longitudinal study of aging. Hypertension 2017;69(2):228–35.

[141] Feig DI, Soletsky B, Johnson RJ. Effect of allopurinol on blood pressure of adolescents with newly diagnosed essential hypertension: a randomized trial. JAMA 2008;300(8):924–32.

[142] MacIsaac RL, Salatzki J, Higgins P, Walters MR, Padmanabhan S, Dominiczak AF, et al. Allopurinol and cardiovascular outcomes in adults with hypertension. Hypertension 2016;67 (3):535–40.

[143] Schalinske KL, Smazal AL. Homocysteine imbalance: a pathological metabolic marker. Adv Nutr 2012;3(6):755–62.

[144] Lim U, Cassano PA. Homocysteine and blood pressure in the Third National Health and Nutrition Examination Survey, 1988–1994. Am J Epidemiol 2002;156(12):1105–13.

[145] Sutton-Tyrrell K, Bostom A, Selhub J, Zeigler-Johnson C. High homocysteine levels are independently related to isolated systolic hypertension in older adults. Circulation 1997;96 (6):1745–9.

[146] Sundstrom J, Sullivan L, D'Agostino RB, Jacques PF, Selhub J, Rosenberg IH, et al. Plasma homocysteine, hypertension incidence, and blood pressure tracking: the Framingham Heart Study. Hypertension 2003;42(6):1100–5.

[147] Bowman TS, Gaziano JM, Stampfer MJ, Sesso HD. Homocysteine and risk of developing hypertension in men. J Hum Hypertens 2006;20(8):631–4.

[148] Skeete J, DiPette DJ. Relationship between homocysteine and hypertension: new data add to the debate. J Clin Hypertens 2017;19(11):1171–2.

[149] Huang Z, Chen C, Li S, Kong F, Shan P, Huang W. Serum markers of endothelial dysfunction and inflammation increase in hypertension with prediabetes mellitus. Genet Test Mol Biomarkers 2016;20(6):322–7.

[150] Creager Mark, Beckman Joshua, Loscalzo Joseph, editors. Vascular medicine: a companion to Braunwald's heart disease, 2nd ed. Philadelphia: WB Saunders; 2012, 880 pp.

[151] Holterman CE, Thibodeau JF, Towaij C, Gutsol A, Montezano AC, Parks RJ, et al. Nephropathy and elevated BP in mice with podocyte-specific NADPH oxidase 5 expression. J Am Soc Nephrol 2014;25(4):784–97.

[152] Jha JC, Watson AMD, Mathew G, de Vos LC, Jandeleit-Dahm K. The emerging role of NADPH oxidase NOX5 in vascular disease. Clin Sci (Lond) 2017;131(10):981–90.

[153] Pratico D, Iuliano L, Mauriello A, Spagnoli L, Lawson JA, Rokach J, et al. Localization of distinct F2-isoprostanes in human atherosclerotic lesions. J Clin Invest 1997;100(8):2028–34.

[154] Speed JS, Pollock DM. Endothelin, kidney disease, and hypertension. Hypertension 2013;61 (6):1142–5.

[155] Liu X, Zhang Q, Wu H, Du H, Liu L, Shi H, et al. Blood neutrophil to lymphocyte ratio as a predictor of hypertension. Am J Hypertens 2015;28(11):1339–46.

[156] Sun X, Luo L, Zhao X, Ye P, Du R. The neutrophil-to-lymphocyte ratio on admission is a good predictor for all-cause mortality in hypertensive patients over 80 years of age. BMC Cardiovasc Disord 2017;17(1):167.

[157] Fülöp T, Tapolyai M. Beauty in simplicity: abnormal neutrophil to lymphocyte ratio in resistant hypertension. J Clin Hypertens 2015;17(7):538–40.

[158] Karagoz A, Vural A, Gunaydin ZY, Bektas O, Gul M, Celik A, et al. The role of neutrophil to lymphocyte ratio as a predictor of diastolic dysfunction in hypertensive patients. Eur Rev Med Pharmacol Sci 2015;19(3):433–40.

[159] Kilicaslan B, Dursun H, Kaymak S, Aydin M, Ekmekci C, Susam I, et al. The relationship between neutrophil to lymphocyte ratio and blood pressure variability in hypertensive and normotensive subjecs. Turk Kardiyol Dern Ars 2015;43(1):18–24.

[160] Bilal A, Farooq JH, Kiani I, Assad S, Ghazanfar H, Ahmed I. Importance of mean red cell distribution width in hypertensive patients. Cureus 2016;8(11):e902.

[161] Tanindi A, Topal FE, Topal F, Celik B. Red cell distribution width in patients with prehypertension and hypertension. Blood Press 2012;21(3):177–81.

[162] Wen Y. High red blood cell distribution width is closely associated with risk of carotid artery atherosclerosis in patients with hypertension. Exp Clin Cardiol 2010;15(3):37–40.

[163] James PA, Oparil S, Carter BL, Cushman WC, Dennison-Himmelfarb C, Handler J, et al. 2014 evidence-based guideline for the management of high blood pressure in adults: report from the panel members appointed to the Eighth Joint National Committee (JNC 8). JAMA 2014;311(5):507–20.

[164] Egan BM, Julius S. Prehypertension: risk stratification and management considerations. Curr Hypertens Rep 2008;10(5):359–66.

[165] Julius S, Nesbitt SD, Egan BM, Weber MA, Michelson EL, Kaciroti N, et al. Feasibility of treating prehypertension with an angiotensin-receptor blocker. N Engl J Med 2006;354 (16):1685–97.

[166] https://www.nice.org.uk/guidance/cg127.

[167] Williams B, MacDonald TM, Morant S, Webb DJ, Sever P, McInnes G, et al. Spironolactone versus placebo, bisoprolol, and doxazosin to determine the optimal treatment for drug-resistant hypertension (PATHWAY-2): a randomised, double-blind, crossover trial. Lancet 2015;386(10008):2059–68.

[168] Preston RA, Materson BJ, Reda DJ, Williams DW, Hamburger RJ, Cushman WC, et al. Age-race subgroup compared with renin profile as predictors of blood pressure response to antihypertensive therapy. Department of Veterans Affairs Cooperative Study Group on antihypertensive agents. JAMA 1998;280(13):1168–72.

[169] Turner ST, Schwartz GL, Chapman AB, Beitelshees AL, Gums JG, Cooper-DeHoff RM, et al. Plasma renin activity predicts blood pressure responses to beta-blocker and thiazide diuretic as monotherapy and add-on therapy for hypertension. Am J Hypertens 2010;23(9):1014–22.

[170] Turner ST, Schwartz GL, Chapman AB, Beitelshees AL, Gums JG, Cooper-DeHoff RM, et al. Power to identify a genetic predictor of antihypertensive drug response using different methods to measure blood pressure response. J Transl Med 2012;10(1):47.

[171] Canzanello VJ, Baranco-Pryor E, Rahbari-Oskoui F, Schwartz GL, Boerwinkle E, Turner ST, et al. Predictors of blood pressure response to the angiotensin receptor blocker candesartan in essential hypertension. Am J Hypertens 2008;21(1):61–6.

[172] Hiltunen TP, Donner KM, Sarin AP, Saarela J, Ripatti S, Chapman AB, et al. Pharmacogenomics of Hypertension: a genome-wide, placebo-controlled cross-over study, using four classes of antihypertensive drugs. J Am Heart Assoc 2015;4(1):e001521.

[173] Hiltunen TP, Rimpelä JM, Mohney RP, Stirdivant SM, Kontula KK. Effects of four different antihypertensive drugs on plasma metabolomic profiles in patients with essential hypertension. PLoS One 2017;12(11):e0187729.

[174] Gong Y, McDonough CW, Wang Z, Hou W, Cooper-DeHoff RM, Langaee TY, et al. Hypertension susceptibility loci and blood pressure response to antihypertensives: results from the pharmacogenomic evaluation of antihypertensive responses study. Circ Cardiovasc Genet 2012;5(6):686–91.

[175] Johnson JA, Boerwinkle E, Zineh I, Chapman AB, Bailey K, Cooper-DeHoff RM, et al. Pharmacogenomics of antihypertensive drugs: rationale and design of the Pharmacogenomic Evaluation of Antihypertensive Responses (PEAR) study. Am Heart J 2009;157(3):442–9.

[176] Magvanjav O, Gong Y, McDonough CW, Chapman AB, Turner ST, Gums JG, et al. Genetic variants associated with uncontrolled blood pressure on thiazide diuretic/beta-blocker combination therapy in the PEAR (Pharmacogenomic Evaluation of Antihypertensive Responses) and INVEST (International Verapamil-SR Trandolapril Study) trials. J Am Heart Assoc 2017;6(11)https://doi.org/10.1161/JAHA.117.006522.

[177] Suonsyrjä T, Hannila-Handelberg T, Paavonen KJ, Miettinen HE, Donner K, Strandberg T, et al. Laboratory tests as predictors of the antihypertensive effects of amlodipine, bisoprolol, hydrochlorothiazide and losartan in men: results from the randomized, double-blind, cross-over GENRES Study. J Hypertens 2008;26(6):1250–6.

[178] Wolf-Maier K, Cooper RS, Banegas JR, Giampaoli S, Hense HW, Joffres M, et al. Hypertension prevalence and blood pressure levels in 6 European countries, Canada, and the United States. JAMA 2003;289(18):2363–9.

[179] Cheng H-M, Chuang S-Y, Sung S-H, Yu W-C, Pearson A, Lakatta EG, et al. Derivation and validation of diagnostic thresholds for central blood pressure measurements based on long-term cardiovascular risks. J Am Coll Cardiol 2013;62(19):1780–7.

[180] Rothwell PM, Howard SC, Dolan E, O'Brien E, Dobson JE, Dahlof B, et al. Effects of beta blockers and calcium-channel blockers on within-individual variability in blood pressure and risk of stroke. Lancet Neurol 2010;9(5):469–80.

[181] Kikuya M, Hozawa A, Ohokubo T, Tsuji I, Michimata M, Matsubara M, et al. Prognostic significance of blood pressure and heart rate variabilities: the Ohasama study. Hypertension 2000;36(5):901–6.

[182] Johansson JK, Niiranen TJ, Puukka PJ, Jula AM. Prognostic value of the variability in home-measured blood pressure and heart rate: the Finn-Home Study. Hypertension 2012;59 (2):212–8.

[183] Parati G, Pomidossi G, Albini F, Malaspina D, Mancia G. Relationship of 24-hour blood pressure mean and variability to severity of target-organ damage in hypertension. J Hypertens 1987;5(1):93–8.

[184] Tatasciore A, Renda G, Zimarino M, Soccio M, Bilo G, Parati G, et al. Awake systolic blood pressure variability correlates with target-organ damage in hypertensive subjects. Hypertension 2007;50(2):325–32.

[185] Matsui Y, Ishikawa J, Eguchi K, Shibasaki S, Shimada K, Kario K. Maximum value of home blood pressure: a novel indicator of target organ damage in hypertension. Hypertension 2011;https://doi.org/10.1161/HYPERTENSIONAHA.111.171645.

[186] Muntner P, Shimbo D, Tonelli M, Reynolds K, Arnett DK, Oparil S. The relationship between visit-to-visit variability in systolic blood pressure and all-cause mortality in the general population: findings from NHANES III, 1988 to 1994. Hypertension 2011; https://doi.org/10.1161/HYPERTENSIONAHA.110.162255.

[187] Kario K, Pickering TG, Umeda Y, Hoshide S, Hoshide Y, Morinari M, et al. Morning surge in blood pressure as a predictor of silent and clinical cerebrovascular disease in elderly hypertensives: a prospective study. Circulation 2003;107(10):1401–6.

[188] McEniery CM, Cockcroft JR, Roman MJ, Franklin SS, Wilkinson IB. Central blood pressure: current evidence and clinical importance. Eur Heart J 2014;35(26):1719–25.

[189] Sharman JE, Laurent S. Central blood pressure in the management of hypertension: soon reaching the goal? J Hum Hypertens 2013;27(7):405–11.

[190] Epstein BJ, Anderson S. Discordant effects of beta-blockade on central aortic systolic and brachial systolic blood pressure: considerations beyond the cuff. Pharmacotherapy 2007; 27(9):1322–33.

[191] Asmar RG, London GM, O'Rourke ME, Safar ME, REASON Project Coordinators and Investigators. Improvement in blood pressure, arterial stiffness and wave reflections with a very-low-dose perindopril/indapamide combination in hypertensive patient: a comparison with atenolol. Hypertension 2001;38(4):922–6.

[192] Middeke M. Central aortic blood pressure: important parameter in diagnosis and therapy. Dtsch Med Wochenschr 2017;142(19):1430–6.

[193] Tomiyama H, O'Rourke MF, Hashimoto H, Matsumoto C, Odaira M, Yoshida M, et al. Central blood pressure: a powerful predictor of the development of hypertension. Hypertens Res 2013;36(1):19–24.

[194] Rinaldi ER, Yannoutsos A, Borghi C, Safar ME, Blacher J. Central hemodynamics for risk reduction strategies: additive value over and above brachial blood pressure. Curr Pharm Des 2015;21(6):730–6.

[195] Zuo J, Chu S, Tan I, Butlin M, Zhao J, Avolio A. Association of haemodynamic indices of central and peripheral pressure with subclinical target organ damage. Pulse 2018;5 (1–4):133–43.

[196] Yang L, Qin B, Zhang X, Chen Y, Hou J. Association of central blood pressure and cardiovascular diseases in diabetic patients with hypertension. Medicine 2017;96(42):e8286.

[197] Huang CM, Wang KL, Cheng HM, Chuang SY, Sung SH, Yu WC, et al. Central versus ambulatory blood pressure in the prediction of all-cause and cardiovascular mortalities. J Hypertens 2011;29(3):454–9.

[198] Mitchell GF, Hwang SJ, Larson MG, Hamburg NM, Benjamin EJ, Vasan RS, et al. Transfer function-derived central pressure and cardiovascular disease events: the Framingham Heart Study. J Hypertens 2016;34(8):1528–34.

[199] Mitchell GF, Hwang SJ, Vasan RS, Larson MG, Pencina MJ, Hamburg NM, et al. Arterial stiffness and cardiovascular events: the Framingham Heart Study. Circulation 2010;121 (4):505–11.

[200] Vlachopoulos C, Aznaouridis K, O'Rourke MF, Safar ME, Baou K, Stefanadis C. Prediction of cardiovascular events and all-cause mortality with central haemodynamics: a systematic review and meta-analysis. Eur Heart J 2010;31(15):1865–71.

[201] Laurent S, Briet M, Boutouyrie P. Arterial stiffness as surrogate end point: needed clinical trials. Hypertension 2012;60(2):518–22.

[202] Laurent S, Sharman J, Boutouyrie P. Central versus peripheral blood pressure: finding a solution. J Hypertens 2016;34(8):1497–9.

[203] Franklin SS, Khan SA, Wong ND, Larson MG, Levy D. Is pulse pressure useful in predicting risk for coronary heart disease? The Framingham Heart Study. Circulation 1999; 100(4):354–60.

[204] Franklin SS, Lopez VA, Wong ND, Mitchell GF, Larson MG, Vasan RS, et al. Single versus combined blood pressure components and risk for cardiovascular disease: the Framingham Heart Study. Circulation 2009;119(2):243–50.

[205] Franklin SS, Gokhale SS, Chow VH, Larson MG, Levy D, Vasan RS, et al. Does low diastolic blood pressure contribute to the risk of recurrent hypertensive cardiovascular disease events? Hypertension 2015;65(2):299–305.

[206] Vaccarino V, Berger AK, Abramson J, Black HR, Setaro JF, Davey JA, et al. Pulse pressure and risk of cardiovascular events in the systolic hypertension in the elderly program. Am J Cardiol 2001;88(9):980–6.

[207] Staessen JA, Thijs L, O'Brien ET, Bulpitt CJ, de Leeuw PW, Fagard RH, et al. Ambulatory pulse pressure as predictor of outcome in older patients with systolic hypertension. Am J Hypertens 2002;15(10 Pt 1):835–43.

[208] Domanski MJ, Davis BR, Pfeffer MA, Kastantin M, Mitchell GF. Isolated systolic hypertension: prognostic information provided by pulse pressure. Hypertension 1999;34(3):375–80.

[209] Cushman WC, Materson BJ, Williams DW, Reda DJ. Pulse pressure changes with six classes of antihypertensive agents in a randomized, controlled trial. Hypertension 2001;38(4):953–7.

[210] Fyhrquist F, Dahlof B, Devereux RB, Kjeldsen SE, Julius S, Beevers G, et al. Pulse pressure and effects of losartan or atenolol in patients with hypertension and left ventricular hypertrophy. Hypertension 2005;45(4):580–5.

[211] Williams B, O'Rourke M. The Conduit Artery Functional Endpoint (CAFE) study in ASCOT. J Hum Hypertens 2001;15(Suppl 1):S69–73.

[212] Dahlof B, Sever PS, Poulter NR, Wedel H, Beevers DG, Caulfield M, et al. Prevention of cardiovascular events with an antihypertensive regimen of amlodipine adding perindopril as required versus atenolol adding bendroflumethiazide as required, in the Anglo-Scandinavian Cardiac Outcomes Trial-Blood Pressure Lowering Arm (ASCOT-BPLA): a multicentre randomised controlled trial. Lancet 2005;366(9489):895–906.

[213] Wang JG, Staessen JA, Franklin SS, Fagard R, Gueyffier F. Systolic and diastolic blood pressure lowering as determinants of cardiovascular outcome. Hypertension 2005;45(5):907–13.

[214] Oparil S, Izzo Jr JL. Pulsology rediscovered: commentary on the Conduit Artery Function Evaluation (CAFE) study. Circulation 2006;113(9):1162–3.

[215] Wilkinson IB, McEniery CM, Cockcroft JR. Atenolol and cardiovascular risk: an issue close to the heart. Lancet 2006;367(9511):627–9.

[216] https://www.uptodate.com/contents/increased-pulse-pressure.

[217] Lewington S, Clarke R, Qizilbash N, Peto R, Collins R. Age-specific relevance of usual blood pressure to vascular mortality: a meta-analysis of individual data for one million adults in 61 prospective studies. Lancet 2002;360(9349):1903–13.

[218] Devereux RB, Pickering TG, Alderman MH, Chien S, Borer JS, Laragh JH. Left ventricular hypertrophy in hypertension. Prevalence and relationship to pathophysiologic variables. Hypertension 1987;9(2 Pt 2):II53–60.

[219] Whelton PK, Carey RM, Aronow WS, Casey Jr DE, Collins KJ, Dennison Himmelfarb C, et al. 2017 ACC/AHA/AAPA/ABC/ACPM/AGS/APhA/ASH/ASPC/NMA/PCNA guideline for the prevention, detection, evaluation, and Management of High Blood Pressure in adults: executive summary: a report of the American College of Cardiology/American Heart Association task force on clinical practice guidelines. Hypertension 2018;71(6):1269–324.

[220] Currie G, McKay G, Delles C. Biomarkers in diabetic nephropathy: present and future. World J Diabetes 2014;5(6):763–76.

[221] Prasad PV, Edelman RR, Epstein FH. Noninvasive evaluation of intrarenal oxygenation with BOLD MRI. Circulation 1996;94(12):3271–5.

[222] Li LP, Vu AT, Li BS, Dunkle E, Prasad PV. Evaluation of intrarenal oxygenation by BOLD MRI at 3.0 T. J Magn Reson Imaging 2004;20(5):901–4.

[223] Gillis KA, McComb C, Foster JE, Taylor AH, Patel RK, Morris ST, et al. Inter-study reproducibility of arterial spin labelling magnetic resonance imaging for measurement of renal perfusion in healthy volunteers at 3 Tesla. BMC Nephrol 2014;15:23.

[224] Pena MJ, Mischak H, Heerspink HJ. Proteomics for prediction of disease progression and response to therapy in diabetic kidney disease. Diabetologia 2016;59(9):1819–31.

[225] Good DM, Zurbig P, Argiles A, Bauer HW, Behrens G, Coon JJ, et al. Naturally occurring human urinary peptides for use in diagnosis of chronic kidney disease. Mol Cell Proteomics 2010;9(11):2424–37.

[226] Lindhardt M, Persson F, Currie G, Pontillo C, Beige J, Delles C, et al. Proteomic prediction and renin angiotensin aldosterone system inhibition prevention of early diabetic nephRopathy in TYpe 2 diabetic patients with normoalbuminuria (PRIORITY): essential study design and rationale of a randomised clinical multicentre trial. BMJ Open 2016; 6(3):e010310.

[227] Drozdz D, Kawecka-Jaszcz K. Cardiovascular changes during chronic hypertensive states. Pediatr Nephrol 2014;29(9):1507–16.

[228] Santos M, Shah AM. Alterations in cardiac structure and function in hypertension. Curr Hypertens Rep 2014;16(5):428.

[229] Muiesan ML, Salvetti M, Monteduro C, Bonzi B, Paini A, Viola S, et al. Left ventricular concentric geometry during treatment adversely affects cardiovascular prognosis in hypertensive patients. Hypertension 2004;43(4):731–8.

[230] ADVANCE Echocardiography Substudy Investigators, ADVANCE Collaborative Group. Effects of perindopril-indapamide on left ventricular diastolic function and mass in patients with type 2 diabetes: the ADVANCE Echocardiography Substudy. J Hypertens 2011; 29(7):1439–47. https://doi.org/10.1097/HJH.0b013e3283480fe9.

[231] Barron AJ, Hughes AD, Sharp A, Baksi AJ, Surendran P, Jabbour RJ, et al. Long-term antihypertensive treatment fails to improve E/e′ despite regression of left ventricular mass: an Anglo-Scandinavian cardiac outcomes trial substudy. Hypertension 2014;63(2):252–8.

[232] Fagard RH, Celis H, Thijs L, Wouters S. Regression of left ventricular mass by antihypertensive treatment: a meta-analysis of randomized comparative studies. Hypertension 2009;54 (5):1084–91.

[233] Solomon SD, Janardhanan R, Verma A, Bourgoun M, Daley WL, Purkayastha D, et al. Effect of angiotensin receptor blockade and antihypertensive drugs on diastolic function in patients with hypertension and diastolic dysfunction: a randomised trial. Lancet 2007;369 (9579):2079–87.

[234] Solomon SD, Verma A, Desai A, Hassanein A, Izzo J, Oparil S, et al. Effect of intensive versus standard blood pressure lowering on diastolic function in patients with uncontrolled hypertension and diastolic dysfunction. Hypertension 2010;55(2):241–8.

[235] Almuntaser I, Mahmud A, Brown A, Murphy R, King G, Crean P, et al. Blood pressure control determines improvement in diastolic dysfunction in early hypertension. Am J Hypertens 2009;22(11):1227–31.

[236] Mettimano M, Specchia ML, Migneco A, Savi L. Microalbuminuria as a marker of cardiac damage in essential hypertension. Eur Rev Med Pharmacol Sci 2001;5(1):31–6.

[237] Dong R, Zhang M, Hu Q, Zheng S, Soh A, Zheng Y, et al. Galectin-3 as a novel biomarker for disease diagnosis and a target for therapy (review). Int J Mol Med 2018;41(2):599–614.

[238] Uri K, Fagyas M, Manyine Siket I, Kertesz A, Csanadi Z, Sandorfi G, et al. New perspectives in the renin-angiotensin-aldosterone system (RAAS) IV: circulating ACE2 as a biomarker of systolic dysfunction in human hypertension and heart failure. PLoS One 2014;9(4):e87845.

[239] Bhandari SK, Batech M, Shi J, Jacobsen SJ, Sim JJ. Plasma renin activity and risk of cardiovascular and mortality outcomes among individuals with elevated and nonelevated blood pressure. Kidney Res Clin Pract 2016;35(4):219–28.

[240] Verma S, Gupta M, Holmes DT, Xu L, Teoh H, Gupta S, et al. Plasma renin activity predicts cardiovascular mortality in the Heart Outcomes Prevention Evaluation (HOPE) study. Eur Heart J 2011;32(17):2135–42.

[241] Sim JJ, Shi J, Al-Moomen R, Behayaa H, Kalantar-Zadeh K, Jacobsen SJ. Plasma renin activity and its association with ischemic heart disease, congestive heart failure, and cerebrovascular disease in a large hypertensive cohort. J Clin Hypertens 2014;16(11):805–13.

[242] Sever PS, Chang CL, Prescott MF, Gupta A, Poulter NR, Whitehouse A, et al. Is plasma renin activity a biomarker for the prediction of renal and cardiovascular outcomes in treated hypertensive patients? Observations from the Anglo-Scandinavian Cardiac Outcomes Trial (ASCOT). Eur Heart J 2012;33(23):2970–9.

[243] Kuznetsova T, Mischak H, Mullen W, Staessen JA. Urinary proteome analysis in hypertensive patients with left ventricular diastolic dysfunction. Eur Heart J 2012;33(18):2342–50.

[244] Zhang Z, Staessen JA, Thijs L, Gu Y, Liu Y, Jacobs L, et al. Left ventricular diastolic function in relation to the urinary proteome: a proof-of-concept study in a general population. Int J Cardiol 2014;176(1):158–65.

[245] Ariesen MJ, Claus SP, Rinkel GJ, Algra A. Risk factors for intracerebral hemorrhage in the general population: a systematic review. Stroke 2003;34(8):2060–5.

[246] Pires PW, Dams Ramos CM, Matin N, Dorrance AM. The effects of hypertension on the cerebral circulation. Am J Physiol Heart Circ Physiol 2013;304(12):H1598–614.

[247] Gasecki D, Kwarciany M, Nyka W, Narkiewicz K. Hypertension, brain damage and cognitive decline. Curr Hypertens Rep 2013;15(6):547–58.

[248] Prins ND, Scheltens P. White matter hyperintensities, cognitive impairment and dementia: an update. Nat Rev Neurol 2015;11(3):157–65.

[249] Kloppenborg RP, Nederkoorn PJ, Geerlings MI, van den Berg E. Presence and progression of white matter hyperintensities and cognition: a meta-analysis. Neurology 2014;82 (23):2127–38.

[250] Ai Q, Pu YH, Sy C, Liu LP, Gao PY. Impact of regional white matter lesions on cognitive function in subcortical vascular cognitive impairment. Neurol Res 2014;36(5):434–43.

[251] Potter GM, Doubal FN, Jackson CA, Chappell FM, Sudlow CL, Dennis MS, et al. Enlarged perivascular spaces and cerebral small vessel disease. Int J Stroke 2015;10(3):376–81.

[252] Yang S, Qin W, Yang L, Fan H, Li Y, Yin J, et al. The relationship between ambulatory blood pressure variability and enlarged perivascular spaces: a cross-sectional study. BMJ Open 2017;7(8):e015719.

[253] Yang S, Yuan J, Zhang X, Fan H, Li Y, Yin J, et al. Higher ambulatory systolic blood pressure independently associated with enlarged perivascular spaces in basal ganglia. Neurol Res 2017;39(9):787–94.

[254] Zhang C, Chen Q, Wang Y, Zhao X, Wang C, Liu L, et al. Risk factors of dilated Virchow-Robin spaces are different in various brain regions. PLoS One 2014;9(8):e105505.

[255] Li Y, Liu N, Huang Y, Wei W, Chen F, Zhang W. Risk factors for silent lacunar infarction in patients with transient ischemic attack. Med Sci Monit 2016;22:447–53.

[256] Tsubota-Utsugi M, Satoh M, Tomita N, Hara A, Kondo T, Hosaka M, et al. Lacunar infarcts rather than white matter hyperintensity as a predictor of future higher level functional decline: the Ohasama study. J Stroke Cerebrovasc Dis 2017;26(2):376–84.

[257] Kherada N, Heimowitz T, Rosendorff C. Antihypertensive therapies and cognitive function: a review. Curr Hypertens Rep 2015;17(10):79.

[258] Scuteri A, Tesauro M, Guglini L, Lauro D, Fini M, Di Daniele N. Aortic stiffness and hypotension episodes are associated with impaired cognitive function in older subjects with subjective complaints of memory loss. Int J Cardiol 2013;169(5):371–7.

[259] Hughes TM, Sink KM. Hypertension and its role in cognitive function: current evidence and challenges for the future. Am J Hypertens 2016;29(2):149–57.

[260] Kitagawa K. Cerebral blood flow measurement by PET in hypertensive subjects as a marker of cognitive decline. J Alzheimers Dis 2010;20(3):855–9.

[261] Pase MP, Himali JJ, Mitchell GF, Beiser A, Maillard P, Tsao C, et al. Association of aortic stiffness with cognition and brain aging in young and middle-aged adults: the framingham third generation cohort study. Hypertension 2016;67(3):513–9.

[262] Wang WJ, Li QQ, Xu JD, Cao XX, Li HX, Tang F, et al. Over-expression of ubiquitin carboxy terminal hydrolase-L1 induces apoptosis in breast cancer cells. Int J Oncol 2008;33(5):1037–45.

[263] González-Quevedo A, González-García S, Peña-Sánchez M, et al. Blood-based biomarkers could help identify subclinical brain damage caused by arterial hypertension. MEDICC Rev 2017;18:46–53.

[264] Sharrett AR, Hubbard LD, Cooper LS, Sorlie PD, Brothers RJ, Nieto FJ, et al. Retinal arteriolar diameters and elevated blood pressure: the Atherosclerosis Risk in Communities Study. Am J Epidemiol 1999;150(3):263–70.

[265] Erden S, Bicakci E. Hypertensive retinopathy: incidence, risk factors, and comorbidities. Clin Exp Hypertens 2012;34(6):397–401.

[266] Wong TY, McIntosh R. Hypertensive retinopathy signs as risk indicators of cardiovascular morbidity and mortality. Br Med Bull 2005;73(1):57–70.

[267] Kabedi NN, Mwanza JC, Lepira FB, Kayembe TK, Kayembe DL. Hypertensive retinopathy and its association with cardiovascular, renal and cerebrovascular morbidity in Congolese patients. Cardiovasc J Afr 2014;25(5):228–32.

[268] Ong YT, Wong TY, Klein R, Klein BE, Mitchell P, Sharrett AR, et al. Hypertensive retinopathy and risk of stroke. Hypertension 2013;62(4):706–11.

[269] Shantha GPS, Kumar AA, Bhaskar E, Sivagnanam K, Srinivasan D, Sundaresan M, et al. Hypertensive retinal changes, a screening tool to predict microalbuminuria in hypertensive patients: a cross-sectional study. Nephrol Dial Transplant 2010;25(6):1839–45.

[270] Bock KD. Regression of retinal vascular changes by antihypertensive therapy. Hypertension 1984;6(6 Pt 2):III158–62.

[271] UK Prospective Diabetes Study Group. Tight blood pressure control and risk of macrovascular and microvascular complications in type 2 diabetes: UKPDS 38. UK Prospective Diabetes Study Group. BMJ 1998;317(7160):703–13.

[272] Fraser-Bell S, Symes R, Vaze A. Hypertensive eye disease: a review. Clin Experiment Ophthalmol 2017;45(1):45–53.

[273] Ikram MK, Xueling S, Jensen RA, Cotch MF, Hewitt AW, Ikram MA, et al. Four novel loci (19q13, 6q24, 12q24, and 5q14) influence the microcirculation in vivo. PLoS Genet 2010;6(10):e1001184.

[274] Sim X, Jensen RA, Ikram MK, Cotch MF, Li X, MacGregor S, et al. Genetic loci for retinal arteriolar microcirculation. PLoS One 2013;8(6):e65804.

[275] Jensen RA, Sim X, Smith AV, Li X, Jakobsdottir J, Cheng CY, et al. Novel genetic loci associated with retinal microvascular diameter. Circ Cardiovasc Genet 2016;9(1):45–54.

[276] Heggermont WA, Heymans S. MicroRNAs are involved in end-organ damage during hypertension. Hypertension 2012;60(5):1088–93.

[277] Ahn SJ, Woo SJ, Park KH. Retinal and choroidal changes with severe hypertension and their association with visual outcome. Invest Ophthalmol Vis Sci 2014;55(12):7775–85.

[278] Gallo A, Mattina A, Rosenbaum D, Koch E, Paques M, Girerd X. Retinal arteriolar remodeling evaluated with adaptive optics camera: relationship with blood pressure levels. Ann Cardiol Angeiol 2016;65(3):203–7.

[279] Rosenbaum D, Mattina A, Koch E, Rossant F, Gallo A, Kachenoura N, et al. Effects of age, blood pressure and antihypertensive treatments on retinal arterioles remodeling assessed by adaptive optics. J Hypertens 2016;34(6):1115–22.

[280] Ritt M, Schmieder RE. Wall-to-lumen ratio of retinal arterioles as a tool to assess vascular changes. Hypertension 2009;54(2):384–7.

[281] Harazny J, Schmieder RE. Reliability of retinal microcirculation measurements by scanning laser Doppler flowmetry in humans. J Hypertens 2012;30(6):1266.

[282] Cheung CY-I, Ikram MK, Sabanayagam C, Wong TY. Retinal microvasculature as a model to study the manifestations of hypertension. Hypertension 2012;60(5):1094–103.

CHAPTER 8

Renal biomarkers of preeclampsia

Sara Faiz
Baylor College of Medicine, Houston, TX, United States

Introduction

Preeclampsia is usually defined as new-onset hypertension (i.e., systolic blood pressures (SBP) $\geq$ 140 mmHg and/or diastolic blood pressures (DBP) $\geq$ 90 mmHg) and proteinuria (>0.3 g/day) arising after 20 weeks of gestation in a previously normotensive woman [1]. Elevated blood pressure (BP) should be recorded on two measurements at least 4 h apart. Proteinuria is defined as the excretion of at least 300 mg of protein in a 24-h urine collection. Alternatively, a urine protein (mg/dL)/creatinine ratio (mg/dL) $\geq$ 0.3 has good sensitivity (98.2%) and specificity (98.8%) as a diagnostic tool [2]. Conversely, a positive qualitative dipstick test for proteinuria provides too variable results to be considered as a reliable diagnostic tool of proteinuria. It can be used if no other method is readily available. In that case only, a 1+ dipstick result is considered as the cut-off for the diagnosis of proteinuria [1]. Since recently, in recognition of the syndromic nature of preeclampsia, proteinuria is no longer considered as mandatory for the diagnosis of preeclampsia [3, 4]. Consequently, in the absence of proteinuria, preeclampsia can be diagnosed as new onset hypertension associated with:

- thrombocytopenia < 100,000/μL,
- elevated liver transaminases (> twice the normal values),
- impaired renal function (with serum creatinine > 1.1 mg/dL or doubling of serum creatinine level in the absence of any other renal disease),
- pulmonary edema,
- new-onset visual or cerebral disturbances [3].

Early and late preeclampsia

Preeclampsia has been categorized by some researchers into early onset (<34 weeks of gestation) versus late onset (>34 weeks of gestation) [5]. These

Seema S. Ahuja and Brian Castillo: Kidney Biomarkers. https://doi.org/10.1016/B978-0-12-815923-1.00009-2

two subtypes seem to have different etiologies and phenotypes. In placental or early-onset preeclampsia, the etiology is abnormal placentation. In maternal preeclampsia or late-onset preeclampsia, the problem arises from the interaction between a presumably normal placenta and maternal factors that are plagued with endothelial dysfunction, making them susceptible to microvascular damage. These commonly used classifications seem to have prognostic value, because placental or early-onset preeclampsia carries a significantly higher risk of maternal and fetal complications [5–7]. Despite the pathophysiologic differences between these subtypes of preeclampsia, one must recognize that the distinction is not always clear cut, because the two subtypes may harbor significant overlap, such as in the older woman with vascular disease who experiences abnormal placentation. Thus, although subtyping may be helpful in the understanding and prognostication of the condition, most patients with preeclampsia have elements of both pathologies [8].

Other hypertensive disorders of pregnancy

Chronic hypertension

Chronic hypertension (CHT) is defined as systolic blood pressure 140 mmHg or diastolic blood pressure 90 mmHg, or both, (at least 2 readings taken 4 h apart) that antedates pregnancy, is present before the 20th week of pregnancy, or persists longer than 12 weeks postpartum and occurs in 5%–8% of pregnancies.

Gestational hypertension

Gestational hypertension is the new onset of hypertension after 20 weeks, often near term with no proteinuria or other end organ effects, with normalization of blood pressure after delivery.

Superimposed preeclampsia

Superimposed preeclampsia (SPE) refers to the development of PE in women with underlying hypertension [9].

Incidence of preeclampsia

It is difficult to obtain accurate estimates of the incidence of preeclampsia because of a lack of standardization of diagnostic criteria in population databases, regional population variation and lack of databases. Abalos et al. estimated the incidence of preeclampsia in the world and in the United States as approximately 4.6% and 2.3%, respectively [10]. In a study using data from the National Hospital Discharge Survey (NHDS) public-use data set, it was reported that preeclampsia rates have increased in the United States, with age-adjusted rates rising from 2.4% between 1987 and 1988 to 2.9% of deliveries between 2003 and 2004 [11]. The incidence of preeclampsia is increasing in the United States and may be related to the higher prevalence of predisposing disorders such as hypertension, diabetes, obesity, delay in child bearing, and the use of artificial reproductive technologies with associated increase in multifetal gestation [11].

Global burden

Worldwide, preeclampsia and related-conditions are among the leading causes of maternal mortality, perinatal death, preterm birth, and low birthweight [12] The only definitive treatment for preeclampsia is termination of pregnancy/delivery of the fetus and placenta, which can cause significant mortality and morbidity for both mother and child, particularly when it occurs remote from term, between 24 and 34 weeks' gestation. Approximately 12%–25% of fetal growth restriction and small for gestational age infants as well as 15%–20% of all preterm births are attributable to preeclampsia; the associated complications of prematurity are substantial, including neonatal deaths and serious long-term neonatal morbidity [12, 13]. One-quarter of stillbirths and neonatal deaths in developing countries are associated with preeclampsia/eclampsia [14]. Preeclampsia does not only have short-term risks, but long term can lead to cardiovascular disease and/or type 2 DM in both mothers and their offspring [15, 16].

A World Health Organization review identified hypertensive disorders of pregnancy (HDP) among the leading causes of maternal mortality, causing nearly 18% of all maternal deaths worldwide, with an estimated 62,000–77,000 deaths per year [4]. They are the single leading cause of maternal mortality in industrialized countries, accounting for 16% of deaths [17]. In Africa and Asia, hypertensive disorders accounted for 9% of maternal deaths, whereas, in Latin America and the Caribbean, the figure was over 25% [17].

Predisposing factors

Obesity, chronic hypertension, and diabetes are among the risk factors for preeclampsia, which also include nulliparity, adolescent pregnancy, and conditions leading to hyperplacentation and large placentas, e.g., twin pregnancy [18].

In the Hyperglycemia and Adverse Pregnancy Outcome (HAPO) Study, a higher risk of preeclampsia was found in obese nongestational diabetic women than in nonobese gestational diabetics. Obese women are more insulin resistant as compared with normal weight women; hence, increased insulin resistance may be relevant to the development of preeclampsia in obese women and women developing gestational diabetes. However, obesity in addition to

Table 1 Common risk factors for preeclampsia [14].

Pregnancy-specific issues	Maternal preexisting conditions
• Nulliparity • Partner-related factors (new paternity, limited sperm exposure (e.g., barrier contraception) • Multifetal gestation • Hydatidiform mole	• Older age • African-American race • Higher body mass index • Pregestational diabetes • Chronic hypertension • Renal disease • Antiphospholipid antibody syndrome • Connective tissue disorder (e.g., systemic lupus erythematosus) • Family history or preeclampsia • Lack of smoking

gestational diabetes was associated with a greater risk of preeclampsia than either factor alone, thereby implicating other potential mechanisms such as inflammation in the development of preeclampsia in this high-risk group [19]. The risk of development of PE in women of Black or South Asian racial origin is higher than in Caucasians [20]. Hypertriglyceridemia is also known to precede preeclampsia. A recent metaanalysis of 24 case-control studies, 2720 women, revealed that high triglyceride levels are associated with a fourfold increased prevalence for PE [21].

Systemic lupus erythematosus (SLE) affects women in their childbearing years and is associated with immune alteration, specifically a reduction in regulatory T cells. Over 20% of pregnant women with SLE have pregnancies complicated with PE [22].

Risk factors can be classified into pregnancy-specific characteristics and maternal preexisting features (Table 1) [14]. The presence of preeclampsia in a first-degree relative increases a woman's risk of severe preeclampsia two- to fourfold [23]. If a woman becomes pregnant by a man who has already fathered a preeclamptic pregnancy in a different woman, her risk of developing preeclampsia is almost doubled [24].

Pathogenesis

Preeclampsia only occurs in the presence of a placenta and almost always remits after its delivery. As in the case of the hydatidiform mole, the presence of a fetus is not necessary for the development of preeclampsia [25]. Similarly, in a case of preeclampsia with an extra-uterine pregnancy, removal of the fetus alone was not sufficient, and symptoms persisted until the placenta was delivered [26]. Cases of postpartum eclampsia have been associated with retained placental fragments, with rapid improvement after uterine curettage [27]. Various pathways, including deficient heme oxygenase expression, genetic factors, oxidative stress, immune factors such as angiotensin receptor autoantibodies or altered natural killer cell signaling and, deficient catechol-*O*-methyl transferase or deficient corin enzymes, have been all proposed to have key roles in inducing placental disease [28–33].

A 2-stage model of preeclampsia has been proposed to address its pathophysiology. Stage 1 of preeclampsia, reduced placental perfusion, is considered the "root cause." This then translates, in some but not all women, into stage 2: the multisystemic maternal syndrome of preeclampsia (Fig. 1) [34]. The most compelling observation supporting reduced placental perfusion in preeclampsia is the failure of the maternal uterine spiral arteries that supply the intervillous space, to undergo the vascular remodeling characteristic of normal pregnancy (Fig. 2) [34].

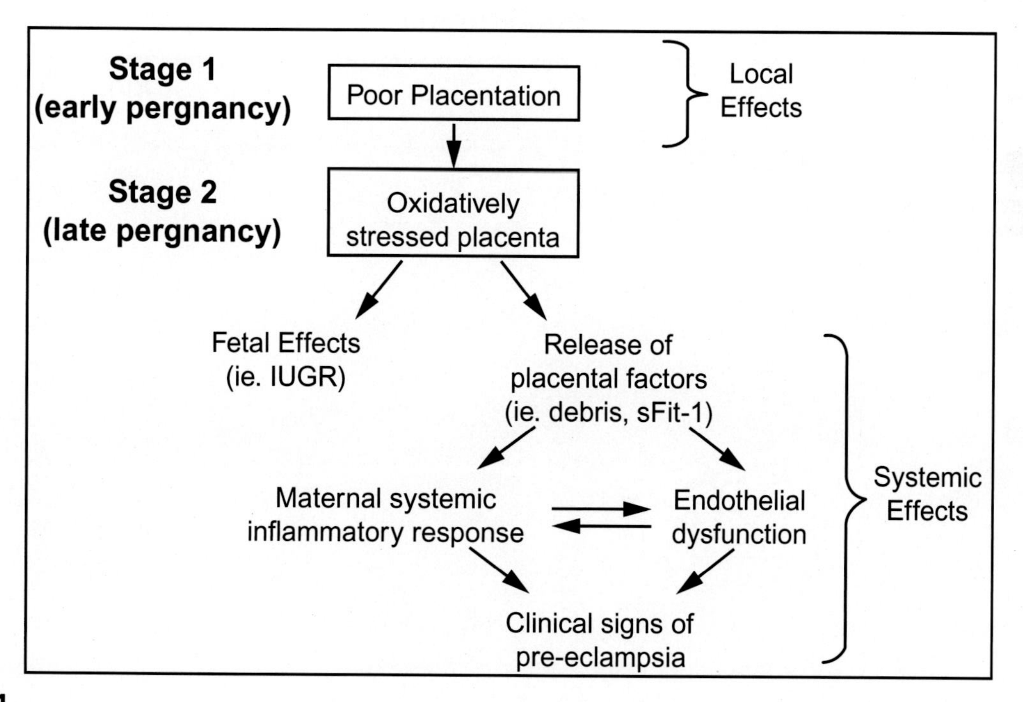

FIG. 1
The two stages in the development of preeclampsia. Preclinical stage 1 preeclampsia occurs in early pregnancy when insufficient trophoblast invasion leads to poor placentation resulting in placental hypoxia. Stage 2 occurs systemically when an oxidatively stressed placenta releases factors into the maternal circulation, which cause the maternal systemic inflammatory response and endothelial dysfunction that lead to the clinical signs of preeclampsia [160].

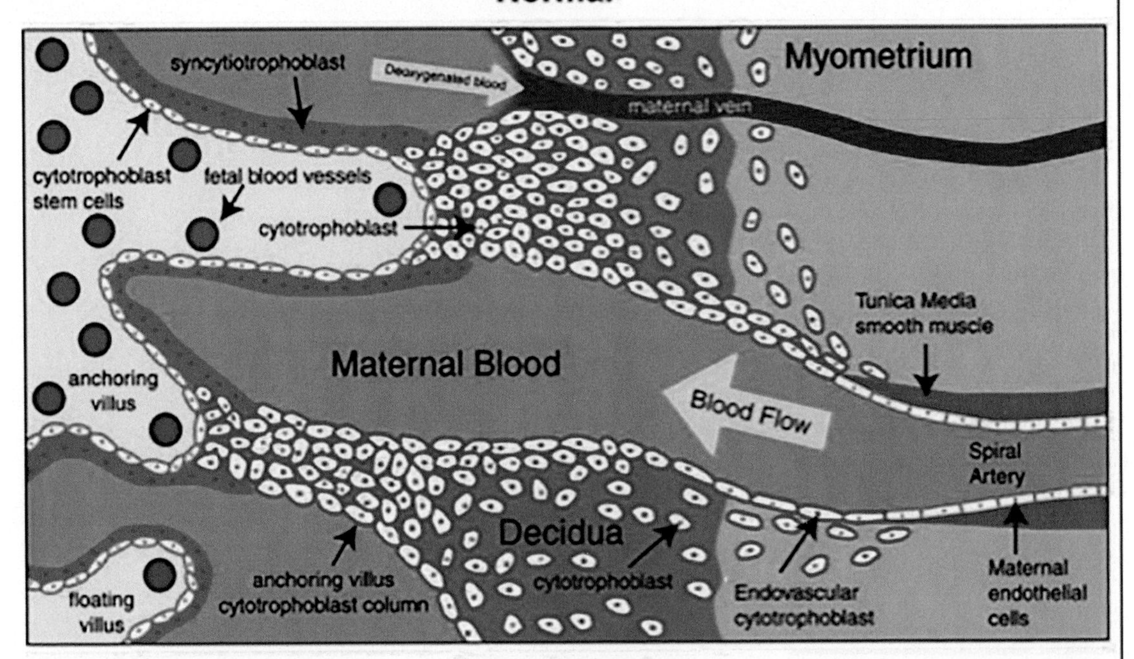

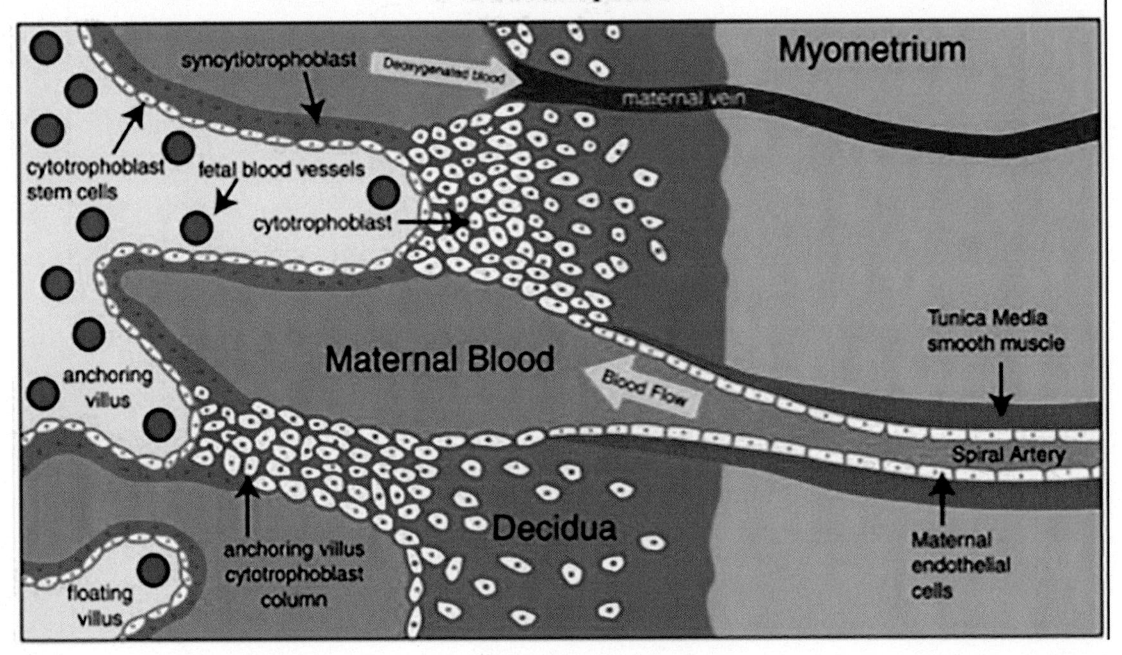

FIG. 2

Abnormal placentation in preeclampsia. In normal placental development, invasive cytotrophoblasts of fetal origin invade the maternal spiral arteries, transforming them from small-caliber resistance vessels to high-caliber capacitance vessels capable of providing placental perfusion adequate to sustain the growing fetus. During the process of vascular invasion, the cytotrophoblasts differentiate from an epithelial phenotype to an endothelial phenotype, a process referred to as "pseudovasculogenesis" or "vascular mimicry" (*upper panel*). In preeclampsia, cytotrophoblasts fail to adopt an invasive endothelial phenotype. Instead, invasion of the spiral arteries is shallow, and they remain small-caliber, resistance vessels (*lower panel*) [161].

Clinical features

The maternal syndrome and clinical features of preeclampsia are unified by the presence of systemic endothelial dysfunction and microangiopathy, in which the target organ may be the brain (seizures or eclampsia), liver (HELLP syndrome), or kidney (glomerular endotheliosis and proteinuria) [35].

Effect on kidney

Women with preeclampsia have a four to five times increased risk of ESRD compared with women without preeclampsia. A large Norwegian registry database showed that the relative risk of developing end-stage renal disease (ESRD) was 4.7 (95% CI, 3.6–6.1) with the risk increasing to 15.5 (95% CI, 7.8–30.8) if 2 or 3 pregnancies were complicated by preeclampsia [36].

The glomerular injury of preeclampsia manifests as a clinical triad, namely, hypertension, albuminuria, and a loss of intrinsic filtration capacity that lowers the GFR. Glomerular endotheliosis, the characteristic renal lesion in preeclampsia, is characterized by occlusion of capillary lumens, glomerular endothelial swelling, and loss of endothelial fenestrations [37]. The glomerular capillary lumina are obscured, which results in a substantial decrease in glomerular filtration. In a study comparing postpartum women with a history of preeclampsia, with healthy, postpartum controls, GFR was depressed by 40% on postpartum day 1, but by only 19% and 8% in the second and fourth postpartum weeks, respectively [38]. Studies have shown that glomerular damage and proteinuria often are extensive during a preeclamptic pregnancy [35] but normalize soon after birth in the majority [38]. In a small subset of patients, the resolution might, however, take years [39].

Effect on cardiovascular system

While it has been traditionally assumed that preeclampsia is a self-limited disease that resolves once the baby and placenta are delivered, some studies have shown that this maternal endothelial dysfunction can last for years after the episode of preeclampsia [40, 41].

Preeclampsia is known to increase the risk of developing a cardiovascular disease later in life by a factor of 2 [11]. In a study of 3658 women with preeclampsia, about 50% of them developed future hypertension with a 3.70 times higher risk compared to women with a normotensive pregnancy [42]. Sibai et al. carried out a 25-year follow-up of patients with a history of preeclampsia versus a control group, and assessed the risk of developing chronic hypertension. The overall incidence of chronic hypertension was significantly higher in patients with preeclampsia (14.8% vs 5.6%; *P*, .0001), which clearly indicates an increase of this risk with time [43]. The risk of fatal ischemic heart disease was increased in women with preeclampsia, and early onset preeclampsia (before week 37) significantly added to this risk, resulting in a relative risk of 7.71 [42].

Liver and coagulation abnormalities

The HELLP (hemolysis, elevated liver enzymes, and low platelet count) syndrome is considered the most severe form of PE, thus requiring a prompt delivery of the newborn to avoid the progression of the disease and complications [44]. The HELLP syndrome has been related to 1% of cases of maternal mortality and about 20% of perinatal mortality [45].

Effects on nervous system

Potential cerebral complications of preeclampsia include ischemic stroke, hemorrhagic stroke, cerebral edema, and seizure (Eclampsia) [46]. Eclamptic seizures contribute substantially to maternal morbidity and mortality. Several prodromal symptoms such as severe headache, altered mental status, blurred vision, and hyperreflexia with clonus may precede the onset of seizures. However, in 40% of eclampsia cases, no prodromal signs are present [1]. In large metaanalyses relative to women with uncomplicated pregnancies, women with a history of preeclampsia/eclampsia had a 1.8–2 times increased risk of cerebrovascular disease [40, 42]. Preeclampsia has been associated with posterior reversible encephalopathy syndrome (PRES) [46]. PRES is characterized by vasogenic cerebral edema that causes focal neurological symptoms. Because the edema of PRES tends to preferentially involve the parietal and occipital lobes, visual symptoms are common. In addition, headache, altered mental status, cerebral ischemia, and hemorrhage can be seen in PRES [47, 48]. The cerebral edema and symptoms are often reversible although permanent parenchymal injury and clinical disability can occur in severe cases [46].

Fetal complications

As delivery of the placenta is the only effective treatment for PE, babies born to women with PE often suffer from intrauterine growth restriction and preterm birth along with some of the associated neonatal comorbidities (i.e., respiratory distress syndrome, intraventricular hemorrhage) and increased fetal proinflammatory profiles [49]. An antiangiogenic state such as preeclampsia may additionally predispose infants to bronchopulmonary dysplasia (BPD). Ozkan et al. showed that the incidence of BPD in preterm infants born to preeclamptic women was significantly higher than in preterm infants born to normotensive women (38.5% vs 19.5%) [50]. ROP was also more severe in infants born to preeclamptic mothers [51]. The onset of NEC was significantly earlier and duration of NEC was longer in these infants [52]. The children born to preeclamptic mothers have also been shown to be at high risk for complications like diabetes mellitus, cardiovascular disease, and hypertension [53].

Biomarkers

Angiogenic proteins

The past decade has seen new insights into pathogenesis of preeclampsia, most notably the role of antiangiogenic proteins in producing preeclampsia phenotypes in pregnant rodents and elevated levels in women with preeclampsia [54]. It has been suggested that the clinical manifestations of preeclampsia result, in part, from an imbalance between circulating proangiogenic and antiangiogenic factors in the maternal circulation. Two important antiangiogenic factors are soluble vascular endothelial growth factor 1 (sVEGFR1) (also referred to as soluble fms-like tyrosine kinase 1 or sFlt1) and soluble endoglin (sEng). Clark et al. suggested that sFlt-1 is produced in the placenta and secreted into the bloodstream, where it is thought to bind and neutralize VEGF, and placental growth factor (PlGF), with high affinity [55].

Placental syncytiotrophoblasts and in particular syncytial knots were identified as a major source of sFlt1 and soluble endoglin production [56, 57]. Syncytial knots are induced by placental hypoxia and are noted predominantly in preeclamptic placentas. Syncytial knots have been shown to release aggregates into the maternal circulation suggesting an additional source of increased sFlt1 in the maternal blood besides secretion by the placenta [57]. It has been suggested that shed syncytial aggregates get trapped in the capillary beds of lung tissue, where they further undergo disaggregation or apoptosis/necrosis to release the smaller microparticles into the systemic circulation [58]. sFlt1 is an endogenous soluble antiangiogenic protein that acts by binding proangiogenic proteins—vascular endothelial and placental growth factors VEGF and PlGF.

Although VEGF is commonly referred to as a single factor, it is actually a family of proteins. This family consists of VEGF-A (often referred to as simply VEGF), VEGF-B, VEGF-C, and VEGF-D (VEGF-C and VEGF-D are both involved in the growth of lymphatic vessels) [59]. PlGF is also a member of the VEGF family [59]. VEGF and its family members are prone to rapid degradation and thus have a relatively short plasma half-life of approximately 30 min [60]. Because endothelial cells are among those that secrete VEGF and it acts in a paracrine fashion, the roles of VEGF are often thought to be limited to endothelial proliferation, migration, and increased vascular permeability [61]. The actions exerted by VEGF family members are elicited through two tyrosine kinase receptors, VEGF receptor-1 (VEGFR-1) and VEGF receptor-2 (VEGFR-2). VEGFR-1 is also known as Flt-1 (fms-like tyrosine kinase 1), and VEGFR-2 is commonly referred to as Flk-1 (fetal liver kinase 1). VEGFR-1 also has a splice variant sVEGFR-1/sFlt-1, which contains only the extracellular domain of the receptor, making it soluble in plasma [62]. Because sFlt-1 contains the binding site for VEGF, it is still able to bind all isoforms of the growth factor, as well as its close relative placental growth factor (PlGF) [63].

The decreased blood flow to the placenta due to failed remodeling leads to a local hypoxic environment in the placenta. In response to hypoxia, vascular endothelial growth factor (VEGF) is transcriptionally upregulated; this growth factor is a particularly effective proangiogenic factor and is important to maintain the health of existing vessels [64].

Different combinations of the VEGF receptors can result in either a pro- or antiangiogenic effect, for example, when PlGF binds to Flt-1, it is considered proangiogenic, because this interaction increases the bioavailability of VEGF to bind to the more proangiogenic Flk-1 [65]. VEGF binding to Flt-1, on the other hand, is considered mildly antiangiogenic [66]; this is because the angiogenic actions of Flt-1 are not as powerful as those of Flk-1. When VEGF binds to Flt-1, there is less VEGF available to bind Flk-1. Similarly, sFlt-1 acts as an antagonist for VEGF, sequestering it in the plasma [67].

One condition, which has been shown to upregulate sFlt-1 expression, is hypoxia; this likely advances the pathogenesis of preeclampsia [68]. The alternative splicing of sFlt-1 is regulated at the level of mRNA. Both sFlt-1 and Flt-1 are transcribed from the FLT1 gene and have the same start site as well as transcriptional regulatory sequences [69]. This splicing regulation has been shown to be acted on by factors such as VEGF. With increased VEGF exposure, sFlt-1 expression has been shown to be upregulated [70]. Increased expression of VEGF in hypoxic conditions could potentially be a mechanism, which promotes sFlt-1 upregulation in preeclampsia.

Fan et al. observed a significant increase in VEGF in the decidua of preeclamptic women, which could be a cause of the significant increase of sFlt-1 in placental trophoblasts observed in these women [71]. Though the hypoxic placenta secretes increased levels of VEGF [72], the expression of sFlt-1 increases to a greater extent. The imbalance in the pro- and antiangiogenic factors produced and released into the maternal circulation is believed to drive the maternal syndrome[73]. The imbalance can be seen as an increase in plasma sFlt-1 levels and a decrease in both free VEGF and free PlGF [54].

It has been suggested that Flt-1 acts as a VEGF decoy receptor to regulate the availability of VEGF for Flk-1, which mediates angiogenesis [74, 75]. VEGF and PlGF bound to sFlt-1 are no longer available to their innate receptors on endothelial cells, and VEGF signaling is disrupted [76]. Later evidence showed, however, that Flt-1 also participates in angiogenesis during adulthood [77]. Activating Flt-1 leads to nitric oxide (NO) release and stimulates organization of endothelial cells into capillary-like tubes [78].

Angiogenic proteins as potential biomarkers

Whereas sFlt-1, with its molecular weight of 110 kDa, is too large to be filtered by the glomerulus; both PlGF and VEGF (30 and 45 kDa, respectively) are easily filtered by the kidneys and excreted in the urine. However, in addition to the VEGF made by the vascular endothelium and other cell types, podocytes surrounding the glomerulus also produce VEGF [79]. Therefore, VEGF measured in the urine may not necessarily represent circulating levels accurately, but PlGF, which can only be found in the circulating blood, is more likely to give an accurate representation of the angiogenic state [80]. Also free VEGF concentrations are not useful for clinical purposes because the circulating concentrations are below the detection limit of most commercially available ELISA kits during pregnancy [76].

First-trimester screening

In a large prospective clinical study involving nearly 8000 women, Poon et al. demonstrated that a combination of angiogenic factors (PlGF), pregnancy-associated plasma protein A, and uterine artery Doppler velocimetry in the first trimester can predict the subsequent development of early onset preeclampsia in a low-risk population with a diagnostic sensitivity of 93% at a 5% false-positive rate [81].

One promising strategy is urine screening with a PlGF assay followed by blood confirmation with sFlt-1/PlGF ratio [82]. In a nested case-control study by Levine et al., urinary PlGF was measured in 120 normotensive controls and women who subsequently developed preeclampsia [66]. This study revealed that a low concentration of PlGF in the urine at midgestation was strongly associated with the subsequent development of preterm preeclampsia. The adjusted odds ratio for the risk of developing preterm preeclampsia in those women who had low urinary PlGF concentrations (118 pg/mL) at 21–32 weeks was 22.5. A prospective study of 69 pregnant women (35 with preeclampsia and 34 normotensive controls) also showed that compared to controls, the urinary PlGF and PlGF/creatinine levels were significantly reduced in women with preeclampsia compared to normotensive controls. The difference was maintained after normalization for urinary creatinine concentrations and gestational age. Reduced urinary PlGF level antedated the diagnosis of preeclampsia by several weeks in three cases in their cohort [80].

As for all screening tests, the positive predictive value of angiogenic factor screening depends on the population tested. [76]. Perni et al. demonstrated in a longitudinal study of women with chronic hypertension, that alterations in sFlt-1, PlGF, sFlt-1/PlGF, and sEng were dramatically altered as early as 20 weeks of gestation in women who developed subsequent early onset preeclampsia [83]. Several recent prospective studies investigated the use of angiogenic factors in a high-risk group identified by abnormal uterine artery dopplers. The combination of early second-trimester ultrasound, including measurements of pulsatility index, with determination of angiogenic factors, sFlt-1 and PlGF, largely improved the predictive value of ultrasound screening alone [84–87].

A recent prospective study of nonintervention screening for PE during gestation in 61,174 singleton pregnancies explored the possibility of carrying out first-stage screening in the whole population by maternal factors and some of the biomarkers and proceeding to second-stage screening by the triple test only for a subgroup of the population selected on the basis of the risk derived from first-stage screening [88]. A 2-stage screening was recommended such that if the method of first-stage screening is maternal factors, then measurement of biomarkers can be reserved for only 70% of the population and if some of the biomarkers are included in first-stage screening then the need for the complete triple test can be reduced to 20%–40% of the population [88]. They had previously reported the maternal factors for increased risk for PE to be: advancing maternal age, increasing weight, Black and South Asian racial origin, medical history of chronic hypertension, diabetes mellitus and SLE or APS, conception by in vitro fertilization, family history of PE and personal history of PE; in the latter group the risk is inversely related to the gestational age at delivery of the previous pregnancy [20]. They had found that the performance of screening by both biophysical and biochemical markers is superior to screening by either method alone [20].

sFlt-1 and sFlt-1/PlGF ratio

In a cross-sectional nested case control study, Levine et al. compared gestational age–matched women with active preeclampsia and those with a normal pregnancy and revealed that concentrations of sFlt-1 were significantly higher in the former group [89]. They also showed that concentrations of this antiangiogenic protein were significantly increased 5–6 weeks before the detection of hypertension and proteinuria [89, 90]. Kim et al. found significantly elevated levels of sFlt-1 and significantly decreased levels of free PlGF at 14–23 weeks of gestation in women who went on to develop preeclampsia. These findings were observed even prior to the development of hypertension. It was found that a plasma log [sFlt-1/PlGF] ratio greater than 1.4 was associated with an increased risk of developing preeclampsia; this cut-off exhibited an 84% sensitivity and a 78% specificity [91].

A few studies [92, 93] have demonstrated that sFlt-1 measurement alone gives diagnostic specificity and sensitivity of up to approximately 80% and approximately 70%, respectively, for the prediction of preeclampsia during the second trimester. Examining the ratio of sFlt-1 (rising in preeclampsia) and PlGF (falling in preeclampsia) improved prediction of the condition remarkably [92–94]. Rana et al. examined the role of angiogenic factors as predictors of adverse outcomes in a cohort of women with twin pregnancies who were being evaluated for suspected preeclampsia. They found that within 2 weeks of evaluation, an adverse outcome occurred in about 65% of women and plasma sFlt1/PlGF ratio was identified as the strongest predictor of adverse outcomes alone and in combination with clinically available measurements. They also found that the sFlt1/PlGF ratio was inversely correlated with duration of pregnancy from time of evaluation to delivery and that 3 in 4 patients with a sFlt-1/PlGF ratio >85 needed delivery within 7 days [95].

These findings were similar to their study evaluating the role of sFlt1/PlGF ratio in singleton pregnancy. Normotensive women carrying twins had approximately twofold higher circulating sFlt1 and threefold higher sFlt1/PlGF ratio than normotensive women with singleton pregnancies [95]. This increased circulating antiangiogenic state may be one mechanism for the increased risk of preeclampsia noted in women carrying twin pregnancies [96].

Clinically, sFlt-1 concentrations have been observed to be directly proportional to severity of proteinuria, but inversely correlated with platelet count, gestational age, and neonatal birth weight adjusted for gestational age [97, 98]. In women with preeclampsia, concentrations of sFlt-1 are higher in those with early onset (<37 weeks), more severe disease [54, 89, 90, 99], and SGA (small for gestational age) neonates [89, 90, 97].

Maternal serum levels of PlGF and sFlt1 can be used to differentiate preeclampsia from chronic hypertension or systemic lupus erythematosus (SLE). In a retrospective case-control study of 52 patients with systemic lupus erythematosus (SLE), Qazi et al. [100] showed that serum concentration of sFlt1 was significantly higher in SLE pregnancies with preeclampsia ($n = 18$) than in those with SLE but without preeclampsia (1768 vs 1177 pg/mL). Rolfo et al. [101] showed that the sFlt1/PlGF ratio was significantly increased in preeclampsia (median: 436) compared with controls (median: 9.4) and CKD (median: 4.0).

Soluble endoglin

Soluble endoglin (sEng), another antiangiogenic protein, has also been implicated in the pathogenesis of preeclampsia [102]. Acting as an antiangiogenic protein, sEng disrupts transforming growth factor-*B* signaling in the vasculature. Overexpression of sFlt1 and sEng through an adenoviral expression system in rats led to a severe preeclampsia-like phenotype with features of HELLP (H: hemolysis (the breakdown of red blood cells); EL: elevated liver enzymes; LP: low platelet count) syndrome and fetal growth restriction [102]. In human pregnancy, alterations in sEng antedated clinical symptoms of preeclampsia by several months [90]. Interestingly, a composite measure incorporating sFlt-1, PlGF, and sEng was more predictive of preterm preeclampsia than the individual biomarkers alone [90]. Women with isolated SGA (small for gestational age) pregnancies were also characterized by early sustained increases in sEng [90], suggesting that sEng may be marker for placental insufficiency.

PP-13 and PAPP-A

Women who develop preeclampsia have significantly reduced first trimester serum levels of peptides produced predominantly by the syncytiotrophoblast of the placenta, such as pregnancy-associated plasma protein A (PAPP-A) and placental protein 13 (PP 13), but none of these markers, taken separately, has sufficient predictive power to be used as screening test [103, 104]. Several studies have suggested the involvement of PP 13 in the trophoblastic invasion in early pregnancy [105–107]. This protein binds extracellular matrix molecules (such as annexin II), which may contribute to the placental implantation. It is believed that its mild lysophospholipase- A activity may modulate these functions [106]. PP13 also stimulates prostacyclin release, which may allow the maternal spiral artery remodeling [107]. Reduced levels of PP13 in early pregnancy can lead to impaired trophoblast invasion. In a nested case-control study of 5867 pregnancies, it was demonstrated that in pregnancies with preeclampsia, first-trimester maternal serum levels of the protein PP-13 and of PAPP-A are reduced and that with more severe or early preeclampsia the levels may be further reduced, pointing toward the potential future use of PP-13 and PAPP-A as useful tools for identifying preeclampsia [104].

Matrix metalloproteinases

Matrix metalloproteinases (MMPs) comprise a family of 23 zinc and calcium-dependent proteases that degrade different components of the extracellular matrix [108]. The profound changes in uterine microarchitecture required to transform the spiral vessels and create an optimum environment for embryonic development involve a grounding transformation in which MMPs are essential [109]. Besides the effect on vascular remodeling, MMP-2 can mediate vascular reactivity by promoting the production of the vasoconstrictor peptide endothelin-1 through cleavage of the vasodilatory calcitonine gene related peptide [110]. Furthermore, the elevation of MMP-1 and MMP-2 has been strongly associated with the endothelial dysfunction observed in preeclamptic women [111, 112]. Expression of different MMPs such as MMP-2, -8, -9, and -11 was shown to be downregulated in placental tissues from pregnancies complicated with preeclampsia at >35 weeks of gestation compared to normal pregnancies [113]. Urine samples from healthy and preeclamptic pregnancies were analyzed in a study that was carried out in order to predict the risk of developing preeclampsia at three different stages of early pregnancy (12-, 16-, and 20-gestational week). From a set of nine MMPs evaluated, only MMP-2 was found to be significantly higher at 12 and 16 weeks [114]. In a different study,

14 biomarkers for preeclampsia, including MMP-2 and MMP-9, were evaluated in urine samples. At delivery, urine concentrations of MMP-2 and MMP-9 were significantly elevated in women with severe preeclampsia compared to normal pregnancy, and this feature persisted 6–8 weeks after delivery [115]. Measurements of the levels of MMPs have not been consistent in preeclampsia, with some studies showing an increase in MMP-2 and MMP-9 [116], while other studies showing a decrease in MMP-9 [117] The discrepancy in the results may be due to the fact that plasma MMPs represents global changes in MMPs in different tissues, and localized changes in MMPs in uteroplacental tissues and fluids may carry more predictive value [118].

Activated MMPs may also contribute to cardiovascular dysfunction in preeclampsia through proteolysis of cell surface receptors, such as VEGFR-2 and β(2)-adrenergic receptor, as previously shown in other animal models of cardiovascular diseases [119]. The animal models have found increased MMP-2 or MMP-9 activity in different tissues, and treatment with doxycycline (a nonspecific MMP inhibitor) ameliorated hypertension, vascular dysfunction, and artery/cardiac remodeling associated with this condition [120]. Doxycycline is a nonspecific MMP inhibitor that was proposed to alleviate hypertension and vascular dysfunction in preeclampsia, but was found to decrease placenta weight and cause IUGR in both normal pregnant and hypertensive pregnant rats, and to reduce trophoblast invasion and placental perfusion in hypertensive pregnant rats [121].

Galectin-7

Members of the galectin family are expressed within the female reproductive tract and have been shown to be involved in multiple biological functions that support the progression of pregnancy [122]. Galectin-7 serum concentration was found to be significantly higher in women with preeclampsia during weeks 10–12 and 17–20 of gestation, compared to uncomplicated gestation matched pregnancies [123].

Copeptin

Santillan et al. suggested a possible use for plasma arginine vasopressin (AVP) measurements in the prediction of preeclampsia [124]. They showed that chronic infusion of AVP during pregnancy in wild-type C57BL/6J mice is sufficient to induce all of the major maternal and fetal phenotypes associated with human preeclampsia, including pregnancy-specific hypertension. Copeptin is an inert prosegment of AVP that is secreted in a 1:1 molar ratio and exhibits a substantially longer biological half-life than AVP, rendering it a clinically useful biomarker of AVP secretion. Copeptin was measured throughout pregnancy in maternal plasma from preeclamptic and control women and was significantly higher throughout preeclamptic pregnancies versus control pregnancies, as early as 6 weeks of gestation.

Podocyte glycoproteins and extracellular vesicles

Over the last few years, there has been a focus on derangements of podocytes and podocyte-specific proteins (such as nephrin, synaptopodin, podocin, and podocalyxin), and their roles in the mechanism(s) of proteinuria in preeclampsia [125]. Studies have shown reduced expressions of nephrin, glomerular *epithelial* protein 1, GLEPP-1, and ezrin in the renal tissue sections of preeclamptic woman compared with those of women with normal pregnancies [126, 127]. However, none of the three urinary markers (testing podocyturia, nephrinuria, or albuminuria) achieved the minimum predictive values required for clinical testing in a prospective study of high-risk pregnant patients [128].

Recently it was suggested that renal injury in preeclampsia is associated with the presence of urinary extracellular vesicles containing immunologically detectable podocyte-specific proteins.

The study showed that, compared with urine from women with normotensive pregnancies, urine from women with preeclamptic pregnancies contained a high ratio of podocin-positive to nephrin-positive urinary extracellular vesicles (podocin$^+$ EVs-to-nephrin$^+$ EVs ratio) and increased nephrinuria, both of which correlated with proteinuria. The potential use of urinary EV RNAs and proteins as diagnostic biomarkers for various kidney and urologic diseases is currently being explored [129].

Cystatin C

Cystatin C is a 13-kDa protein, a member of the cysteine proteinase inhibitors family, which is produced at a constant rate in all nucleated cells, then freely filtered across the glomerular membrane, and finally reabsorbed and thoroughly metabolized in proximal tubules [130]. Unlike creatinine, changes of cystatin C concentrations in serum are less susceptible to the influence of nonrenal factors such as gender, age, muscular mass, and inflammation [131, 132]. Apart from being marker of renal dysfunction, the hypothesis is that cystatin C could have a direct role in etiology of PE, since placental expression of cystatin C is increased in patients with PE [133].

Serum level of CysC, a 13.3-kDa cysteine protease inhibitor, has been reported to be significantly higher in late stages of pregnancy than in nonpregnant women and even higher in PE [134–136]. A nested case-control study from a prospectively assembled cohort during the first trimester revealed that a cystatin C value of >0.85 mg/L together with normal creatinine can identify women with altered glomerular filtration quality and was a strong independent risk factor for the development of PE [137]. Kolialexi et al. also reported in a retrospective case control study that first trimester CysC at mean of 12.1 weeks gestational age in PE cases is higher than in controls (median: 0.91 vs 0.48 mg/L) [138]. The data from prospectively conducted studies showed that the accuracy of CysC in predicting PE increases with increasing duration of pregnancy [137–139].

Serum uric acid

Uric acid (UA) is an oxidation end product of purine metabolism. Serum UA levels depend on the balance of endogenous production, dietary intake and elimination via filtration and active secretion in the kidneys, approximately two-thirds, and via the gastroenteric tract, approximately one-third [140]. Laughon et al. [141] and Risch et al. [137] found higher mean first trimester UA concentrations in cases with PE compared to the concentrations without PE. Together, they supported the hypothesis that higher first trimester UA in unselected women with singleton pregnancies is associated with the later development of PE. Another study seemed to reflect that there were higher levels of serum UA and serum creatinine in the third trimester not only in patients with severe preeclampsia, but also in the normal pregnancies, with no changes in their second trimesters either in normal pregnancies or in patients with severe preeclampsia [142].

Urinary kallikreins

Tissue kallikrein is a serine protease that cleaves low-molecular-weight kininogen to generate kallidin and bradykinin. In human placenta, kallikrein has been shown in syncytiotrophoblast, intravascular trophoblast, fetal endothelium, and in the basal and chorionic plate [143]. Urinary tissue kallikrein levels were found to be significantly decreased in women with severe preeclampsia compared with those of gestation matched normotensive pregnant women at 28 weeks of gestation and at near delivery date [144, 145]. However, it has not been shown to be useful as a screening test for preeclampsia.

Cell-free DNA

It has been observed that fetal DNA concentration in maternal plasma (cfDNA) increases in pregnancy complications associated with placental abnormalities, such as hypertension and preeclampsia. In a recent study, it was observed that total and fetal cfDNA levels were significantly different among healthy, mild-PE, severe-PE, and HELLP syndrome pregnant women, with higher levels in those women with a more severe disease, suggesting that total cfDNA levels may be a promising novel biomarker to identify women with PE and assess its severity in clinical practice [146]. The identification of the methylated *RASSF1A* promoter gene, which is elevated in early-PE pregnancies, offers a promising alternative to the quantification of cfDNA, using Y-chromosome-specific sequences [147].

Clinical utility of biomarkers in disease management

The use of biomarkers in early pregnancy would help in identifying high- and low-risk pregnancies for surveillance and to administer interventions. Identification of pregnancies at high-risk of developing preeclampsia at 11–13 weeks' gestation is beneficial because in such cases prophylactic use of aspirin (150 mg/day from 11–14 weeks' gestation to 36 weeks) reduces the rate of early-PE, with delivery at <32 weeks, by about 90% and preterm-PE, with delivery at <37 weeks, by about 60%; but there is little evidence of a reduction in incidence of PE with delivery at term [148, 149]. In pregnancies with impaired placentation, the use of low-dose aspirin at >16 weeks gestation does not prevent the subsequent development of preeclampsia [150, 151]. Randomized trials on the use of aspirin have reported that the drug is not associated with increased risk of adverse events and in the case of antepartum hemorrhage the risk may actually be reduced [152].

In addition, compounds that upregulate pro-angiogenic factors such as statins have been used to ameliorate preeclampsia in animal models [153]. Recently pravastatin was successfully tested as a therapeutic option in the prevention of preeclampsia in high-risk pregnant women [154].

Extracorporeal apheresis to lower circulating sFlt1 has also been attempted as a treatment modality in women with preeclampsia. In exciting studies, Thadhani et al. using dextran sulfate apheresis were able to extend three preeclamptic pregnancies by 2–4 weeks, all of which resulted in healthy deliveries with no neonatal or maternal morbidity [155]. In another study, lipid apheresis via the heparin-mediated extracorporeal LDL-precipitation (H.E.L.P) technique resulted in six pregnancies being prolonged on average by 15 days after patients were admitted to hospital, and by 9 days when compared to a control group. This may have translated into a calculated 20% lower fetal mortality risk compared to untreated pregnancies. However, independent from the transient heparin-induced effect H.E.L.P.-apheresis did not result in a significant immediate change of sFlt-1 levels below baseline, hence lowering lipids or other yet undefined factors appeared to be of more relevance than reducing sFlt-1 [156].

Wu et al. found in their meta-analysis that PlGF was best at predicting EOPE as a single biomarker; however, a combination model performed better than a single biomarker if studying PE as a single entity [157].

The diagnosis of preeclampsia when superimposed on CKD and CHT is challenging because it may be clinically indistinguishable from benign gestational progression of preexisting hypertension and proteinuria, which often coexist [158]. Superimposed preeclampsia is frequently associated with poor maternal and fetal outcomes; therefore, early and accurate diagnosis is essential to allow timely intervention, whereas misdiagnosis may also lead to unnecessary admissions and iatrogenic preterm delivery. In an analysis of potential biomarkers for superimposed preeclampsia, Bramham et al. concluded that from 20 and up to 42 weeks of gestation, lower maternal PlGF concentrations had high diagnostic accuracy for superimposed preeclampsia requiring delivery within 14 days [159].

References

[1] Lambert G, Brichant JF, Hartstein G, Bonhomme V, Dewandre PY. Preeclampsia: an update. Acta Anaesthesiol Belg 2014;65(4):137–49.

[2] Leanos-Miranda A, Marquez-Acosta J, Romero-Arauz F, Cardenas-Mondragon GM, Rivera-Leanos R, Isordia-Salas I, et al. Protein:creatinine ratio in random urine samples is a reliable marker of increased 24-hour protein excretion in hospitalized women with hypertensive disorders of pregnancy. Clin Chem 2007;53(9):1623–8.

[3] American College of Obstetricians and Gynecologists, Task Force on Hypertension in Pregnancy. Hypertension in pregnancy. Report of the American College of Obstetricians and Gynecologists' Task Force on Hypertension in Pregnancy. Obstet Gynecol 2013;122(5): 1122–31.

[4] Homer CS, Brown MA, Mangos G, Davis GK. Non-proteinuric pre-eclampsia: a novel risk indicator in women with gestational hypertension. J Hypertens 2008;26(2):295–302.

[5] von Dadelszen P, Magee LA, Roberts JM. Subclassification of preeclampsia. Hypertens Pregnancy 2003;22(2):143–8.

[6] Mongraw-Chaffin ML, Cirillo PM, Cohn BA. Preeclampsia and cardiovascular disease death: prospective evidence from the child health and development studies cohort. Hypertension 2010;56(1):166–71.

[7] Paruk F, Moodley J. Maternal and neonatal outcome in early- and late-onset pre-eclampsia. Semin Neonatol 2000;5(3):197–207.

[8] Phipps E, Prasanna D, Brima W, Jim B. Preeclampsia: updates in pathogenesis, definitions, and guidelines. Clin J Am Soc Nephrol 2016;11(6):1102–13.

[9] Acharya A. Promising biomarkers for superimposed pre-eclampsia in pregnant women with established hypertension and chronic kidney disease. Kidney Int 2016;89(4):743–6.

[10] Abalos E, Cuesta C, Grosso AL, Chou D, Say L. Global and regional estimates of preeclampsia and eclampsia: a systematic review. Eur J Obstet Gynecol Reprod Biol 2013;170(1):1–7.

[11] Wallis AB, Saftlas AF, Hsia J, Atrash HK. Secular trends in the rates of preeclampsia, eclampsia, and gestational hypertension, United States, 1987-2004. Am J Hypertens 2008;21(5):521–6.

[12] Duley L. The global impact of pre-eclampsia and eclampsia. Semin Perinatol 2009;33(3): 130–7.

[13] Goldenberg RL, Rouse DJ. Prevention of premature birth. N Engl J Med 1998;339(5):313–20.

[14] Jeyabalan A. Epidemiology of preeclampsia: impact of obesity. Nutr Rev 2013;71(Suppl 1): S18–25.

[15] Manten GT, Sikkema MJ, Voorbij HA, Visser GH, Bruinse HW, Franx A. Risk factors for cardiovascular disease in women with a history of pregnancy complicated by preeclampsia or intrauterine growth restriction. Hypertens Pregnancy 2007;26(1):39–50.

[16] Berg CJ, Mackay AP, Qin C, Callaghan WM. Overview of maternal morbidity during hospitalization for labor and delivery in the United States: 1993-1997 and 2001-2005. Obstet Gynecol 2009;113(5):1075–81.

[17] Khan KS, Wojdyla D, Say L, Gulmezoglu AM, Van Look PF. WHO analysis of causes of maternal death: a systematic review. Lancet 2006;367(9516):1066–74.

[18] WHO. WHO recommendations for prevention and treatment of pre-eclampsia and eclampsia. WHO Guidelines Approved by the Guidelines Review Committee, Geneva: WHO; 2011.

[19] Catalano PM, McIntyre HD, Cruickshank JK, McCance DR, Dyer AR, Metzger BE, et al. The hyperglycemia and adverse pregnancy outcome study. Associations of GDM and obesity with pregnancy outcomes. Diabetes Care 2012;35(4):780–6.

[20] Wright D, Syngelaki A, Akolekar R, Poon LC, Nicolaides KH. Competing risks model in screening for preeclampsia by maternal characteristics and medical history. Am J Obstet Gynecol 2015;213(1):62 e1–e10.

[21] Ray JG, Diamond P, Singh G, Bell CM. Brief overview of maternal triglycerides as a risk factor for pre-eclampsia. BJOG 2006;113(4):379–86.

[22] Clowse ME, Jamison M, Myers E, James AH. A national study of the complications of lupus in pregnancy. Am J Obstet Gynecol 2008;199(2). 127 e1-6.

[23] Carr DB, Epplein M, Johnson CO, Easterling TR, Critchlow CW. A sister's risk: family history as a predictor of preeclampsia. Am J Obstet Gynecol 2005;193(3 Pt 2):965–72.

[24] Lie RT, Rasmussen S, Brunborg H, Gjessing HK, Lie-Nielsen E, Irgens LM. Fetal and maternal contributions to risk of pre-eclampsia: population based study. BMJ 1998;316(7141):1343–7.

[25] Wang A, Rana S, Karumanchi SA. Preeclampsia: the role of angiogenic factors in its pathogenesis. Physiology 2009;24:147–58.

[26] Shembrey MA, Noble AD. An instructive case of abdominal pregnancy. Aust N Z J Obstet Gynaecol 1995;35(2):220–1.

[27] Matsuo K, Kooshesh S, Dinc M, Sun CC, Kimura T, Baschat AA. Late postpartum eclampsia: report of two cases managed by uterine curettage and review of the literature. Am J Perinatol 2007;24(4):257–66.

[28] Cudmore M, Ahmad S, Al-Ani B, Fujisawa T, Coxall H, Chudasama K, et al. Negative regulation of soluble Flt-1 and soluble endoglin release by heme oxygenase-1. Circulation 2007;115(13):1789–97.

[29] Cui Y, Wang W, Dong N, Lou J, Srinivasan DK, Cheng W, et al. Role of corin in trophoblast invasion and uterine spiral artery remodelling in pregnancy. Nature 2012;484(7393):246–50.

[30] Hiby SE, Walker JJ, O'Shaughnessy KM, Redman CW, Carrington M, Trowsdale J, et al. Combinations of maternal KIR and fetal HLA-C genes influence the risk of preeclampsia and reproductive success. J Exp Med 2004;200(8):957–65.

[31] Kanasaki K, Palmsten K, Sugimoto H, Ahmad S, Hamano Y, Xie L, et al. Deficiency in catechol-O-methyltransferase and 2-methoxyoestradiol is associated with pre-eclampsia. Nature 2008;453(7198):1117–21.

[32] Zhou A, Carrell RW, Murphy MP, Wei Z, Yan Y, Stanley PL, et al. A redox switch in angiotensinogen modulates angiotensin release. Nature 2010;468(7320):108–11.

[33] Zhou CC, Zhang Y, Irani RA, Zhang H, Mi T, Popek EJ, et al. Angiotensin receptor agonistic autoantibodies induce pre-eclampsia in pregnant mice. Nat Med 2008;14(8):855–62.

[34] Roberts JM, Gammill HS. Preeclampsia: recent insights. Hypertension 2005;46(6):1243–9.

[35] Karumanchi SA, Maynard SE, Stillman IE, Epstein FH, Sukhatme VP. Preeclampsia: a renal perspective. Kidney Int 2005;67(6):2101–13.

[36] Vikse BE, Irgens LM, Leivestad T, Skjaerven R, Iversen BM. Preeclampsia and the risk of end-stage renal disease. N Engl J Med 2008;359(8):800–9.

[37] Stillman IE, Karumanchi SA. The glomerular injury of preeclampsia. J Am Soc Nephrol 2007;18(8):2281–4.

[38] Hladunewich MA, Myers BD, Derby GC, Blouch KL, Druzin ML, Deen WM, et al. Course of preeclamptic glomerular injury after delivery. Am J Physiol Ren Physiol 2008;294(3): F614–20.

[39] Berks D, Steegers EA, Molas M, Visser W. Resolution of hypertension and proteinuria after preeclampsia. Obstet Gynecol 2009;114(6):1307–14.

[40] McDonald SD, Malinowski A, Zhou Q, Yusuf S, Devereaux PJ. Cardiovascular sequelae of preeclampsia/eclampsia: a systematic review and meta-analyses. Am Heart J 2008;156(5): 918–30.

[41] Smith GC, Pell JP, Walsh D. Pregnancy complications and maternal risk of ischaemic heart disease: a retrospective cohort study of 129,290 births. Lancet 2001;357(9273):2002–6.

[42] Bellamy L, Casas JP, Hingorani AD, Williams DJ. Pre-eclampsia and risk of cardiovascular disease and cancer in later life: systematic review and meta-analysis. BMJ 2007;335 (7627):974.

[43] Sibai B, Dekker G, Kupferminc M. Pre-eclampsia. Lancet 2005;365(9461):785–99.

[44] Vallejo Maroto I, Miranda Guisado ML, Stiefel Garcia-Junco P, Pamies Andreu E, Marenco ML, Castro de Gavilan D, et al. Clinical and biological characteristics of a group of 54 pregnant women with HELLP syndrome. Med Clin 2004;122(7):259–61.

[45] Miranda ML, Vallejo-Vaz AJ, Cerrillo L, Marenco ML, Villar J, Stiefel P. The HELLP syndrome (hemolysis, elevated liver enzymes and low platelets): clinical characteristics and maternal-fetal outcome in 172 patients. Pregnancy Hypertens 2011;1(2):164–9.

[46] Hammer ES, Cipolla MJ. Cerebrovascular dysfunction in preeclamptic pregnancies. Curr Hypertens Rep 2015;17(8):64.

[47] Fugate JE, Claassen DO, Cloft HJ, Kallmes DF, Kozak OS, Rabinstein AA. Posterior reversible encephalopathy syndrome: associated clinical and radiologic findings. Mayo Clin Proc 2010;85(5):427–32.

[48] Liman TG, Bohner G, Heuschmann PU, Endres M, Siebert E. The clinical and radiological spectrum of posterior reversible encephalopathy syndrome: the retrospective Berlin PRES study. J Neurol 2012;259(1):155–64.

[49] Guillemette L, Lacroix M, Allard C, Patenaude J, Battista MC, Doyon M, et al. Preeclampsia is associated with an increased pro-inflammatory profile in newborns. J Reprod Immunol 2015;112:111–4.

[50] Ozkan H, Cetinkaya M, Koksal N. Increased incidence of bronchopulmonary dysplasia in preterm infants exposed to preeclampsia. J Matern Fetal Neonatal Med 2012;25(12):2681–5.

[51] Ozkan H, Cetinkaya M, Koksal N, Ozmen A, Yildiz M. Maternal preeclampsia is associated with an increased risk of retinopathy of prematurity. J Perinat Med 2011;39(5):523–7.

[52] Cetinkaya M, Ozkan H, Koksal N. Maternal preeclampsia is associated with increased risk of necrotizing enterocolitis in preterm infants. Early Hum Dev 2012;88(11):893–8.

[53] Lawlor DA, Macdonald-Wallis C, Fraser A, Nelson SM, Hingorani A, Davey Smith G, et al. Cardiovascular biomarkers and vascular function during childhood in the offspring of

mothers with hypertensive disorders of pregnancy: findings from the Avon Longitudinal Study of Parents and Children. Eur Heart J 2012;33(3):335–45.

[54] Maynard SE, Min JY, Merchan J, Lim KH, Li J, Mondal S, et al. Excess placental soluble fms-like tyrosine kinase 1 (sFlt1) may contribute to endothelial dysfunction, hypertension. J Clin Invest 2003;111(5):649–58.

[55] Clark DE, Smith SK, He Y, Day KA, Licence DR, Corps AN, et al. A vascular endothelial growth factor antagonist is produced by the human placenta and released into the maternal circulation. Biol Reprod 1998;59(6):1540–8.

[56] Sela S, Itin A, Natanson-Yaron S, Greenfield C, Goldman-Wohl D, Yagel S, et al. A novel human-specific soluble vascular endothelial growth factor receptor 1: cell-type-specific splicing and implications to vascular endothelial growth factor homeostasis and preeclampsia. Circ Res 2008;102(12):1566–74.

[57] Rajakumar A, Cerdeira AS, Rana S, Zsengeller Z, Edmunds L, Jeyabalan A, et al. Transcriptionally active syncytial aggregates in the maternal circulation may contribute to circulating soluble fms-like tyrosine kinase 1 in preeclampsia. Hypertension 2012;59(2):256–64.

[58] Naljayan MV, Karumanchi SA. New developments in the pathogenesis of preeclampsia. Adv Chronic Kidney Dis 2013;20(3):265–70.

[59] Shibuya M, Claesson-Welsh L. Signal transduction by VEGF receptors in regulation of angiogenesis and lymphangiogenesis. Exp Cell Res 2006;312(5):549–60.

[60] Eppler SM, Combs DL, Henry TD, Lopez JJ, Ellis SG, Yi JH, et al. A target-mediated model to describe the pharmacokinetics and hemodynamic effects of recombinant human vascular endothelial growth factor in humans. Clin Pharmacol Ther 2002;72(1):20–32.

[61] Maharaj AS, D'Amore PA. Roles for VEGF in the adult. Microvasc Res 2007;74(2-3):100–13.

[62] Kendall RL, Thomas KA. Inhibition of vascular endothelial cell growth factor activity by an endogenously encoded soluble receptor. Proc Natl Acad Sci U S A 1993;90(22):10705–9.

[63] Kendall RL, Wang G, Thomas KA. Identification of a natural soluble form of the vascular endothelial growth factor receptor, FLT-1, and its heterodimerization with KDR. Biochem Biophys Res Commun 1996;226(2):324–8.

[64] Ahmed A, Li XF, Dunk C, Whittle MJ, Rushton DI, Rollason T. Colocalisation of vascular endothelial growth factor and its Flt-1 receptor in human placenta. Growth Factors 1995;12(3):235–43.

[65] Eddy AC, Bidwell III GL, George EM. Pro-angiogenic therapeutics for preeclampsia. Biol Sex Differ 2018;9(1):36.

[66] Olsson AK, Dimberg A, Kreuger J, Claesson-Welsh L. VEGF receptor signalling—in control of vascular function. Nat Rev Mol Cell Biol 2006;7(5):359–71.

[67] Ambati BK, Nozaki M, Singh N, Takeda A, Jani PD, Suthar T, et al. Corneal avascularity is due to soluble VEGF receptor-1. Nature 2006;443(7114):993–7.

[68] Thomas CP, Andrews JI, Raikwar NS, Kelley EA, Herse F, Dechend R, et al. A recently evolved novel trophoblast-enriched secreted form of fms-like tyrosine kinase-1 variant is up-regulated in hypoxia and preeclampsia. J Clin Endocrinol Metab 2009;94(7):2524–30.

[69] Thomas CP, Andrews JI, Liu KZ. Intronic polyadenylation signal sequences and alternate splicing generate human soluble Flt1 variants and regulate the abundance of soluble Flt1 in the placenta. FASEB J 2007;21(14):3885–95.

[70] Saito T, Takeda N, Amiya E, Nakao T, Abe H, Semba H, et al. VEGF-A induces its negative regulator, soluble form of VEGFR-1, by modulating its alternative splicing. FEBS Lett 2013;587(14):2179–85.

[71] Fan X, Rai A, Kambham N, Sung JF, Singh N, Petitt M, et al. Endometrial VEGF induces placental sFLT1 and leads to pregnancy complications. J Clin Invest 2014;124(11):4941–52.

[72] Forsythe JA, Jiang BH, Iyer NV, Agani F, Leung SW, Koos RD, et al. Activation of vascular endothelial growth factor gene transcription by hypoxia-inducible factor 1. Mol Cell Biol 1996;16(9):4604–13.

[73] Hollegaard B, Lykke JA, Boomsma JJ. Time from pre-eclampsia diagnosis to delivery affects future health prospects of children. Evol Med Public Health 2017;2017(1):53–66.

[74] Hiratsuka S, Minowa O, Kuno J, Noda T, Shibuya M. Flt-1 lacking the tyrosine kinase domain is sufficient for normal development and angiogenesis in mice. Proc Natl Acad Sci U S A 1998;95(16):9349–54.

[75] Kearney JB, Kappas NC, Ellerstrom C, DiPaola FW, Bautch VL. The VEGF receptor flt-1 (VEGFR-1) is a positive modulator of vascular sprout formation and branching morphogenesis. Blood 2004;103(12):4527–35.

[76] Hagmann H, Thadhani R, Benzing T, Karumanchi SA, Stepan H. The promise of angiogenic markers for the early diagnosis and prediction of preeclampsia. Clin Chem 2012;58(5): 837–45.

[77] Carmeliet P, Moons L, Luttun A, Vincenti V, Compernolle V, De Mol M, et al. Synergism between vascular endothelial growth factor and placental growth factor contributes to angiogenesis and plasma extravasation in pathological conditions. Nat Med 2001;7(5):575–83.

[78] Bussolati B, Dunk C, Grohman M, Kontos CD, Mason J, Ahmed A. Vascular endothelial growth factor receptor-1 modulates vascular endothelial growth factor-mediated angiogenesis via nitric oxide. Am J Pathol 2001;159(3):993–1008.

[79] Brown LF, Berse B, Tognazzi K, Manseau EJ, Van de Water L, Senger DR, et al. Vascular permeability factor mRNA and protein expression in human kidney. Kidney Int 1992; 42(6):1457–61.

[80] Aggarwal PK, Jain V, Sakhuja V, Karumanchi SA, Jha V. Low urinary placental growth factor is a marker of pre-eclampsia. Kidney Int 2006;69(3):621–4.

[81] Poon LC, Kametas NA, Maiz N, Akolekar R, Nicolaides KH. First-trimester prediction of hypertensive disorders in pregnancy. Hypertension 2009;53(5):812–8.

[82] Levine RJ, Thadhani R, Qian C, Lam C, Lim KH, Yu KF, et al. Urinary placental growth factor and risk of preeclampsia. JAMA 2005;293(1):77–85.

[83] Perni U, Sison C, Sharma V, Helseth G, Hawfield A, Suthanthiran M, et al. Angiogenic factors in superimposed preeclampsia: a longitudinal study of women with chronic hypertension during pregnancy. Hypertension 2012;59(3):740–6.

[84] Crispi F, Llurba E, Dominguez C, Martin-Gallan P, Cabero L, Gratacos E. Predictive value of angiogenic factors and uterine artery Doppler for early- versus late-onset pre-eclampsia and intrauterine growth restriction. Ultrasound Obstet Gynecol 2008;31(3):303–9.

[85] Espinoza J, Romero R, Nien JK, Gomez R, Kusanovic JP, Goncalves LF, et al. Identification of patients at risk for early onset and/or severe preeclampsia with the use of uterine artery Doppler velocimetry and placental growth factor. Am J Obstet Gynecol 2007;196(4). 326 e1-13.

[86] Stepan H, Geipel A, Schwarz F, Kramer T, Wessel N, Faber R. Circulatory soluble endoglin and its predictive value for preeclampsia in second-trimester pregnancies with abnormal uterine perfusion. Am J Obstet Gynecol 2008;198(2). 175 e1-6.

[87] Stepan H, Unversucht A, Wessel N, Faber R. Predictive value of maternal angiogenic factors in second trimester pregnancies with abnormal uterine perfusion. Hypertension 2007;49(4): 818–24.

[88] Wright A, Wright D, Syngelaki A, Georgantis A, Nicolaides KH. Two-stage screening for preterm preeclampsia at 11-13 weeks' gestation. Am J Obstet Gynecol 2018.

[89] Levine RJ, Maynard SE, Qian C, Lim KH, England LJ, Yu KF, et al. Circulating angiogenic factors and the risk of preeclampsia. N Engl J Med 2004;350(7):672–83.

[90] Levine RJ, Lam C, Qian C, Yu KF, Maynard SE, Sachs BP, et al. Soluble endoglin and other circulating antiangiogenic factors in preeclampsia. N Engl J Med 2006;355(10):992–1005.

[91] Kim SY, Ryu HM, Yang JH, Kim MY, Han JY, Kim JO, et al. Increased sFlt-1 to PlGF ratio in women who subsequently develop preeclampsia. J Korean Med Sci 2007;22(5):873–7.

[92] De Vivo A, Baviera G, Giordano D, Todarello G, Corrado F, D'Anna R. Endoglin, PlGF and sFlt-1 as markers for predicting pre-eclampsia. Acta Obstet Gynecol Scand 2008; 87(8):837–42.

[93] Lim JH, Kim SY, Park SY, Yang JH, Kim MY, Ryu HM. Effective prediction of preeclampsia by a combined ratio of angiogenesis-related factors. Obstet Gynecol 2008;111(6):1403–9.

[94] Verlohren S, Galindo A, Schlembach D, Zeisler H, Herraiz I, Moertl MG, et al. An automated method for the determination of the sFlt-1/PIGF ratio in the assessment of preeclampsia. Am J Obstet Gynecol 2010;202(2). 161 e1-e11.

[95] Rana S, Powe CE, Salahuddin S, Verlohren S, Perschel FH, Levine RJ, et al. Angiogenic factors and the risk of adverse outcomes in women with suspected preeclampsia. Circulation 2012;125(7):911–9.

[96] Rana S, Hacker MR, Modest AM, Salahuddin S, Lim KH, Verlohren S, et al. Circulating angiogenic factors and risk of adverse maternal and perinatal outcomes in twin pregnancies with suspected preeclampsia. Hypertension 2012;60(2):451–8.

[97] Chaiworapongsa T, Romero R, Espinoza J, Bujold E, Mee Kim Y, Goncalves LF, et al. Evidence supporting a role for blockade of the vascular endothelial growth factor system in the pathophysiology of preeclampsia. Young Investigator Award. Am J Obstet Gynecol 2004;190(6). 1541–7; discussion 7–50.

[98] Veas CJ, Aguilera VC, Munoz IJ, Gallardo VI, Miguel PL, Gonzalez MA, et al. Fetal endothelium dysfunction is associated with circulating maternal levels of sE-selectin, sVCAM1, and sFlt-1 during pre-eclampsia. J Matern Fetal Neonatal Med 2011;24(11):1371–7.

[99] Wikstrom AK, Larsson A, Eriksson UJ, Nash P, Norden-Lindeberg S, Olovsson M. Placental growth factor and soluble FMS-like tyrosine kinase-1 in early-onset and late-onset preeclampsia. Obstet Gynecol 2007;109(6):1368–74.

[100] Qazi U, Lam C, Karumanchi SA, Petri M. Soluble Fms-like tyrosine kinase associated with preeclampsia in pregnancy in systemic lupus erythematosus. J Rheumatol 2008;35(4): 631–4.

[101] Rolfo A, Attini R, Nuzzo AM, Piazzese A, Parisi S, Ferraresi M, et al. Chronic kidney disease may be differentially diagnosed from preeclampsia by serum biomarkers. Kidney Int 2013; 83(1):177–81.

[102] Venkatesha S, Toporsian M, Lam C, Hanai J, Mammoto T, Kim YM, et al. Soluble endoglin contributes to the pathogenesis of preeclampsia. Nat Med 2006;12(6):642–9.

[103] Khalil A, Cowans NJ, Spencer K, Goichman S, Meiri H, Harrington K. First trimester maternal serum placental protein 13 for the prediction of pre-eclampsia in women with a priori high risk. Prenat Diagn 2009;29(8):781–9.

[104] Spencer K, Cowans NJ, Chefetz I, Tal J, Meiri H. First-trimester maternal serum PP-13, PAPP-A and second-trimester uterine artery Doppler pulsatility index as markers of pre-eclampsia. Ultrasound Obstet Gynecol 2007;29(2):128–34.

[105] Than NG, Abdul Rahman O, Magenheim R, Nagy B, Fule T, Hargitai B, et al. Placental protein 13 (galectin-13) has decreased placental expression but increased shedding and maternal serum concentrations in patients presenting with preterm pre-eclampsia and HELLP syndrome. Virchows Arch 2008;453(4):387–400.

[106] Than NG, Pick E, Bellyei S, Szigeti A, Burger O, Berente Z, et al. Functional analyses of placental protein 13/galectin-13. Eur J Biochem 2004;271(6):1065–78.

[107] Huppertz B, Sammar M, Chefetz I, Neumaier-Wagner P, Bartz C, Meiri H. Longitudinal determination of serum placental protein 13 during development of preeclampsia. Fetal Diagn Ther 2008;24(3):230–6.

[108] Bonnans C, Chou J, Werb Z. Remodelling the extracellular matrix in development and disease. Nat Rev Mol Cell Biol 2014;15(12):786–801.

[109] Pollheimer J, Fock V, Knofler M. Review: the ADAM metalloproteinases—novel regulators of trophoblast invasion? Placenta 2014;35(Suppl):S57–63.

[110] Myers JE, Merchant SJ, Macleod M, Mires GJ, Baker PN, Davidge ST. MMP-2 levels are elevated in the plasma of women who subsequently develop preeclampsia. Hypertens Pregnancy 2005;24(2):103–15.

[111] Estrada-Gutierrez G, Cappello RE, Mishra N, Romero R, Strauss 3rd JF, Walsh SW. Increased expression of matrix metalloproteinase-1 in systemic vessels of preeclamptic women: a critical mediator of vascular dysfunction. Am J Pathol 2011;178(1):451–60.

[112] Nugent WH, Mishra N, Strauss III JF, Walsh SW. Matrix metalloproteinase 1 causes vasoconstriction and enhances vessel reactivity to angiotensin II via protease-activated receptor 1. Reprod Sci 2016;23(4):542–8.

[113] Zhu J, Zhong M, Pang Z, Yu Y. Dysregulated expression of matrix metalloproteinases and their inhibitors may participate in the pathogenesis of pre-eclampsia and fetal growth restriction. Early Hum Dev 2014;90(10):657–64.

[114] Martinez-Fierro ML, Perez-Favila A, Garza-Veloz I, Espinoza-Juarez MA, Avila-Carrasco L, Delgado-Enciso I, et al. Matrix metalloproteinase multiplex screening identifies increased MMP-2 urine concentrations in women predicted to develop preeclampsia. Biomarkers 2018;23(1):18–24.

[115] Wang Y, Gu Y, Loyd S, Jia X, Groome LJ. Increased urinary levels of podocyte glycoproteins, matrix metallopeptidases, inflammatory cytokines, and kidney injury biomarkers in women with preeclampsia. Am J Physiol Ren Physiol 2015;309(12):F1009–17.

[116] Eleuterio NM, Palei AC, Rangel Machado JS, Tanus-Santos JE, Cavalli RC, Sandrim VC. Positive correlations between circulating adiponectin and MMP2 in preeclampsia pregnant. Pregnancy Hypertens 2015;5(2):205–8.

[117] Montagnana M, Lippi G, Albiero A, Scevarolli S, Salvagno GL, Franchi M, et al. Evaluation of metalloproteinases 2 and 9 and their inhibitors in physiologic and pre-eclamptic pregnancy. J Clin Lab Anal 2009;23(2):88–92.

[118] Chen J, Khalil RA. Matrix metalloproteinases in normal pregnancy and preeclampsia. Prog Mol Biol Transl Sci 2017;148:87–165.

[119] Rodrigues SF, Tran ED, Fortes ZB, Schmid-Schonbein GW. Matrix metalloproteinases cleave the beta2-adrenergic receptor in spontaneously hypertensive rats. Am J Physiol Heart Circ Physiol 2010;299(1):H25–35.

[120] Rizzi E, Castro MM, Prado CM, Silva CA, Fazan Jr R, Rossi MA, et al. Matrix metalloproteinase inhibition improves cardiac dysfunction and remodeling in 2-kidney, 1-clip hypertension. J Card Fail 2010;16(7):599–608.

[121] Palei AC, Granger JP, Tanus-Santos JE. Matrix metalloproteinases as drug targets in preeclampsia. Curr Drug Targets 2013;14(3):325–34.

[122] Blois SM, Barrientos G. Galectin signature in normal pregnancy and preeclampsia. J Reprod Immunol 2014;101-102:127–34.

[123] Menkhorst E, Koga K, Van Sinderen M, Dimitriadis E. Galectin-7 serum levels are altered prior to the onset of pre-eclampsia. Placenta 2014;35(4):281–5.

[124] Santillan MK, Santillan DA, Scroggins SM, Min JY, Sandgren JA, Pearson NA, et al. Vasopressin in preeclampsia: a novel very early human pregnancy biomarker and clinically relevant mouse model. Hypertension 2014;64(4):852–9.

[125] Craici IM, Wagner SJ, Weissgerber TL, Grande JP, Garovic VD. Advances in the pathophysiology of pre-eclampsia and related podocyte injury. Kidney Int 2014;86(2):275–85.

[126] Zhao S, Gu X, Groome LJ, Wang Y. Decreased nephrin and GLEPP-1, but increased VEGF, Flt-1, and nitrotyrosine, expressions in kidney tissue sections from women with preeclampsia. Reprod Sci 2009;16(10):970–9.

[127] Garovic VD, Wagner SJ, Petrovic LM, Gray CE, Hall P, Sugimoto H, et al. Glomerular expression of nephrin and synaptopodin, but not podocin, is decreased in kidney sections from women with preeclampsia. Nephrol Dial Transplant 2007;22(4):1136–43.

[128] Jim B, Mehta S, Qipo A, Kim K, Cohen HW, Moore RM, et al. A comparison of podocyturia, albuminuria and nephrinuria in predicting the development of preeclampsia: a prospective study. PLoS One 2014;9(7).

[129] Gilani SI, Anderson UD, Jayachandran M, Weissgerber TL, Zand L, White WM, et al. Urinary extracellular vesicles of podocyte origin and renal injury in preeclampsia. J Am Soc Nephrol 2017;28(11):3363–72.

[130] Chew JS, Saleem M, Florkowski CM, George PM. Cystatin C—a paradigm of evidence based laboratory medicine. Clin Biochem Rev 2008;29(2):47–62.

[131] Hojs R, Bevc S, Ekart R, Gorenjak M, Puklavec L. Serum cystatin C as an endogenous marker of renal function in patients with mild to moderate impairment of kidney function. Nephrol Dial Transplant 2006;21(7):1855–62.

[132] Stevens LA, Coresh J, Schmid CH, Feldman HI, Froissart M, Kusek J, et al. Estimating GFR using serum cystatin C alone and in combination with serum creatinine: a pooled analysis of 3,418 individuals with CKD. Am J Kidney Dis 2008;51(3):395–406.

[133] Thilaganathan B, Ralph E, Papageorghiou AT, Melchiorre K, Sheldon J. Raised maternal serum cystatin C: an early pregnancy marker for preeclampsia. Reprod Sci 2009;16(8): 788–93.

[134] Strevens H, Wide-Swensson D, Torffvit O, Grubb A. Serum cystatin C for assessment of glomerular filtration rate in pregnant and non-pregnant women. Indications of altered filtration process in pregnancy. Scand J Clin Lab Invest 2002;62(2):141–7.

[135] Franceschini N, Qiu C, Barrow DA, Williams MA. Cystatin C and preeclampsia: a case control study. Ren Fail 2008;30(1):89–95.

[136] Novakov Mikic A, Cabarkapa V, Nikolic A, Maric D, Brkic S, Mitic G, et al. Cystatin C in preeclampsia. J Matern Fetal Neonatal Med 2012;25(7):961–5.

[137] Risch M, Purde MT, Baumann M, Mohaupt M, Mosimann B, Renz H, et al. High first-trimester maternal blood cystatin C levels despite normal serum creatinine predict pre-eclampsia in singleton pregnancies. Scand J Clin Lab Invest 2017;77(8):634–43.

[138] Kolialexi A, Gourgiotis D, Daskalakis G, Marmarinos A, Lykoudi A, Mavreli D, et al. Validation of serum biomarkers derived from proteomic analysis for the early screening of preeclampsia. Dis Markers 2015;2015.

[139] Thilaganathan B, Wormald B, Zanardini C, Sheldon J, Ralph E, Papageorghiou AT. Early-pregnancy multiple serum markers and second-trimester uterine artery Doppler in predicting preeclampsia. Obstet Gynecol 2010;115(6):1233–8.

[140] Lam C, Lim KH, Kang DH, Karumanchi SA. Uric acid and preeclampsia. Semin Nephrol 2005;25(1):56–60.

[141] Laughon SK, Catov J, Powers RW, Roberts JM, Gandley RE. First trimester uric acid and adverse pregnancy outcomes. Am J Hypertens 2011;24(4):489–95.

[142] Guo HX, Wang CH, Li ZQ, Gong SP, Zhou ZQ, Leng LZ, et al. The application of serum cystatin C in estimating the renal function in women with preeclampsia. Reprod Sci 2012;19(7): 712–7.

[143] Valdes G, Chacon C, Corthorn J, Figueroa CD, Germain AM. Tissue kallikrein in human placenta in early and late gestation. Endocrine 2001;14(2):197–204.

[144] Khedun SM, Naicker T, Moodley J. Tissue kallikrein activity in pregnancy. Aust N Z J Obstet Gynaecol 2000;40(4):451–4.

[145] Khedun SM, Naicker T, Moodley J, Naidoo S. Urinary tissue kallikrein excretion in black African women with severe pre-eclampsia. Acta Obstet Gynecol Scand 1999;78(4):316–20.

[146] Munoz-Hernandez R, Medrano-Campillo P, Miranda ML, Macher HC, Praena-Fernandez JM, Vallejo-Vaz AJ, et al. Total and fetal circulating cell-free DNA, angiogenic, and antiangiogenic factors in preeclampsia and HELLP syndrome. Am J Hypertens 2017;30(7):673–82.

[147] Salvianti F, Inversetti A, Smid M, Valsecchi L, Candiani M, Pazzagli M, et al. Prospective evaluation of RASSF1A cell-free DNA as a biomarker of pre-eclampsia. Placenta 2015;36(9): 996–1001.

[148] Rolnik DL, Wright D, Poon LC, O'Gorman N, Syngelaki A, de Paco Matallana C, et al. Aspirin versus placebo in pregnancies at high risk for preterm preeclampsia. N Engl J Med 2017; 377(7):613–22.

[149] Roberge S, Bujold E, Nicolaides KH. Aspirin for the prevention of preterm and term preeclampsia: systematic review and metaanalysis. Am J Obstet Gynecol 2018;218(3): 287–93. e1.

[150] Bujold E, Roberge S, Lacasse Y, Bureau M, Audibert F, Marcoux S, et al. Prevention of preeclampsia and intrauterine growth restriction with aspirin started in early pregnancy: a meta-analysis. Obstet Gynecol 2010;116(2 Pt 1):402–14.

[151] Roberge S, Villa P, Nicolaides K, Giguere Y, Vainio M, Bakthi A, et al. Early administration of low-dose aspirin for the prevention of preterm and term preeclampsia: a systematic review and meta-analysis. Fetal Diagn Ther 2012;31(3):141–6.

[152] Roberge S, Bujold E, Nicolaides KH. Meta-analysis on the effect of aspirin use for prevention of preeclampsia on placental abruption and antepartum hemorrhage. Am J Obstet Gynecol 2018;218(5):483–9.

[153] Kumasawa K, Ikawa M, Kidoya H, Hasuwa H, Saito-Fujita T, Morioka Y, et al. Pravastatin induces placental growth factor (PGF) and ameliorates preeclampsia in a mouse model. Proc Natl Acad Sci U S A 2011;108(4):1451–5.

[154] Costantine MM, Cleary K, Hebert MF, Ahmed MS, Brown LM, Ren Z, et al. Safety and pharmacokinetics of pravastatin used for the prevention of preeclampsia in high-risk pregnant women: a pilot randomized controlled trial. Am J Obstet Gynecol 2016;214(6). 720 e1-e17.

[155] Thadhani R, Kisner T, Hagmann H, Bossung V, Noack S, Schaarschmidt W, et al. Pilot study of extracorporeal removal of soluble fms-like tyrosine kinase 1 in preeclampsia. Circulation 2011;124(8):940–50.

[156] Winkler K, Contini C, Konig B, Krumrey B, Putz G, Zschiedrich S, et al. Treatment of very preterm preeclampsia via heparin-mediated extracorporeal LDL-precipitation (H.E.L.P.) apheresis: the Freiburg preeclampsia H.E.L.P.-Apheresis study. Pregnancy Hypertens 2018;12:136–43.

[157] Wu P, van den Berg C, Alfirevic Z, O'Brien S, Rothlisberger M, Baker PN, et al. Early pregnancy biomarkers in pre-eclampsia: a systematic review and meta-analysis. Int J Mol Sci 2015;16(9): 23035–56.

[158] Crews DC, Plantinga LC, Miller III ER, Saran R, Hedgeman E, Saydah SH, et al. Prevalence of chronic kidney disease in persons with undiagnosed or prehypertension in the United States. Hypertension 2010;55(5):1102–9.

[159] Bramham K, Seed PT, Lightstone L, Nelson-Piercy C, Gill C, Webster P, et al. Diagnostic and predictive biomarkers for pre-eclampsia in patients with established hypertension and chronic kidney disease. Kidney Int 2016;89(4):874–85.

[160] Borzychowski AM, Sargent IL, Redman CW. Inflammation and pre-eclampsia. Semin Fetal Neonatal Med 2006;11(5):309–16.

[161] Lam C, Lim KH, Karumanchi SA. Circulating angiogenic factors in the pathogenesis and prediction of preeclampsia. Hypertension 2005;46(5):1077–85.

Further reading

Brosens JJ, Pijnenborg R, Brosens IA. The myometrial junctional zone spiral arteries in normal and abnormal pregnancies: a review of the literature. Am J Obstet Gynecol 2002;187(5):1416–23.

Dechend R, Muller DN, Wallukat G, Homuth V, Krause M, Dudenhausen J, et al. AT1 receptor agonistic antibodies, hypertension, and preeclampsia. Semin Nephrol 2004;24(6):571–9.

Espino YSS, Flores-Pliego A, Espejel-Nunez A, Medina-Bastidas D, Vadillo-Ortega F, Zaga-Clavellina V, et al. New insights into the role of matrix metalloproteinases in preeclampsia. Int J Mol Sci 2017;18(7).

Greer IA, Lyall F, Perera T, Boswell F, Macara LM. Increased concentrations of cytokines interleukin-6 and interleukin-1 receptor antagonist in plasma of women with preeclampsia: a mechanism for endothelial dysfunction? Obstet Gynecol 1994;84(6):937–40.

Hubel CA. Dyslipidemia, iron, and oxidative stress in preeclampsia: assessment of maternal and feto-placental interactions. Semin Reprod Endocrinol 1998;16(1):75–92.

Jadli A, Ghosh K, Satoskar P, Damania K, Bansal V, Shetty S. Combination of copeptin, placental growth factor and total annexin V microparticles for prediction of preeclampsia at 10-14 weeks of gestation. Placenta 2017;58:67–73.

James JL, Whitley GS, Cartwright JE. Pre-eclampsia: fitting together the placental, immune and cardiovascular pieces. J Pathol 2010;221(4):363–78.

Konijnenberg A, Stokkers EW, van der Post JA, Schaap MC, Boer K, Bleker OP, et al. Extensive platelet activation in preeclampsia compared with normal pregnancy: enhanced expression of cell adhesion molecules. Am J Obstet Gynecol 1997;176(2):461–9.

LaMarca BB, Bennett WA, Alexander BT, Cockrell K, Granger JP. Hypertension produced by reductions in uterine perfusion in the pregnant rat: role of tumor necrosis factor-alpha. Hypertension 2005;46(4):1022–5.

LaMarca B, Wallace K, Granger J. Role of angiotensin II type I receptor agonistic autoantibodies (AT1-AA) in preeclampsia. Curr Opin Pharmacol 2011;11(2):175–9.

LaMarca B, Wallukat G, Llinas M, Herse F, Dechend R, Granger JP. Autoantibodies to the angiotensin type I receptor in response to placental ischemia and tumor necrosis factor alpha in pregnant rats. Hypertension 2008;52(6):1168–72.

Lok CA, Van Der Post JA, Sargent IL, Hau CM, Sturk A, Boer K, et al. Changes in microparticle numbers and cellular origin during pregnancy and preeclampsia. Hypertens Pregnancy 2008;27(4):344–60.

Maitre JL. Mechanics of blastocyst morphogenesis. Biol Cell 2017;109(9):323–38.

Mause SF, Weber C. Microparticles: protagonists of a novel communication network for intercellular information exchange. Circ Res 2010;107(9):1047–57.

McNally R, Alqudah A, Obradovic D, McClements L. Elucidating the pathogenesis of pre-eclampsia using in vitro models of spiral uterine artery remodelling. Curr Hypertens Rep 2017;19(11):93.

Nomura Y, John RM, Janssen AB, Davey C, Finik J, Buthmann J, et al. Neurodevelopmental consequences in offspring of mothers with preeclampsia during pregnancy: underlying biological mechanism via imprinting genes. Arch Gynecol Obstet 2017;295(6):1319–29.

Perry Jr KG, Martin Jr JN. Abnormal hemostasis and coagulopathy in preeclampsia and eclampsia. Clin Obstet Gynecol 1992;35(2):338–50.

Redman CW, Sacks GP, Sargent IL. Preeclampsia: an excessive maternal inflammatory response to pregnancy. Am J Obstet Gynecol 1999;180(2 Pt 1):499–506.

Sacks GP, Studena K, Sargent K, Redman CW. Normal pregnancy and preeclampsia both produce inflammatory changes in peripheral blood leukocytes akin to those of sepsis. Am J Obstet Gynecol 1998;179(1):80–6.

Shah DM. The role of RAS in the pathogenesis of preeclampsia. Curr Hypertens Rep 2006;8(2): 144–52.

Terrone DA, Rinehart BK, May WL, Moore A, Magann EF, Martin Jr JN. Leukocytosis is proportional to HELLP syndrome severity: evidence for an inflammatory form of preeclampsia. South Med J 2000;93(8):768–71.

Vince GS, Starkey PM, Austgulen R, Kwiatkowski D, Redman CW. Interleukin-6, tumour necrosis factor and soluble tumour necrosis factor receptors in women with pre-eclampsia. Br J Obstet Gynaecol 1995;102(1):20–5.

Whitley GS, Cartwright JE. Cellular and molecular regulation of spiral artery remodelling: lessons from the cardiovascular field. Placenta 2010;31(6):465–74.

Zhou Y, Damsky CH, Fisher SJ. Preeclampsia is associated with failure of human cytotrophoblasts to mimic a vascular adhesion phenotype. One cause of defective endovascular invasion in this syndrome? J Clin Invest 1997;99(9):2152–64.

Zhou Y, Fisher SJ, Janatpour M, Genbacev O, Dejana E, Wheelock M, et al. Human cytotrophoblasts adopt a vascular phenotype as they differentiate. A strategy for successful endovascular invasion? J Clin Invest 1997;99(9):2139–51.

Index

Note: Page numbers followed by *f* indicate figures and *t* indicate tables.

F

M

N

Q

R

Printed in the United States
By Bookmasters